AF365183

Digital Signal Processing

Digital
Signal
Processing

Dr. G. Jagadeeswar Reddy

Principal
Narayana Engineering College, Nellore
Andhra Pradesh

BS Publications

<u>A unit of **BSP Books Pvt. Ltd.**</u>

4-4-309/316, Giriraj Lane, Sultan Bazar,
Hyderabad - 500 095
Phone : 040 - 23445605, 23445688

Digital Signal Processing by *Dr. G. Jagadeeswar Reddy*

© 2015, *by Publisher*

Published by :

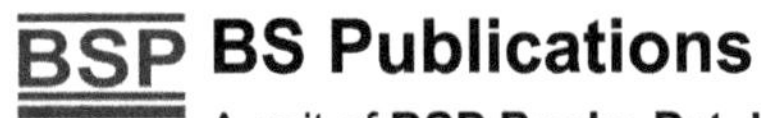 **BS Publications**

A unit of **BSP Books Pvt. Ltd.**

4-4-309/316, Giriraj Lane, Sultan Bazar,
Hyderabad – 500 095
Phone : 040 – 23445605, 23445688
e-mail : info@bspbooks.net

ISBN : 978-93-52300-82-2 (HB)

Dedicated to

My Wife
Sunitha

&

My Daughters
Nikitha and Namitha

Preface

The goal of DSP is usually to measure, filter and/or compress continuous real-world analog signals. The first step is usually to convert the signal from an analog to digital form, by sampling and then digitizing it using an analog-to-digital converter (ADC), which turns the analog signal into a stream of numbers. However, often, the required output signal is another analog output signal, which requires a digital-to analog converter (DAC). Even if this process is more complex than analog processing and has a discrete value range, the application of computational power to digital signal processing allows for many advantages over analog processing in many applications, such as error detection and correction in transmission as well as data compression.

This volume is mainly the expansion of the concepts of Signals and Systems. A key feature of this book is the presentation of the aspects of DSP in a crystal clear manner. Secondly, it provides concept oriented information and every concept is accompanied with relevant example, a problem and corresponding solution. As it addresses the various problems and caters to diversified needs of the students in view of examinations, one need not refer to different books; instead one can get everything in this volume.

The book is divided into six chapters. Chapter one is concerned with the back drop of Digital Signal Processing and it's co-relation with Analog Signal Processing. Second chapter discusses various elements like Discrete Time System, Discrete Fourier Series, Linear Convolution and Circular Convolution. Chapter three is devoted to the study of Z-Transforms, Stability Criterion, Discrete Filters etc. The fourth chapter concentrates on the applications of IIR Digital Filters. The fifth chapter discusses various aspects of FIR Digital Filters and related windows. The last chapter concentrates on Multirate Digital Signal Processing and Applications.

Though the subject is tough, it is made easy with all the above features.

- Author

Contents

CHAPTER 3

Z-transform

CHAPTER 4

IIR Digital Filters

FIR Digital Filters

CHAPTER 6

Multirate Digital Signal Processing

Chapter 1

Introduction to
Digital Signal Processing

1.1 Introduction

1.1.1 Signal and Signal Processing

A signal is defined as "any physical quantity which varies with one or more independent variables like time, space". Mathematically it can be represented as a function of one or more independent variables. For example, the function

$$s(t) = 2t \qquad \qquad \text{....(1.1.1)}$$

describes a signal, which varies linearly with the independent variable t (time). 'Speech' signal is an example, which varies with single independent variable.

Consider the function

$$s(x, y) = 2x + 3y + 5xy \qquad \qquad \text{.....(1.1.2)}$$

This function describes a signal, which varies with two independent variables x and y. 'Image' is a signal which varies with two independent variables.

Most of the signals encountered are analog in nature i.e., they vary with continuous variable, such as time or space. "Processing of these signals by analog systems such as a filters or frequency analyzers or frequency multipliers for the purpose of changing their characteristics or extracting some desired information is called Signal Processing". In this case both the input signal and the output signal are in analog form, which is shown in Fig. 1.1.

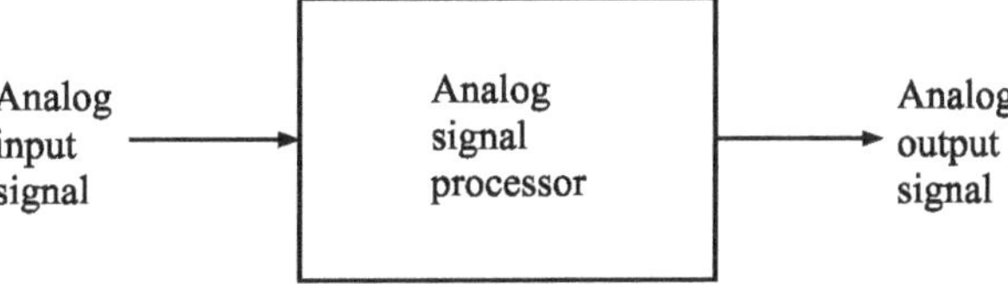

Fig. 1.1 Analog signal processing.

1.1.2 Basic Elements of Digital Signal Processing Systems

Digital Signal Processing is an alternative method to process an analog signal. It requires an interface between an analog input and digital signal processor, called an analog-to-digital (A/D) converter. We should provide another interface between digital signal processor and an analog output signal called a digital-to-analog (D/A) converter as shown in the Fig. 1.2.

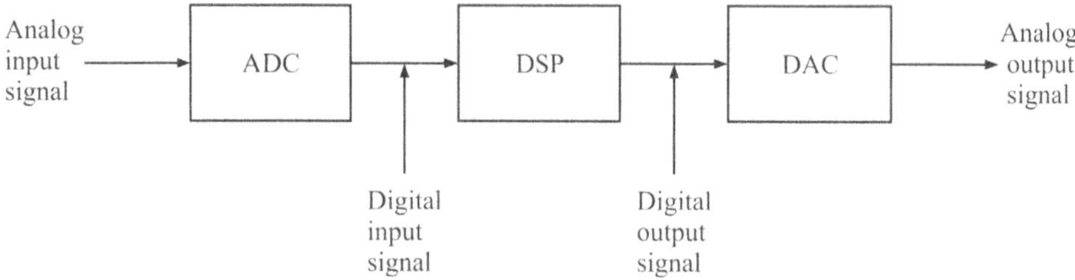

Fig. 1.2 Block diagram of basic DSP system.

The ADC or analog-to-digital (A/D) converter contains a sampler, quantizer and an encoder. Sampler takes an analog signal, samples it with a predefined sampling period and gives out discrete-time signal, which is discrete in time domain and continuous (varying) in amplitude. This signal contains different number of amplitude levels. Quantizer approximates these different levels with fixed number of levels by rounding or truncating the values. For example, if the allowable signal values in the digital signal are integers, say 0 to 7, the continuous-amplitude signal will be quantized into these integer values. Thus the signal value 5.63 will be approximated by the value 6 if the quantization process is performed by rounding to the nearest integer or by 5 if the quantization process is performed by truncation. The encoder converts these set of integer into digital form (i.e., binary form).

The digital signal processor may be a large programmable digital computer or a small microprocessor or a hardwired digital processor to perform the operation on the digital signal i.e., on the output of encoder. Hardwired digital processor performs a specified set of operations on the digital signal i.e., reconfiguring is difficult with the hardwired machines, where as programmable machines provide flexibility to change the operations through a change in the software.

The output of DSP block is digital signal. Digital to analog converter converts digital signal into an analog signal, which may not be required on some applications like extracting information from the radar signal, such as the position of the air-craft and its speed, may simply be printed on paper.

1.1.3 Advantages of Digital Signal Processing over Analog Signal Processing

There are many reasons why digital systems are preferred over an analog system. Some of the advantages are:

(a) *Noise Immunity:* Digital systems are more immune to noise compared to an analog systems.

(b) *Arbitrarily High Accuracy:* Tolerances in analog circuit components make difficult to control the accuracy of an analog systems, where as digital systems provide much better control of accuracy.

(c) *High Reliability:* Digital systems are more reliable compared to an analog systems.

(d) *Software Manipulation:* Digital signal processing operations can be changed by changing the program in digital programmable system, i.e., these are flexible systems.

(e) *Integration of Digital Systems:* Digital systems can be cascaded or integrated easily without any loading problems.

(f) *Storage of Digital Signals:* Digital signals are easily stored on magnetic media such as magnetic tape without loss of quality of reproduction of signal.

(g) *Transportable:* As digital signals can be stored on magnetic tapes these can be processed off time i.e., these are easily transported.

(h) *Digital systems* are independent of temperature, ageing and other external parameters.

(i) *Cheaper:* Cost of processing per signal in DSP is reduced by time-sharing of given processor among a number of signals.

Disadvantage of digital systems is that they are not faster compared to analog systems.

1.2 Discrete Time Signals and Sequences

Discrete-time signals or sequences, which are discrete in time domain and continuous in amplitude, can be obtained by sampling continuous time or analog signals as shown in Fig. 1.3.

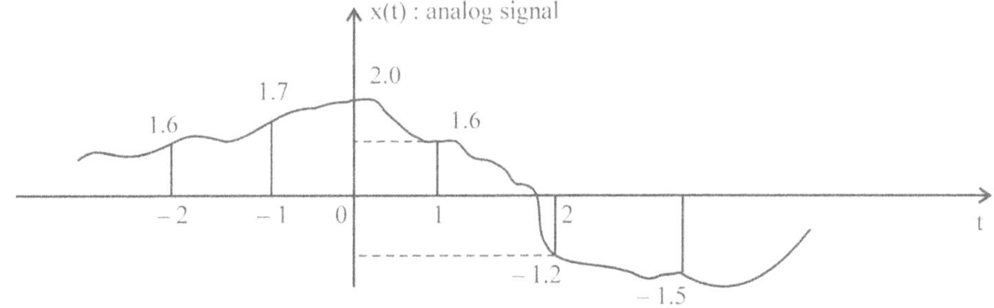

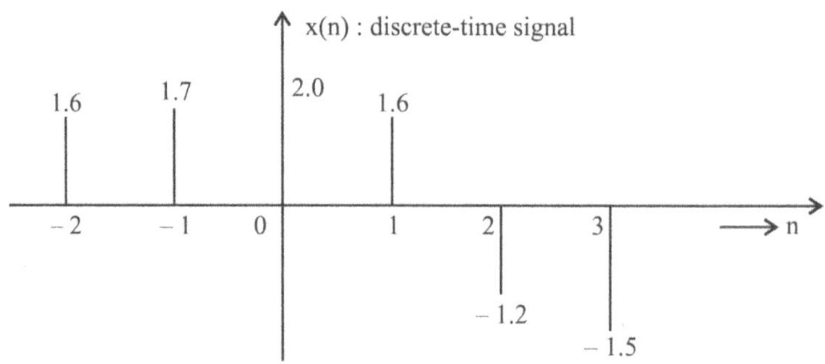

Fig. 1.3 Obtaining discrete-time signal from analog signal.

1.2.1 Representation of Discrete-Time Signals

Discrete-time signals can be represented as follows by using the four methods

    (i)   Graphical Representation

    (ii)   Functional Representation

    (iii)   Tabular Representation

    (iv)   Sequence Representation

(i)   *Graphical Representation:* Discrete-time signals can be represented by a graph when the signal is defined for every integer value of n for $-\infty < n < \infty$. This is shown in Fig. 1.4.

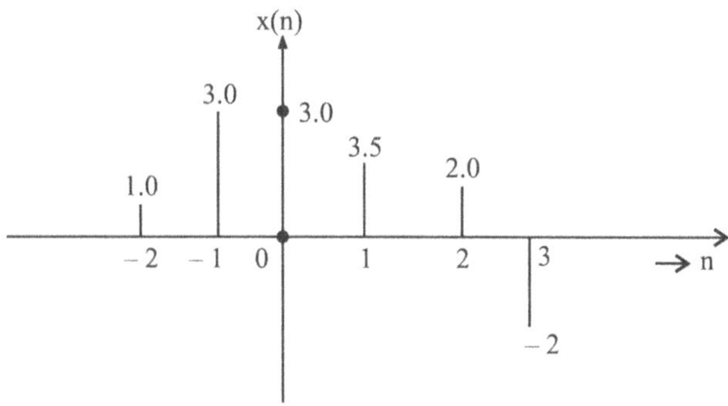

Fig. 1.4 Graphical representation of a discrete-time signal.

(ii) *Functional Representation:* Discrete-time signals can be represented functionally as given below

$$x(n) = \begin{cases} 1, & \text{for } n = 0, 1 \\ 2, & \text{for } n = 2 \\ 3, & \text{for } n = -1 \\ 0, & \text{elsewhere} \end{cases}$$

(iii) *Tabular Representation:* Discrete-time signals can be represented by a table as

n	.....	–2	–1	0	1	2	....
x(n)		1	2	3	1	2	

(iv) *Sequence Representation:* An infinite-duration $(-\infty \leq n \leq \infty)$ signal with the time as origin (n = 0) and indicated by the symbol ↑, if symbol is not shown in representation, origin is at the beginning of the sequence.

$$x(n) = \left\{ ....1,\ \underset{\uparrow}{2},\ 3,\ 4,\ 2,.... \right\}$$

$$x(n) = \{\ 2,\ 3,\ 1,\ 4\}$$

Here origin is the first position i.e., x(0) = 2.

1.2.2 Elementary Discrete-Time Signals

There are some basic signals which play an important role in the study of discrete-time signals and systems. These are:

   (i)   Unit-sample (Cronekar) Sequence, $\delta(n)$ (or) Impulse sequence

   (ii) Unit-step Sequence, u(n)

  (iii) Unit-ramp Sequence, r(n)

  (iv) Exponential Sequence

(i)   *Unit-sample Sequence:* This is illustrated in Fig. 1.5, it is denoted with $\delta(n)$ and is defined as

$$\delta(n) = \begin{cases} 1, & n = 0 \\ 0, & n \neq 0 \end{cases}$$

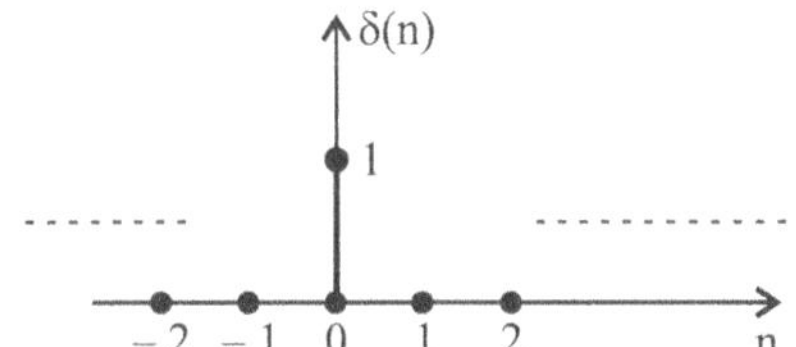

Fig. 1.5 Graphical representation of $\delta(n)$.

(ii) *Unit-step Sequence*: It is denoted by u(n) and is defined as

$$u(n) = \begin{cases} 1, & n \geq 0 \\ 0, & n < 0 \end{cases}$$

This is shown in Fig. 1.6.

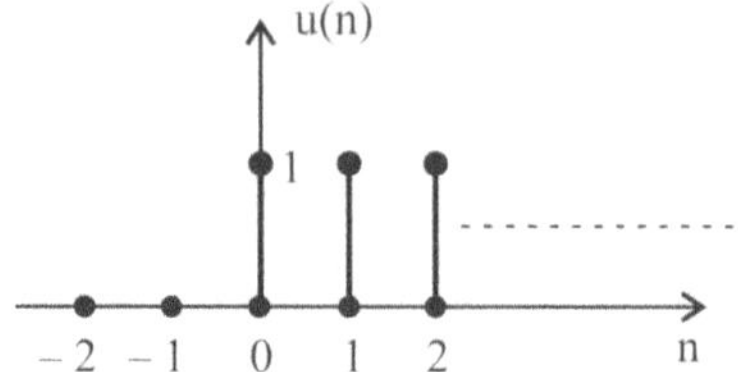

Fig. 1.6 Graphical representation of u(n).

(iii) *Unit-ramp Sequence:* It is denoted by r(n) and is defined as

$$r(n) = \begin{cases} n, & \text{for } n \geq 0 \\ 0, & \text{for } n < 0 \end{cases}$$

This is shown in Fig. 1.7.

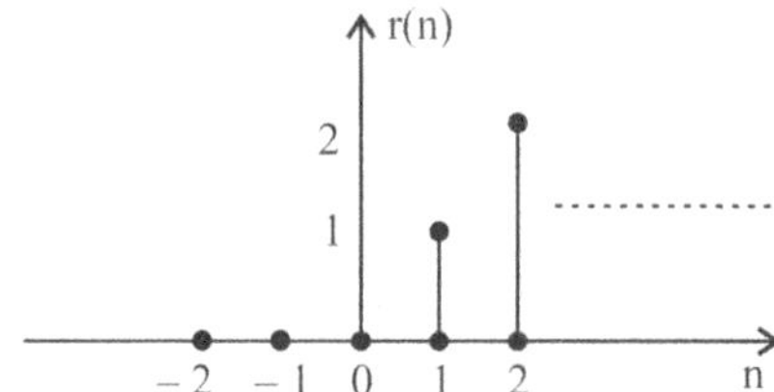

Fig. 1.7 Graphical representation of r(n).

(iv) *Exponential sequence*: It is defined as $x(n) = a^n$ for all values of n.

If the parameter 'a' is real, then x(n) is a real sequence. Fig. 1.8 illustrates this sequence.

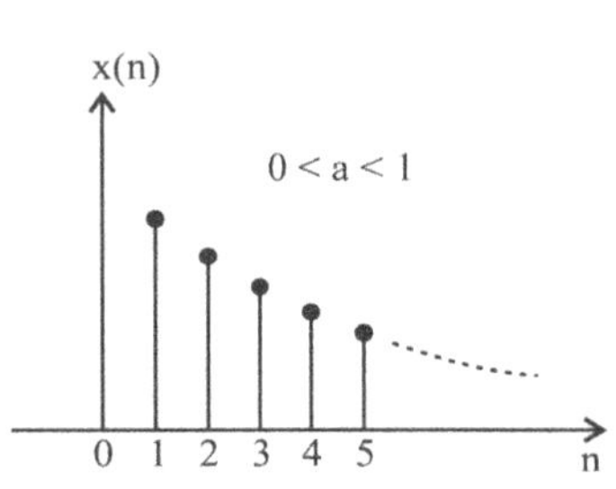

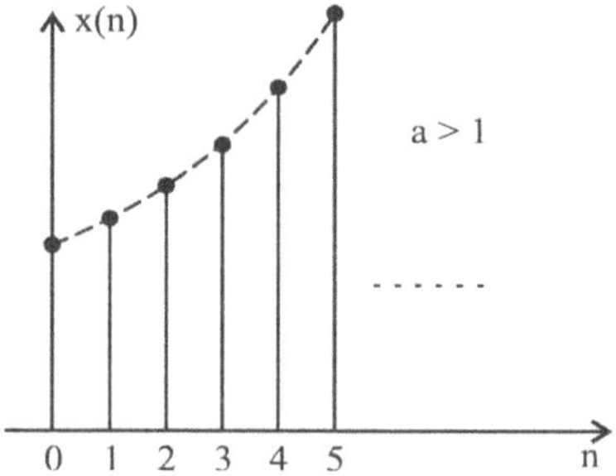

Fig. 1.8 Graphical representation of exponential sequence.

Example 1.1: Let e(n) be an exponential sequence and let x(n) and y(n) denote two arbitrary sequences. Show that

$$\left[e(n) . x(n) \right] * \left[e(n) . y(n) \right] = e(n) . \left[x(n) * y(n) \right]$$

Solution: Given e(n) is an exponential sequence

$$\therefore \qquad e(n) = a^n$$

We know that linear convolution of $x_1(n)$ and $x_2(n)$ as (* is symbol for linear convolution)

$$x_1(n) * x_2(n) = \sum_{k=-\infty}^{\infty} x_1(k).x_2(n-k)$$

In our problem $x_1(n) = e(n) . x(n)$
$$= a^n x(n)$$
and $\qquad x_2(n) = e(n) . y(n)$
$$= a^n y(n)$$
$\therefore \qquad x_1(n) * x_2(n) = \left[a^n x(n)\right] * \left[a^n\ y(n)\right]$

$$= \sum_{k=-\infty}^{\infty} a^k x(k).a^{n-k} y(n-k)$$

$$= \sum_{k=-\infty}^{\infty} x(k).a^k.a^{-k}.a^n y(n-k)$$

$$= a^n \sum_{k=-\infty}^{\infty} x(k) y(n-k)$$

$$\left[a^n\ x(n)\right] * \left[a^n y(n)\right] = a^n . \left[x(n) * y(n)\right]$$

Hence proved

1.2.3 Manipulation of Discrete-Time Signals

Here we study some simple modifications on independent variable (time) and dependent variable (amplitude of signal). Such modifications are required in DSP techniques. *Modification of the Independent variable (time)*: This can be done in three ways.

   (i) Time shifting
   (ii) Folding
   (iii) Time scaling

  (i) *Time Shifting*: A signal can be shifted right side or delayed by replacing n by $n - k$. and is shifted left side or advanced by replacing n by $n + k$, where k is integer and n is a discrete-time index. This is shown in Fig. 1.9.

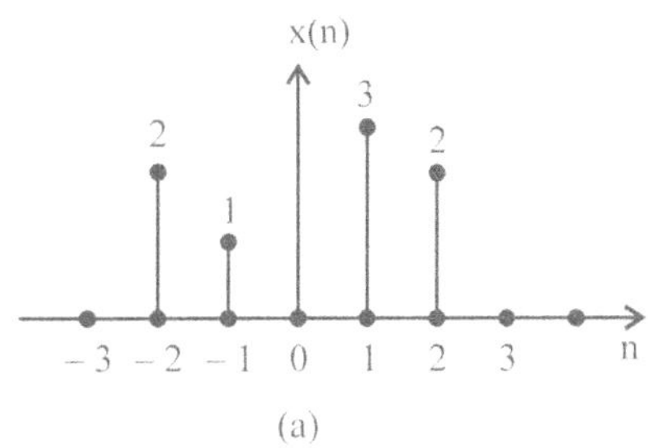

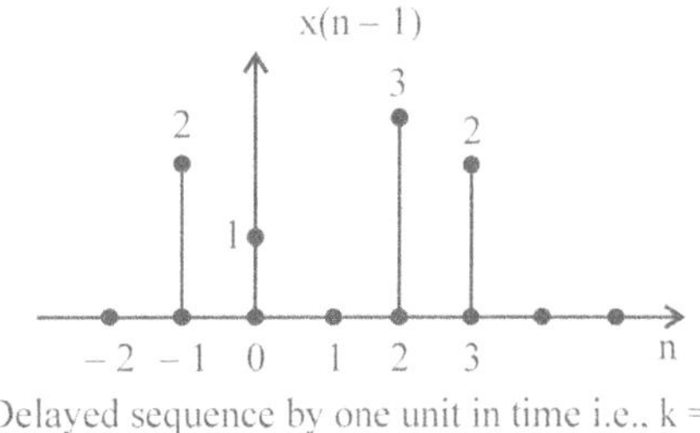

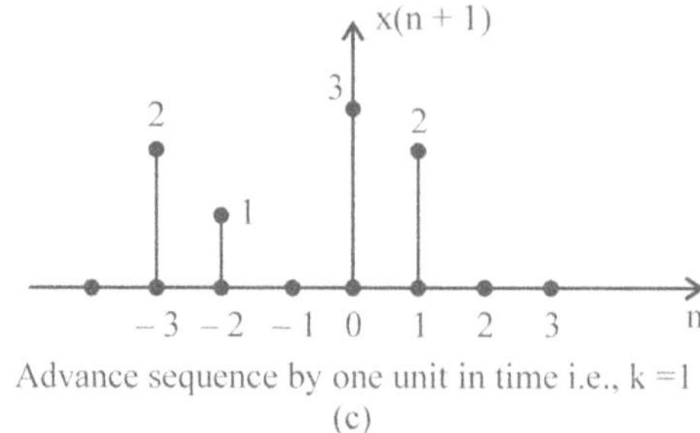

Fig.1.9 (a) Original sequence (b) Delayed by one unit version of original sequence
(c) Advanced sequence by one unit version of original sequence.

(ii) *Folding:* If independent variable (time) n is replaced by $-$ n, then signal folding (mirror image) about the time origin (n = 0) takes place. This is shown in Fig. 1.10.

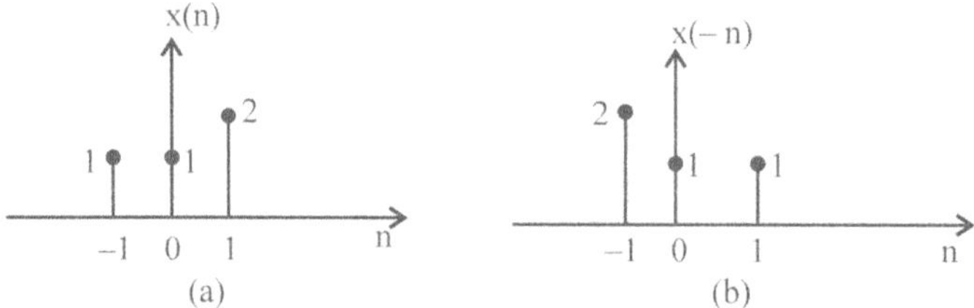

Fig. 1.10 (a) Original sequence (b) Folded version of original sequence.

(iii) *Time Scaling:* Time scaling is performed by replacing independent variable n by kn, where k is an integer. This is shown in Fig. 1.11.

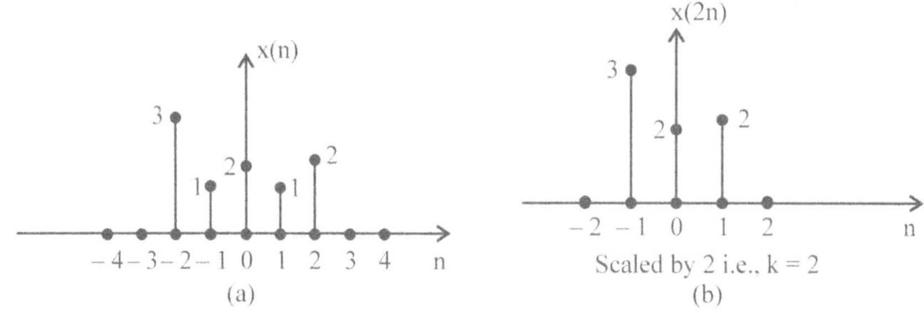

Fig. 1.11 (a) Original sequence (b) Scaled version of original sequence.

Modification of the Dependent Variable (Signal Amplitude):

Signal amplitude can be modified in three ways.

   (i)   Addition of sequences

  (ii)   Multiplication of sequences

 (iii)   Amplitude scaling of sequence

(i)   *Addition of Sequences:* The sum of two discrete time sequences is given by

$$y(n) = x_1(n) + x_2(n), \qquad \infty < n < \infty$$

This is shown in Fig. 1.12 (a)

(ii) *Multiplication of Sequences:* The product of two discrete time sequences is given by

$$y(n) = x_1(n) \cdot x_2(n), \qquad -\infty < n < \infty$$

This is shown in Fig. 1.12 (b)

(iii) *Amplitude Scaling of Sequences:* Amplitude scaling of a signal by a constant A is accomplished by multiplying the value of every signal sample by A.

$$y(n) = A\, x(n), \qquad -\infty < n < \infty$$

where A is real constant quantity

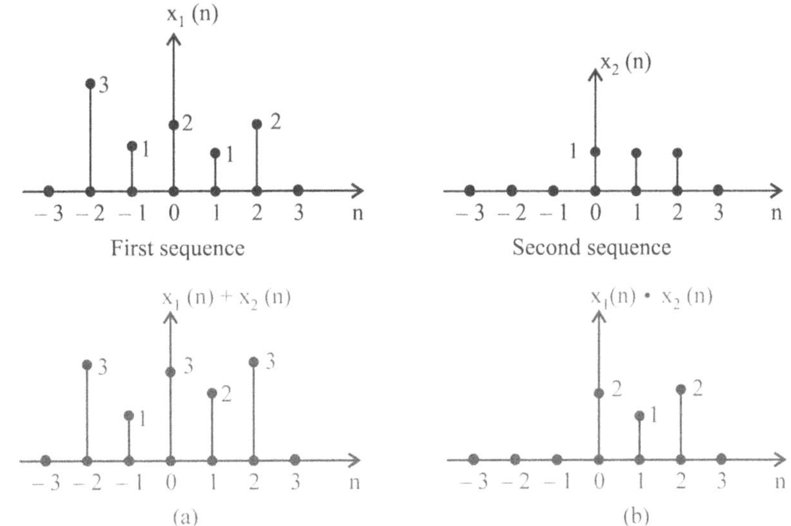

Fig. 1.12 (a) Addition of sequences (b) Multiplication of sequences.

1.2.4 Classification of Discrete-Time Signals

Discrete-time signals are classified based on number of different characteristics as follows:

(i)   Energy signals and power signals

(ii) Periodic signals and Aperiodic signals

(iii) Symmetric (Even) and Antisymmetric (odd) signals.

(i)   *Energy Signals and Power Signals*

The energy E of a signal x(n) is defined as $E = \sum\limits_{n=-\infty}^{\infty} |x(n)|^2$ .....(1.2.1)

Here x(n) may be either complex or real valued signal; E may be finite or infinite.

If E is finite, then x(n) is called energy signal. Many signals that posses infinite energy have a finite average power.

Average power of a signal x(n) is defined as

$$P_{av} = \lim_{N\to\infty} \frac{1}{2N+1} \sum_{n=-N}^{N} |x(n)|^2 \qquad .....(1.2.2)$$

If E is finite, $P_{av} = 0$

If E is infinite, P_{av} may be either finite or infinite if P_{av} is finite, then x(n) is called a power signal.

(ii)   *Periodic Signals and Aperiodic Signals*

If a signal x(n) satisfies the condition x(n) = x(n + N), where N is period then the signals is periodic signal, otherwise it is nonperiodic or aperiodic signal.

Consider a sinusoidal signal $\cos(\omega_0 n + \phi)$, it will be periodic only if $\dfrac{2\pi}{\omega_0}$ is an integral number. If $\dfrac{2\pi}{\omega_0}$ is a rational, then the function will have a period longer than $\dfrac{2\pi}{\omega_0}$.

If $\dfrac{2\pi}{\omega_0}$ is not a rational number, it will not be periodic at all.

The energy of a periodic signal over a single period, say $0 \leq n \leq N - 1$, is finite, but energy of periodic signal for $-\infty \leq n \leq \infty$ is infinite. On the other hand, the average power of the periodic signal is finite.

$\therefore$ Periodic signals are power signals.

(iii)   *Even signals and Odd signals*

A real valued signal x(n) is called symmetric (even)

If   x(n) = x(–n)   .....(1.2.3)

On the other hand, a signal x(n) is called anti symmetric (odd),

       if x(–n) = –x(n)   .....(1.2.4)

Any real sequence can be written as

$$x(n) = x_o(n) + x_e(n)$$

where $x_o(n)$ is odd part of $x(n)$

and $x_e(n)$ is even part of $x(n)$

$x_e(n)$ can be written as

$$x_e(n) = \frac{1}{2}\left[x(n) + x(-n)\right] \qquad\qquad .....(1.2.5)$$

and

$$x_o(n) = \frac{1}{2}\left[x(n) - x(-n)\right]$$

Similarly for complex sequence

Condition for symmetry is $x(n) = x^*(-n)$

where * denotes conjugation

And for Anti-symmetry is $x(n) = -x^*(-n)$

Any complex sequence can be written as

$$x(n) = x_o(n) + x_e(n)$$

where

$$x_e(n) = \frac{1}{2}\left[x(n) + x^*(-n)\right] \qquad\qquad ......(1.2.6)$$

and

$$x_o(n) = \frac{1}{2}\left[x(n) - x^*(-n)\right]$$

Example 1.2: Show that the even and odd parts of a real sequence are, respectively, even and odd sequences.   [JNTU 2002]

Solution: Let $x(n)$ be real sequence, which can be written as

$$x(n) = x_o(n) + x_e(n)$$

where $x_o(n)$ is odd part of $x(n)$ and $x_e(n)$ is even part of $x(n)$

We know that $x_e(n) = \dfrac{1}{2}\left[x(n) + x(-n)\right]$ and $x_o(n) = \dfrac{1}{2}\left[x(n) - x(-n)\right]$

We have to show that even part $(x_e(n))$ of $x(n)$ is a even sequence i.e., it should satisfy $x_e(n) = x_e(-n)$

Consider

$$x_e(n) = \frac{1}{2}\left[x(n) + x(-n)\right]$$

take
$$x_e(-n) = \frac{1}{2}\left[x(-n) + x(n)\right]$$

$$= \frac{1}{2}\left[x(n) + x(-n)\right]$$

$$x_e(-n) = x_e(n)$$

Hence proved.

Similarly we have to show that odd part $(x_o(n))$ of $x(n)$ is a odd sequence i.e., it should satisfy $x_o(n) = -x_o(-n)$

Consider
$$x_o(n) = \frac{1}{2}\left[x(n) - x(-n)\right]$$

take
$$x_o(-n) = \frac{1}{2}\left[x(-n) - x(n)\right]$$

$$= -\frac{1}{2}\left[x(n) - x(-n)\right]$$

$$x_o(-n) = -x_o(n)$$

Hence proved.

1.3 Linear Shift Invariant System, Stability and Causality

Before going to discuss linear shift invariant systems, stability and causality, let us define discrete time system. A discrete time system is a device or an algorithm that operates on a discrete time signal, called the input or excitation, according to some well-defined rule, to produce another discrete time signal called the output or response of the system.

We say that the input signal $x(n)$ is transformed by the system into a signal $y(n)$. These two can be related as

$$y(n) \equiv H[x(n)] \qquad\qquad .....(1.3.1)$$

This is shown graphically in Fig. 1.13.

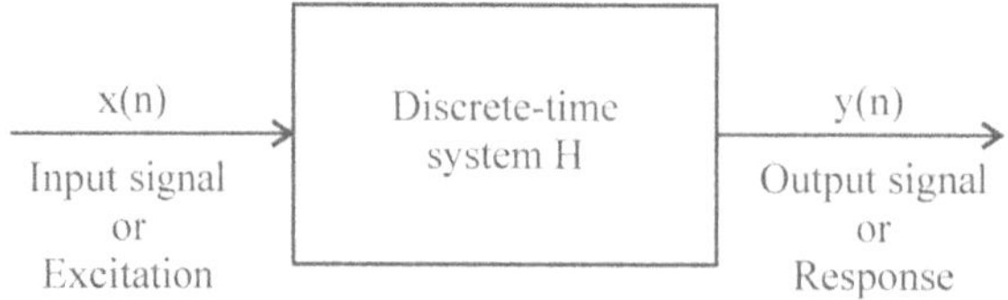

Fig. 1.13 Discrete time system.

1.3.1 Basic Building Blocks of a Discrete Time System

There are three basic building blocks of a discrete time system.

  (i)   Adders or summing element
 (ii)   Multipliers
(iii)   Delay Elements

  (i)   *Adders:* It performs addition of two or more discrete-time signals as shown in Fig. 1.14(a).

 (ii)   *Multipliers*: There are two types of multipliers (a) constant multiplier (b) signal multiplier. A signal multiplier performs multiplication of two or more discrete-time signals as shown in Fig. 1.14(b).

       A constant multiplier performs multiplication of a discrete-time signal with a scalar quantity as shown in Fig 1.14(c).

(iii)   *Delay Elements:* There are two types of delay elements

  (a)   positive delay element, which is indicated by Z^{-1} (b) negative delay elements, which is indicated by Z^{+1} (or) advance element. Positive and negative delay elements provide delay as shown in Fig. 1.14(d) and (e) respectively.

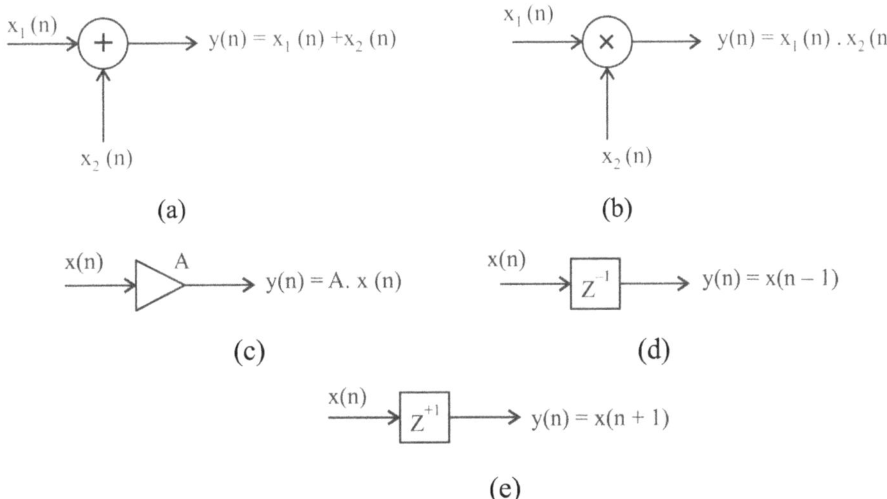

Fig. 1.14 (a) Adder (b) A signal multiplier (c) A constant multiplier (d) A unit delay element (e) A unit advance element.

1.3.2 Classification of Discrete-Time Systems

Discrete-time systems can be classified into

  (i)   Static (memory less) systems and Dynamic (systems with memory)systems.

(ii) Time-invariant and Time-varying systems

(iii) Linear systems and Non-linear systems

(iv) Causal systems and Non-causal systems

(v) Stable systems and Unstable systems.

(i) *Static and Dynamic Systems*: Static systems are also called memory less systems. A discrete-time system is called memory less system if its output at any instant n depends at most on the input at the same instant, but not on past or future values of input samples.

Example:

$$y(n) = A.x(n)$$
$$y(n) = n\ x(n) + Ax^2(n)$$
$$\left.\right\} \text{ Both systems are static systems}$$

On the other hand, output of a system which depends on past or future samples of the input signal is called dynamic system. It is also called a system with memory. These systems require memory for storage for future and past samples of input signal.

Example:

$$y(n) = x(n) + x(n+1) + x(n-1)$$

is a dynamic system.

(ii) *Time-invariant and Time-varying systems*: A system is called time-invariant if its input-output characteristics do not change with time.

If the response to a delayed input, and the delayed response are equal then the system is called time-invariant system (or) shift invariant system.

The response to a delayed input is denoted by $y(n,\ k)$ and the delayed response is denoted by $y(n-k)$. If both responses $y(n,k)$ and $y(n-k)$ are equal then the system is called time-invariant system. If both responses are not equal then the system is called time-varying system.

Example 1.3: Check the following system for Time-invariance
$$y(n) = x(n) + n\ x(n-1)$$

Solution: The response to a delayed input is
$$y(n,\ k) = x(n-k) + n\ x(n-k-1)$$

The delayed response is
$$y(n-k) = x(n-k) + (n-k)x(n-k-1)$$

both responses are not equal

i.e., $$y(n,k) \neq y(n-k)$$

Therefore the given system is not a Time-invariant system. It is a Time-varying system.

Example 1.4: Check the following system for Time-invariance
$$y(n) = x(n) + x(n-1)$$

Solution: The response to a delayed input is
$$y(n, k) = x(n-k) + x(n-1-k)$$

The delayed response is
$$y(n-k) = x(n-k) + x(n-k-1)$$

both responses are equal. Hence given system is a Time-invariant system.

(iii)   *Linear and Non-linear Systems*: A system which satisfies superposition principle is called a linear system. A system which does not satisfy superposition principle is termed as a non-linear system.

Superposition principle is stated as

"Response of the system to a weighted sum of input signals be equal to the corresponding weighted sum of responses of the system to each of the individual input signals".

A system is linear if and only if
$$H[ax_1(n) + bx_2(n)] = aH[x_1(n)] + b\, H[x_2(n)] \qquad .....(1.3.2)$$

where $x_1(n)$ and $x_2(n)$ are arbitrary input signals and a and b are arbitrary constants.

Example 1.5: Check the following systems for Linearity.

(i) $y(n) = x(n^2)$ (ii) $y(n) = e^{x(n)}$

Solution:

(i)   The corresponding outputs for two discrete-time sequences $x_1(n)$ and $x_2(n)$ are
$$y_1(n) = x_1(n^2)$$
$$y_2(n) = x_2(n^2)$$

A linear combination of two input sequences results in the output
$$y_3(n) = H[x_3(n)] = H[a\, x_1(n) + b\, x_2(n)]$$
$$= x_3(n^2) = a\, x_1(n^2) + b\, x_2(n^2) \qquad .....(i)$$

A linear combination of the two outputs results in the output
$$a\, y_1(n) + b\, y_2(n) = a\, x_1(n^2) + b\, x_2(n^2) \qquad .....(ii)$$

Since both outputs are equal, the system is linear.

(ii)   The corresponding outputs for two discrete-time sequences $x_1(n)$ and $x_2(n)$ are

$$y_1(n) = e^{x_1(n)}$$

$$y_2(n) = e^{x_2(n)}$$

A linear combination of $x_1(n)$ and $x_2(n)$ results in the output.

$$y_3(n) = H\,[x_3(n)] = H[a\,x_1(n) + b\,x_2(n)] = e^{ax_1(n)+bx_2(n)} \qquad .....(iii)$$

Linear combination of the two outputs results in the output

$$a\,y_1(n) + b\,y_2(n) = ae^{x_1(n)} + b\,e^{x_2(n)} \qquad .....(iv)$$

Here both outputs i.e., equations (iii) and (iv) are not equal, hence the system is non-linear.

(iv) *Causal and Non-causal Systems*: A system whose present output depends only on present and past inputs, but not on future inputs is called a causal system. If a system response depends on future values, then it is a non-causal system.

Example:

    (i) $y(n) = x(n) + x(n+1)$

Since response depends upon a future value $(x(n+1))$, it is a non-causal system.

    (ii) $y(n) = x(n) + x(n-1)$

Since response does not depend upon future values, it is a causal system.

    (iii) $y(n) = x(-n)$

Take $n = -1$, then $y(-1) = x(1)$, which is a future value i.e., it depends on future values. Hence the system is a non causal system.

    (iv) $y(n) = x(n^2)$ and $y(n) = x(2n)$

Take $n = 2$, then $y(2) = x(4)$ in both the systems which is a future value. Hence both systems are non causal systems.

(v) *Stable and Unstable Systems:* A system which produces bounded (finite) output for a bounded (finite) input is called as stable system, otherwise it is called as an unstable system. Examples will be discussed in the section 1.3.5.

1.3.3 Representation of Discrete Time Signal as Summation of Impulses

Graphical representations of impulse sequences and its shifted versions are shown in Fig. 1.15 (a) (b) and (c).

Let us consider a sequence

$$x(n) = \left\{1, \underset{\uparrow}{1}, 2, 3\right\}$$

which is shown graphically in fig 1.15 (d)

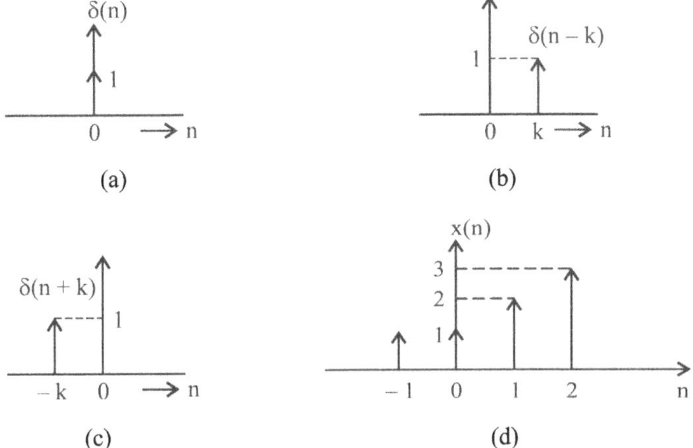

Fig. 1.15 (a) Impulse sequence (b) Shifted towards right (c) Shifted towards left
(d) A general sequence.

product of $x(n)$ and $\delta(n)$ gives $x(0)$

i.e., $x(n) \, \delta(n) = x(0) \rightarrow x(0) \, \delta(n) = x(0)$

similarly $x(n) \, \delta(n-1) = x(1) \rightarrow x(1) \, \delta(n-1) = x(1)$

$\qquad\qquad\quad x(n) \, \delta(n-2) = x(2) \rightarrow x(2) \, \delta(n-2) = x(2)$

$\qquad\qquad\quad x(n) \, \delta(n+1) = x(-1) \rightarrow x(-1) \, \delta(n+1) = x(-1)$

$\therefore$ This can be written as

$$x(n) = \sum_{k=-\infty}^{\infty} x(k) \, \delta(n-k) \qquad\qquad .....(1.3.3)$$

Example: Represent the sequence $x(n) = \left\{2, \underset{\uparrow}{3}, 5\right\}$ as sum of impulse sequences

Solution is $x(n) = 2 \, \delta(n+1) + 3 \, \delta(n) + 5 \, \delta(n-1)$

1.3.4 Response of Linear Time Invariant (LTI) System

Fig. 1.16 shows an LTI system with an excitation $x(n)$ and response $y(n)$.

$$x(n) \longrightarrow \boxed{H} \longrightarrow y(n)$$

Fig. 1.16 A discrete time LTI system.

Unit sample response (or) Impulse response: It is defined as "response of a system when input signal is impulse sequence", impulse response is denoted with h (n)

i.e., when $x(n) = \delta(n) \rightarrow y(n) = h(n)$

Since it is a time invariant system, when impulse sequence is delayed by k, response i.e., h (n) should be delayed by k.

$\therefore$ $\delta(n-k) \rightarrow h(n-k)$

This also can be written as $h(n-k) = H[\delta(n-k)]$

Response of an LTI (Linear Time Invariant) system is

$$y(n) = H[x(n)] \qquad\qquad \dots\dots(1.3.4)$$

we known that x(n) can be represented as sum of impulse sequence as in eqn. (1.3.3).

i.e., $$x(n) = \sum_{k=-\infty}^{\infty} x(k)\delta(n-k)$$

Substitute this in the equation (1)

$\therefore$ $$y(n) = H\left[\sum_{k=-\infty}^{\infty} x(k)\delta(n-k)\right]$$

$$= \sum_{k=-\infty}^{\infty} x(k)\, H\left[\delta(n-k)\right]$$

$$y(n) = \sum_{k=-\infty}^{\infty} x(k)h(n-k) \qquad\qquad \dots\dots(1.3.5)$$

which is an equation for convolution of x (n) and h (n).

Hence the response of an LTI system is convolution of input sequence and impulse response.

Example 1.6: Consider a discrete linear shift invariant system with unit sample response h(n). If the input x(n) is a periodic sequence with period N i.e., $x(n) = x(n + N)$, show that the output y(n) is also a periodic sequence with period N.

Solution: Given $x(n) = x(n + N)$

We known the response of an LTI (or) Liner shift invariant system as

$$y(n) = \sum_{k=-\infty}^{\infty} x(k)h(n-k)$$

also $$y(n) = \sum_{k=-\infty}^{\infty} h(x)x(n-k)$$

consider $$y(n+N) = \sum_{k=-\infty}^{\infty} h(k)x(n+N-k)$$

$$\text{given } x(n) = x(n+N)$$

delay it by k

then $\quad x(n-k) = x(n+N-k)$

$$\therefore \quad y(n+N) = \sum_{k=-\infty}^{\infty} h(x)x(n-k)$$

$$= y(n)$$

Hence the output is also periodic when input is periodic

1.3.5 Stability of an LTI System

Let x(n) is input sequence, assume that it is finite with a value M_x.

Response of an LTI system is

$$y(n) = \sum_{k=-\infty}^{\infty} x(k)h(n-k)$$

$$= \sum_{k=-\infty}^{\infty} h(k)x(n-k)$$

Take absolute on both sides

$$|y(n)| \leq \sum_{k=-\infty}^{\infty} |h(k)| \, |x(n-k)|$$

$$\leq M_x \sum_{k=-\infty}^{\infty} |h(k)| \qquad\qquad .....(1.3.6)$$

According to definition for stability, for a finite input sequence, system should produce finite output.

From eqn (1.3.6), to get finite output, $\displaystyle\sum_{k=-\infty}^{\infty} |h(k)|$ must be finite

$$\therefore \quad \sum_{k=-\infty}^{\infty} |h(k)| < \infty$$

Hence an LTI system is stable if its impulse response (or) unit sample response is absolutely summable.

Example 1.7: Test the stability of the following systems

$$\text{(i)} \quad y(n) = x(-n-2) \qquad\qquad \text{(ii)} \quad y(n) = n\,x(n)$$

Solution: We know that when $x(n) = \delta(n)$, the output $y(n) = h(n)$

(i) $\therefore$ $h(n) = \delta(-n-2)$

$$n = 0 \rightarrow \delta(-2) = 0 = h(0); \ n = -1 \rightarrow \delta(-1) = 0 = h(-1)$$

$$n = 1 \rightarrow \delta(-3) = 0 = h(1); \ n = -2 \rightarrow \delta(0) = 1 = h(-2)$$

$$\vdots \qquad\qquad\qquad \vdots$$

$$n = \infty \rightarrow \delta(\infty) = 0 = h(\infty); \ n = -\infty \rightarrow \delta(-\infty) = 0 = h(-\infty)$$

$\therefore$

$$\sum_{n=-\infty}^{\infty} |h(n)| = 0 + 0 + \cdots\cdots + 1 + 0 + \cdots\cdots + 0 = 1$$

$\therefore$

$$\sum_{n=-\infty}^{\infty} |h(n)| < \infty$$

$\therefore$ System is stable

(ii) $h(n) = n \, \delta(n)$

$$n = 0 \rightarrow h(0) = 0(1) = 0; \ n = -1 \rightarrow h(-1) = -1(0) = 0$$

$$n = 1 \rightarrow h(1) = 1(0) = 0; \ n = -2 \rightarrow h(-2) = -2(0) = 0$$

$$\vdots \qquad\qquad\qquad \vdots$$

$$n = \infty \rightarrow h(\infty) = \infty(0) = 0; \ n = -\infty \rightarrow h(-\infty) = -\infty(0) = 0$$

$\therefore$

$$\sum_{n=-\infty}^{\infty} |h(n)| = 0 + 0 + \cdots\cdots + 0 = 0$$

System is stable

Example 1.8: Determine the range of values of the parameter 'a' for which the LTI system with impulse response $h(n) = a^n \, u(n)$ is stable.

Solution: Condition for a system to be stable is

$$\sum_{n=-\infty}^{\infty} |h(n)| < \infty$$

$\therefore$

$$\sum_{n=0}^{\infty} |a^n| = 1 + |a| + |a|^2 + \cdots\cdots$$

This infinite series converges to

$$= \frac{1}{1-|a|} \ \text{if} \ |a| < 1$$

$\therefore$ Range of values of parameter 'a' is $|a| < 1$

Example 1.9: Determine the range of values of 'a' and 'b' for which LTI system with

impulse response $h(n) = \begin{cases} a^n, & n \geq 0 \\ b^n, & n < 0 \end{cases}$ is stable

Solution:
$$\sum_{n=-\infty}^{\infty} |h(n)| = \sum_{n=-\infty}^{-1} b^n + \sum_{n=0}^{\infty} a^n$$

Put $n = -1$ in first series

$$= \sum_{n=1}^{\infty} \frac{1}{b^n} + \sum_{n=0}^{\infty} a^n$$

$$= \left(\frac{1}{b} + \frac{1}{b^2} + \cdots \right) + \left(1 + a + a^2 + \cdots \right)$$

$$= \frac{1}{b} \left(1 + \frac{1}{b} + \frac{1}{b^2} + \cdots \right) + \left(1 + a + a^2 + \cdots \right)$$

$$= \frac{1}{b} \left(\frac{1}{1 - \frac{1}{b}} \right) + \frac{1}{1-a} \quad \text{if} \quad \begin{cases} \dfrac{1}{b} < 1 \text{ and } a < 1 \\ b > 1 \text{ and } a < 1 \end{cases}$$

$\therefore$ Range of values of 'a' and 'b' are $b > 1$ and $a < 1$

Example 1.10: A unit sample response of a linear system is given by $h(n) = (n+b)a^n$, $n \geq 0$

$= 0,\ n < 0$

For what values of 'a' and 'b' the system will be stable?

Solution:
$$\sum_{n=0}^{\infty} (n+b)a^n = \sum_{n=0}^{\infty} n\, a^n + \sum_{n=0}^{\infty} b\, a^n$$

$$= \left(0 + a + 2a^2 + 3a^3 + \cdots \right) + b\left(1 + a + a^2 + \cdots \right)$$

$$= a\left(1 + 2a + 3a^2 + \cdots \right) + b\frac{1}{1-a}$$

$$= a\frac{1}{(1-a)^2} + \frac{b}{1-a} \quad \text{if } a < 1 \text{ and } b \text{ must finite to become series}$$

finite

$\therefore$ Values of 'a' and 'b' are a < 1 and b < ∞

1.4 Linear-Constant Coefficient Difference Equations

We know that continuous time systems are described by differential equations. But discrete-time systems are described by difference equations.

Input-output relation of Nth order discrete-time system can be written as

$$\sum_{k=0}^{N} a_k\, y(n-k) = \sum_{k=0}^{M} b_x\, x(n-k) \qquad\qquad .....(1.4.1)$$

where   y(n) is output

                  x(n) is input

and a_k and b_k are constant coefficients. Order of the system is determined by L.H.S summation since input-output relation is linear with constant coefficients, this equation is called "Linear-constant coefficient difference equation" for N^{th} order.

   There are two methods by which difference equations can be solved

1. **Direct Method:** This method is directly applicable in the time domain. We are not discussing this method

2. **Indirect Method:** It is also called z-transform method. This method will be discussed in the chapter 3.

1.5 Frequency Domain Representation of Discrete-Time Systems and Signals

1.5.1 Frequency Domain Representation of Discrete-Time System

System function (or) transfer function of a system can be obtained by taking Z-transform (for Z-transform refer chapter 3) of impulse response h(n)

$$\text{i.e., system function} = H(z) = \sum_{n=-\infty}^{\infty} h(n)z^{-n} \qquad\qquad ....(1.5.1)$$

system function can also be defined as ratio of z-transform of response to z-transform of input with zero initial conditions.

$$\text{i.e.,} \qquad H(z) = \frac{Y(z)}{X(z)} \qquad\qquad ....(1.5.2)$$

   Frequency response of a system can be obtained just by putting $z = e^{j\omega}$ in equation (1.5.1).

i.e., $\qquad H(e^{j\omega}) = H(\omega) = \sum\limits_{n=-\infty}^{\infty} h(n)\, e^{-j\omega n}$ $\qquad\qquad$ ....(1.5.3)

Magnitude spectrum of a system is obtained by taking modulus of $H(e^{j\omega})$ i.e., $|H(e^{j\omega})|$
Phase spectrum of a system is obtained by

$$\theta = \tan^{-1}\left(\frac{H_i(\omega)}{H_r(\omega)}\right) \qquad\qquad ....(1.5.4)$$

where $\qquad\qquad H_i(\omega) = $ Imaginary part of $H(\omega)$

$\qquad\qquad\qquad H_r(\omega) = $ Real part of $H(\omega)$

1.5.2 Frequency Domain Representation of Discrete-Time Signals

Let us consider any discrete-time sequence say $x(n)$

Frequency domain representation of this sequence can be obtained by taking z-transform of $x(n)$ and putting $z = e^{j\omega}$.

i.e., $\qquad\qquad X(z) = \sum\limits_{n=-\infty}^{\infty} x(n)\, z^{-n}$

where $x(z)$ is z-transform of $x(n)$

put $\qquad\qquad z = e^{j\omega}$

$\therefore \qquad\qquad X(e^{j\omega}) = X(\omega) = \sum\limits_{n=-\infty}^{\infty} x(n)\, e^{-j\omega n}$

which is frequency domain representation of $x(n)$.

Example 1.11: An LTI system has unit sample response $h(n) = u(n) - u(n{-}N)$. Find the amplitude and phase spectra.

Solution:

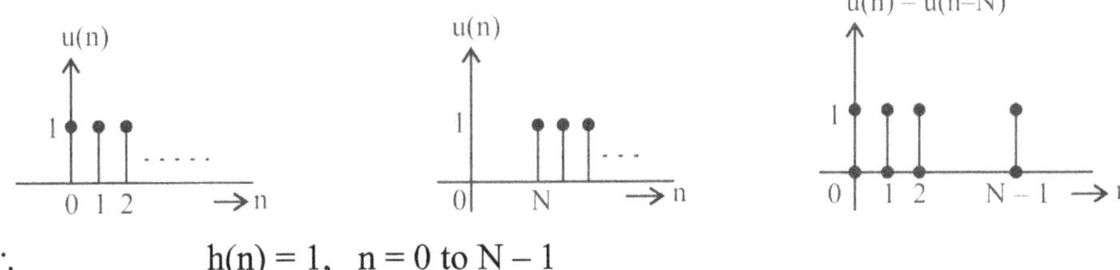

$\therefore \qquad\qquad h(n) = 1, \quad n = 0 \text{ to } N - 1$

from the figures shown

$$H(z) = \sum\limits_{n=0}^{N-1} 1.\, z^{-n}$$

$$= \sum_{n=0}^{N-1} z^{-n}$$

$$= \frac{1 - z^{-N}}{1 - z^{-1}}$$

$$\boxed{\because \sum_{n=0}^{N-1} a^n = \frac{1 - a^N}{1 - a} \qquad \text{Finite Geometric Series}}$$

Frequency response, $z = e^{j\omega}$

$$H(\omega) = \frac{1 - e^{-j\omega N}}{1 - e^{-j\omega}}$$

$$= \frac{e^{\frac{-j\omega N}{2}} \left(e^{\frac{j\omega N}{2}} - e^{\frac{-j\omega N}{2}} \right)}{e^{\frac{-j\omega}{2}} \left(e^{\frac{j\omega}{2}} - e^{\frac{-j\omega}{2}} \right)}$$

$$= e^{\frac{-j\omega}{2}(N-1)} \frac{\sin\left(\dfrac{\omega N}{2}\right)}{\sin\left(\dfrac{\omega}{2}\right)}$$

Magnitude spectra is

$$|H(\omega)| = \frac{\sin\left(\dfrac{\omega N}{2}\right)}{\sin\left(\dfrac{\omega}{2}\right)}$$

Phase spectra is

$$\angle H(\omega) = \theta = -\frac{\omega}{2}(N-1)$$

Review Questions

1. Pick the signal which varies with single independent variable.

 (a) Speech (b) Image (c) $S = 2x + 6xy$ (d) $S = 3x^2 y$ *Ans*: [b]

2. Pick the signal which varies with two independent variables.

 (a) Image (b) $S = 2t^2$ (c) Speech (d) $S = 2t + 3t^3$

                         Ans: [c]

3. Which of the following is not an analog system?

    (a) Frequency analyzers   (b) Analog filters

    (c) Frequency multipliers   (d) Programmable machines

    Ans: [d]

4. Which of the following is not a block in basic DSP system?

    (a) ADC   (b) Digital signal processor

    (c) Analog signal processor   (d) DAC   *Ans*: [c]

5. Which of the following is not a part of an analog-to-digital converter?

    (a) Sampler   (b) Decoder   (c) Encoder   (d) Quantizer *Ans*: [b]

6. The following is the disadvantage of digital systems

    (a) Cost   (b) Speed   (c) Transportability (d) Noise immunity

    Ans: [b]

7. Discrete-time signal is

    (a) Discrete both in time and amplitude

    (b) Discrete in time and continuous in amplitude

    (c) Continuous in time and discrete in amplitude

    (d) All the above   *Ans*: [b]

8. Digital signal is

    (a) Discrete both in time and amplitude

    (b) Continuous in time and discrete in amplitude

    (c) Discrete in time and continuous in amplitude

    (d) All the above   *Ans*: [a]

9. Discrete-time signal can be represented by

    (a) Graphical method   (b) Functional method

    (c) Sequence method   (d) All the above   *Ans*: [d]

10. The other name of unit impulse sequence

    (a) Unit-sample sequence   (b) Unit-step sequence

    (c) Unit ramp sequence   (d) All the above   *Ans*: [a]

11. Unit step sequence is defined as

$$\text{(a)} \quad x(n) = \begin{cases} 1 & n = 0 \\ 0 & n \neq 0 \end{cases} \qquad \text{(b)} \quad x(n) = \begin{cases} 1 & n \geq 0 \\ 0 & n < 0 \end{cases}$$

$$\text{(c)} \quad x(n) = \begin{cases} n & n \geq 0 \\ 0 & n < 0 \end{cases} \qquad \text{(d)} \quad x(n) = \begin{cases} 0 & n \geq 0 \\ 1 & n < 0 \end{cases} \qquad \textit{Ans}: [b]$$

12. Unit impulse is defined as

   (a)　$x(n) = \begin{cases} 1 & n = 0 \\ 0 & n \neq 0 \end{cases}$　　　　(b)　$x(n) = \begin{cases} 1 & n \geq 0 \\ 0 & n < 0 \end{cases}$

   (c)　$x(n) = \begin{cases} 1 & n \geq 0 \\ 0 & n < 0 \end{cases}$　　　　(d)　$x(n) = \begin{cases} 0 & n \geq 0 \\ 1 & n < 0 \end{cases}$　　　*Ans:* [a]

13. Unit ramp is defined as

   (a)　$x(n) = \begin{cases} 1 & n = 0 \\ 0 & n \neq 0 \end{cases}$　　　　(b)　$x(n) = \begin{cases} 1 & n \geq 0 \\ 0 & n < 0 \end{cases}$

   (c)　$x(n) = \begin{cases} 1 & n \geq 0 \\ 0 & n < 0 \end{cases}$　　　　(d)　$x(n) = \begin{cases} 0 & n \geq 0 \\ 1 & n < 0 \end{cases}$　　　*Ans:* [c]

14. Exponential sequence a^n is decaying when

   (a)　$0 < a < 1$　　(b)　$a > 1$　　(c)　$a < 1$　　(d)　$-1 < a < 1$　*Ans:* [a]

15. Discrete-time signal can be modified by modifying independent variable using the following methods.

   (a)　Time shifting　　　　　　(b)　Folding

   (c)　Time-scaling　　　　　　(d)　All the above　　　　　　*Ans:* [d]

16. A signal $x(n)$ can be shifted right side by replacing n with

   (a)　$n + k$　　(b)　$n - k$　　(c)　$n \div k$　(d)　nk　　　*Ans:* [b]

17. A signal $x(n)$ can be shifted left side by replacing n with

   (a)　$n + k$　　(b)　$n - k$　　(c)　$n \div k$　(d)　nk　　　*Ans:* [a]

18. A signal $x(n)$ will be folded if n is replaced by

   (a)　$-n$　　(b)　$\div n$　　(c)　$n + k$　(d)　$n - k$　　　*Ans:* [a]

19. Discrete-time signal can be modified by modifying dependant variable using the following methods

   (a)　Addition of sequences　　　　　　(b)　Multiplication of sequences

   (c)　Amplitude scaling of sequence　　　(d)　All the above　　*Ans:* [d]

20. Energy of a signal is

   (a)　$E = \sum_{n=-\infty}^{\infty} |x(n)|^2$　　　　　　(b)　$E = \dfrac{1}{N} \sum_{n=-\infty}^{\infty} |x(n)|^2$

   (c)　$E = \underset{N \to \infty}{\text{Lim}} \dfrac{1}{2N+1} \sum_{n=-N}^{N} |x(n)|^2$　　(d)　$E = \dfrac{1}{2N} \sum_{n=-N}^{N} |x(n)|^2$　*Ans:* [a]

21. Average power of a signal is

   (a) $P_{av} = \underset{N}{Lim} \dfrac{1}{2N+1} \sum\limits_{n=-N}^{N} |x(n)|^2$ (b) $P_{av} = \dfrac{1}{N} \sum\limits_{n} |x(n)|^2$

   (c) $P_{av} = \sum\limits_{n} |x(n)|^2$ (d) $P_{av} = \dfrac{1}{N} \sum\limits_{n=-N}^{N} |x(n)|^2$ *Ans*: [a]

22. Energy singal's average power is
   (a)   Infinite   (b)   Zero
   (c)   Cannot be determined   (d)   None   *Ans*: [b]

23. Power signal's is energy is
   (a)   Infinite   (b)   Zero
   (c)   Cannot be determined   (d)   None   *Ans*: [a]

24. A periodic signal satisfies the condition
   (a) $x(n) = x(n+N)$ (b) $x(n) = x(n-N)$
   (c) $x(n) = x(-n)$ (d) $x(n) = -x(-n)$ *Ans*: [a]

25. A sinusoidal signal $\cos \omega_0 n$ will be periodic only if $\dfrac{2\pi}{\omega_0}$ is an/a

   (a)   Integer   (b)   Irrational   (c)   Infinite   (d)   Zero

                                                              Ans: [a]

26. A real signal x(n) is called symmetric if
   (a) $x(n) = x(n+N)$ (b) $x(n) = x(-n)$
   (c) $x(n) = -x(-n)$ (d) $x(n) \ne x(n+N)$ *Ans*: [b]

27. A real signal x(n) is called anti symmetric if
   (a) $x(n) = x(n+N)$ (b) $x(n) = x(-n)$
   (c) $x(n) = -x(-n)$ (d) $x(n) \ne x(n+N)$ *Ans*: [c]

28. A complex signal x(n) is called symmetric if
   (a) $x(n) = x(n+N)$ (b) $x(n) = x^*(-n)$
   (c) $x(n) = x(-n)$ (d) $x(n) = -x(-n)$ *Ans*: [b]

29. A complex signal x(n) is called odd signal if
   (a) $x(n) = -x^*(-n)$ (b) $x(n) = -x(-n)$
   (c) $x(n) = x^*(-n)$ (d) $x(n) = x(-n)$ *Ans*: [a]

30. The other name of LTI system is

    (a) LTV   (b)   LSI   (c) TV   (d) None

    Ans: [b]

31. Positive delay element is

    (a) Z^{-1} (b) Z^{+1} (c) Z   (d) None

    Ans: [a]

32. Negative delay element is

    (a) Z^{+1} (b) Z^{-1} (c) $\dfrac{1}{Z}$ (d) None

    Ans: [a]

33. The other name of static system

    (a) Causal   (b) Memory less

    (c) Time variant   (d) Stable   *Ans*: [b]

34. The other name of Dynamic system

    (a) Causal   (b) Memory less

    (c) Time variant   (d) System with memory   *Ans*: [d]

35. Linear system should satisfy

    (a) Stability condition   (b) $y(n-k) \leftrightarrow x(n-k)$

    (c) Superposition principle   (d) None   *Ans*: [c]

36. Time invariant system should satisfy

    (a) Stability condition   (b) $y(n-k) \leftrightarrow x(n-k)$

    (c) Superposition principle   (d) None   *Ans*: [b]

37. Causal system response depends upon

    (a) Future input

    (b) Present input, past input, future input

    (c) Past input, future input

    (d) Present input, past input   *Ans*: [d]

38. Pick a causal system

    (a) $y(n) = x(-n)$ (b) $y(n) = x(2n)$

    (c) $y(n) = x(n)$ (d) $y(n) = x(n^2)$ *Ans*: [c]

39. Pick non causal system
    (a) $\quad y(n) = x(-n)$
    (b) $\quad y(n) = x(n) + x(n-1)$
    (c) $\quad y(n) = 2x(n)$
    (d) $\quad y(n) = x(n-1)$ *Ans*: [a]

40. Pick non causal system
    (a) $\quad y(n) = x(n)$
    (b) $\quad y(n) = x(n) + x(n+1)$
    (c) $\quad y(n) = x(n) + x(n-1)$
    (d) $\quad y(n) = x(n-2)$ *Ans*: [b]

41. A system is said to be unstable if it gives ….. output, for a finite input
    (a) Finite   (b) Zero   (c) Infinite   (d) One
        Ans: [c]

42. Condition for a system to be stable is
    (a) Impulse response should be absolutely summable
    (b) $\displaystyle\sum_{n=-\infty}^{\infty} |h(n)| = \infty$
    (c) $\displaystyle\sum_{n=-N}^{N} |h(n)| < \infty$
    (d) $\displaystyle\sum_{n=-N}^{N} |h(n)| = 0$ *Ans*: [a]

43. Represent the sequence $x(n) = \{1, 2, 3, 4\}$ as sum of impulses
    (a) $\delta(n) + 2\delta(n-1) + 3\delta(n-2) + 4\delta(n-3)$
    (b) $\delta(n+1) + 2\delta(n) + 3\delta(n-1) + 4\delta(n-2)$
    (c) $\delta(n-1) + 2\delta(n) + 3\delta(n+1) + 4\delta(n+2)$
    (d) $\delta(n+1) + \delta(n) + \delta(n-1) + \delta(n-2)$ *Ans*: [b]

44. Response of an LTI systems is
    (a) Multiplication of input and impulse response
    (b) Subtraction of input and impulse response
    (c) Addition of input and impulse response
    (d) Convolution of input and impulse response
        Ans: [d]

45. Discrete-time systems are described by
    (a) Differential equations
    (b) Difference equations
    (c) Linear equations with variable coefficients
    (d) None   *Ans*: [b]

46. Continuous-time systems are described by
    (a)   Differential equations
    (b)   Difference equations
    (c)   Linear equations with variable coefficient
    (d)   None   *Ans*: [a]

47. Solution of difference equations can be obtained by
    (a)   Laplace transform   (b)   Fourier transform
    (c)   Z-transform   (d)   None   *Ans*: [c]

48. Solution of differential equations can be obtained by
    (a)   Laplace transform   (b)   Fourier transform
    (c)   Z-transform   (d)   None   *Ans*: [a]

49. Relation between system function and impulse response is
    (a) $H(z) = h(n)$

    (b) $H(z) = \sum\limits_{n=-\infty}^{\infty} h(n)z^{-n}$

    (c) $H(z)$ = Laplace Transform of h (n)
    (d)   None   *Ans*: [b]

50. Finite Geometric series $\sum\limits_{n=0}^{N-1} a^n$ is

    (a) $\dfrac{1-a^{N-1}}{1-a}$ (b) $\dfrac{1-a^{N}}{1-a}$

    (c) $\dfrac{1-a^{N+1}}{1-a}$ (d) $\dfrac{1-a}{1-a^{N}}$ *Ans*: [b]

Chapter 2

Discrete Fourier Series

2.1 Discrete Fourier Series

A Fourier series is an expansion of a periodic continuous-time signal $x(t)$ in terms of an infinite sum of **sines** and **cosines** or in terms of exponentials. Fourier series make use of the orthogonality relationships of the sine and cosine or exponential functions. The computation and study of Fourier series is known as harmonic analysis and is extremely useful as a way to break up an *arbitrary* periodic signal into a set of simple terms that can be solved individually, and then recombined to obtain the solution to the original problem or an approximation to it to whatever accuracy is desired or practical. A periodic continuous-time signal $x(t)$ can be represented with Fourier series as

$$x(t) = \sum_{k=-\infty}^{\infty} X(k)\, e^{jk\omega_0 t} \qquad .....(2.1.1)$$

where $X(k)$ are Fourier coefficients

Sample the continuous-time (analog) signal with sampling period T, which results a discrete-time signal $x(n)$ with a period N.

put $t = nT$ in eq. (2.1.1)

$$\therefore \qquad x(nT) = x(n) = \sum_{k=-\infty}^{\infty} X(k)\, e^{jk\omega_0 nT}$$

where

$$\omega_0 = \frac{2\pi}{NT}$$

$$x(n) = \sum_{k=-\infty}^{\infty} X(k)\, e^{j2\pi nk/N} \qquad .....(2.1.2)$$

To see that this sequence is periodic, consider

$$x(n+N) = \sum_{k=-\infty}^{\infty} X(k)\, e^{j2\pi(n+N)k/N}$$

$$= \sum_{k=-\infty}^{\infty} X(k)\, e^{j2\pi\, nk/N} \cdot e^{j2\pi k}$$

$$= \sum_{k=-\infty}^{\infty} X(k)\, e^{j2\pi kn/N} \qquad \because \quad e^{j2\pi k} = 1$$

$$= x(n)$$

$\therefore$ This infinite series contains only N distinct exponential components i.e., e^J_o, $e^{j2\pi/N}$, $e^{j4\pi/N}$, ... , $e^{j2\pi(N-1)/N}$. Hence summation can be written from $k = 0$ to $N - 1$, instead of from $-\infty$ to ∞.

$$\therefore \qquad x(n) = \sum_{k=0}^{N-1} X(k)\, e^{j2\pi kn/N} \qquad\qquad .....(2.1.3)$$

To get equation for X(k), multiply both sides of equation (2.1.3) by $e^{-j2\pi mn/N}$ and sum from $n = 0$ to $N-1$.

$$\therefore \qquad \sum_{n=0}^{N-1} x(n)\, e^{-j2\pi mn/N} = \sum_{n=0}^{N-1} \sum_{k=0}^{N-1} X(k)\, e^{j2\pi n(k-m)/N}$$

interchange summations on right side

$$\sum_{n=0}^{N-1} x(n)\, e^{-j2\pi mn/N} = \sum_{k=0}^{N-1} X(k) \sum_{n=0}^{N-1} e^{j2\pi n(k-m)/N}$$

put $k = m$ or $k-m = 0, \pm N, \pm 2N, ...$

$$\sum_{n=0}^{N-1} x(n)\, e^{-j2\pi kn/N} = \sum_{k=0}^{N-1} X(k) \sum_{n=1}^{N-1} 1$$

$$\sum_{n=0}^{N-1} x(n)\, e^{-j2\pi kn/N} = \sum_{k=0}^{N-1} X(k).N \quad \because \quad \sum_{n=0}^{N-1} a^n = \begin{cases} N, & a = 1 \\ \dfrac{1-a^N}{1-a}, & [a \neq 1 \end{cases}$$

$$\therefore \qquad X(k) = \frac{1}{N} \sum_{n=0}^{N-1} x(n)\, e^{-j2\pi n/N} \qquad\qquad .....(2.1.4)$$

$$\therefore \qquad x(n) = \sum_{k=0}^{N-1} X(k)\, e^{j2\pi kn/N} \text{ is known as synthesis equation and it is}$$

also called Discrete–Time Fourier Series (DTFS) or Discrete Fourier Series (DFS).

$$X(k) = \frac{1}{N}\sum_{n=0}^{N-1} x(n)\, e^{-j2\pi kn/N} \text{ is known as Analysis equation.}$$

$\dfrac{1}{N}$ is a scaling factor, it doesn't affect whether it is in synthesis equation or in analysis equation.

Equations (2.1.3) and (2.1.4) can also be expressed as $x(n) = \displaystyle\sum_{k=0}^{N-1} X(k)\, W_N^{-kn}$ and

$$X(k) = \frac{1}{N}\sum_{n=0}^{N-1} x(n) W_N^{kn} \text{ respectively where } W_N = e^{-\frac{j2\pi}{N}}$$

2.2 Properties of Discrete Fourier Series (DFS)

1. ***Linearity*:**

 Consider two periodic sequences $x_1(n)$ and $x_2(n)$, both with period N, such that

$$x_1(n) \xleftrightarrow{\ \text{DFS}\ } X_1(k)$$

 and $\qquad x_2(n) \xleftrightarrow{\ \text{DFS}\ } X_2(k)$

 then $ax_1(n) + bx_2(n) \xleftrightarrow{\ \text{DFS}\ } aX_1(k) + bX_2(k)$

2. ***Shift of a sequence:***

 If $\qquad\qquad x(n) \xleftrightarrow{\ \text{DFS}\ } X(k)$ is a DFS pair, then

$$x(n-m) \xleftrightarrow{\ \text{DFS}\ } e^{-j2\pi km/N} \cdot X(k)$$

 Proof: $\qquad \text{DFS}\,\{x(n)\} = \dfrac{1}{N}\displaystyle\sum_{n=0}^{N-1} x(n)\, e^{-j2\pi kn/N}$

 Consider

$$\text{DFS}\,\{x(n-m)\} = \frac{1}{N}\sum_{n=0}^{N-1} x(n-m)\, e^{-j2\pi kn/N}$$

 put $n - m = \ell \rightarrow n = \ell + m$ and replace n with ℓ in summation.

$$= \frac{1}{N}\sum_{\ell=0}^{N-1} x(\ell)\, e^{-j2\pi k(\ell+m)/N}$$

$$= \frac{1}{N}\sum_{\ell=0}^{N-1} x(\ell)\, e^{-j2\pi k\ell/N} \cdot e^{-j2\pi km/N}$$

$$DFS\{x(n-m)\} = X(k) . e^{-j2\pi km/N}$$

Hence proved.

Similarly $\quad e^{j2\pi n\ell/N} . x(n) \longleftrightarrow X(k-\ell)$ is a pair.

3. *Duality*:

If $\quad x(n) \xleftrightarrow{\text{DFS}} X(k)$, then

$$Nx(n) \xleftrightarrow{\text{DFS}} X(-k)$$

Proof:
$$X(k) = \frac{1}{N} \sum_{n=0}^{N-1} x(n) \, e^{-j2\pi kn/N} \qquad .....(2.2.1)$$

$$x(n) = \sum_{k=0}^{N-1} X(k) \, e^{j2\pi kn/N} \qquad .....(2.2.2)$$

Consider

$$x(-n) = \sum_{k=0}^{N-1} X(k) \, e^{-j2\pi kn/N} \qquad .....(2.2.3)$$

interchange roles of n and k

$$X(-k) = \sum_{n=0}^{N-1} x(n) \, e^{-j2\pi kn/N} \qquad .....(2.2.4)$$

$$X(-k) = \frac{1}{N} \sum_{n=0}^{N-1} N \, x(n) \, e^{-j2\pi kn/N} \qquad .....(2.2.5)$$

(2.2.5) is similar to (2.2.1)

$$\therefore \qquad N \, x(n) \longleftrightarrow X(-k)$$

4. *Symmetry properties*:

If $x(n) \xleftrightarrow{\text{DFS}} X(k)$ is a pair.

Then, when x(n) is complex

(i) $\quad x^{*}(n) \xleftrightarrow{\text{DFS}} X^{*}(-k)$

Proof: We know analysis equation

$$X(k) = \frac{1}{N} \sum_{n=0}^{N-1} x(n) \, e^{-j2\pi kn/N}$$

Consider

$$X(-k) = \frac{1}{N} \sum_{n=0}^{N-1} x(n)\ e^{+j2\pi kn/N}$$

$$X^*(-k) = \frac{1}{N} \sum_{n=0}^{N-1} x^*(n)\ e^{-j2\pi kn/N}$$

$\therefore \qquad X^*(-k) \xleftrightarrow{\ \text{DFS}\ } x^*(n)$ is DFS pair.

(ii) $\quad x^*(-n) \xleftrightarrow{\ \text{DFS}\ } X^*(k)$

Proof: We know synthesis equation

$$x(n) = \sum_{k=0}^{N-1} X(k)\ e^{j2\pi kn/N}$$

$$x(-n) = \sum_{k=0}^{N-1} X(k)\ e^{-j2\pi kn/N}$$

$$x^*(-n) = \sum_{k=0}^{N-1} X^*(k)\ e^{j2\pi kn/N}$$

$\therefore \qquad x^*(-n) \xleftrightarrow{\ \text{DFS}\ } X^*(k)$ is DFS pair.

(iii) $\quad x_e(n) \xleftrightarrow{\ \text{DFS}\ } X_R(k)$ and $x_o(n) \xleftrightarrow{\ \text{DFS}\ } jX_I(k)$

Proof: We know $\qquad x_e(n) = \dfrac{1}{2}\Big[x(n) + x^*(-n)\Big]$

where $\qquad x_e(n)$ is even part of $x(n)$

Synthesis equation is

$$x(n) = \sum_{k=0}^{N-1} X(k)\ e^{j2\pi kn/N}$$

$$x(-n) = \sum_{k=0}^{N-1} \Big[X_R(k) + jX_I(k)\Big] e^{-j2\pi kn/N}$$

where $\qquad X_R(k)$ is real part of $X(k)$

$X_I(k)$ is imaginary part of $X(k)$

$$x^*(-n) = \sum_{k=0}^{N-1} \Big[X_R(k) - jX_I(k)\Big] e^{j2\pi kn/N}$$

$$\therefore \quad x_e(n) = \frac{1}{2}\left[\sum_{k=0}^{N-1}\left[X_R(k)+jX_I(k)\right]e^{j2\pi kn/N} + \sum_{k=0}^{N-1}\left[X_R(k)-jX_I(k)\right]e^{j2\pi kn/N}\right]$$

$$x_e(n) = \sum_{k=0}^{N-1} X_R(k)\, e^{j2\pi kn/N}$$

$$\therefore \quad x_e(n) \xleftrightarrow{\text{DFS}} X_R(k)$$

We know $x_o(n) = \dfrac{1}{2}\left[x(n) - x^*(-n)\right]$

where $x_o(n)$ is odd part of $x(n)$

$$\therefore \quad x_o(n) = \frac{1}{2}\left[\sum_{k=0}^{N-1}[X_R(k)+jX_I(k)]\,e^{j2\pi kn/N} - \sum_{k=0}^{N-1}[X_R(k)-jX_I(k)]\,e^{j2\pi kn/N}\right]$$

$$x_o(n) = \sum_{k=0}^{N-1} jX_I(k)\, e^{j2\pi kn/N}$$

$$\therefore \quad x_o(n) \xleftrightarrow{\text{DFS}} jX_I(k) \text{ is a DFS pair.}$$

when $x(n)$ is real

(i) $X(k) = X^*(-k)$ DFS is conjugate–symmetric

Proof: $X(k) = \dfrac{1}{N}\displaystyle\sum_{n=0}^{N-1} x(n)\, e^{-j2\pi kn/N}$

$$X(-k) = \frac{1}{N}\sum_{n=0}^{N-1} x(n)\, e^{j2\pi kn/N}$$

$$X^*(-k) = \frac{1}{N}\sum_{n=0}^{N-1} x(n)\, e^{-j2\pi kn/N}$$

$$\therefore \quad X(k) = X^*(-k)$$

(ii) $X_R(k) = X_R(-k)$ Real part is even and $X_I(k) = -X_I(-k)$ Imaginary part is odd.

Proof: $X(k) = X_R(k) + jX_I(k)$; $X(k) = \dfrac{1}{N}\displaystyle\sum_{n=0}^{N-1} x(n)\, e^{-j2\pi kn/N}$

$$\therefore \quad X_R(k) + jX_I(k) = \frac{1}{N}\sum_{n=0}^{N-1} x(n)\left[\cos(2\pi kn/N) - j\sin(2\pi kn/N)\right]$$

$$\therefore \quad X_R(k) = \frac{1}{N}\sum_{n=0}^{N-1} x(n)\cos(2\pi kn/N)$$

$$X_R(-k) = \frac{1}{N}\sum_{n=0}^{N-1} x(n)\cos(-2\pi kn/N)$$

$$= \frac{1}{N}\sum_{n=0}^{N-1} x(n)\cos(2\pi kn/N) = X_R(k)$$

$$\therefore \quad X_R(-k) = X_R(k)$$

and $\quad X_I(k) = -\dfrac{1}{N}\displaystyle\sum_{n=0}^{N-1} x(n)\sin(2\pi kn/N)$

$$X_I(-k) = -\frac{1}{N}\sum_{n=0}^{N-1} x(n)\sin(-2\pi kn/N)$$

$$= +\frac{1}{N}\sum_{n=0}^{N-1} x(n)\sin(2\pi kn/N)$$

$$\therefore \quad X_I(k) = -X_I(-k)$$

5. *Periodic convolution (or) circular convolution:*

If $x_1(n) \xleftrightarrow{\text{DFS}} X_1(k)$ and $x_1(n) \xleftrightarrow{\text{DFS}} X_2(k)$ are DFS pairs, then

$$\sum_{m=0}^{N-1} x_1(m)\, x_2(n-m) \xleftrightarrow{\text{DFS}} N\, X_1(k)\, X_2(k)$$

(or)

$$x_1(n) \otimes x_2(n) \xleftrightarrow{\text{DFS}} N\, X_1(k)\, X_2(k)$$

Proof: Let $x_1(n) \otimes x_2(n) = x(n)$

$$x(n) \xleftrightarrow{\text{DFS}} X(k)$$

$$X(k) = \frac{1}{N}\sum_{n=0}^{N-1} x(n)\, e^{-j2\pi kn/N}$$

$$= \frac{1}{N}\sum_{n=0}^{N-1} x_1(n) \otimes x_2(n)\, e^{-j2\pi kn/N}$$

$$= \frac{1}{N}\sum_{n=0}^{N-1}\sum_{m=0}^{N-1} x_1(m) x_2(n-m)\, e^{-j2\pi kn/N}$$

$$= \frac{1}{N} \sum_{m=0}^{N-1} x_1(m) \sum_{n=0}^{N-1} x_2(n-m) \; e^{-j2\pi kn/N}$$

$$= \sum_{m=0}^{N-1} x_1(m) \; e^{-j2\pi kn/N} \cdot X_2(k) \qquad\qquad \text{by shifting property}$$

$$DFS \; \{x_1(n) \otimes x_2(n)\} = N \, X_1(k) \, X_2(k)$$

2.3 DFS Representation of Periodic Sequences

Any periodic sequence x(n) can be represented by synthesis equation

$$x(n) = \sum_{k=0}^{N-1} X(k) \; e^{j2\pi kn/N}$$

where X(k) are DFS coefficients.

X(k) can be obtained by analysis equation

$$X(k) = \frac{1}{N} \sum_{n=0}^{N-1} x(n) \; e^{-j2\pi kn/N}$$

Example 2.1

Determine DFS coefficients (spectra) of the signals

    (i) $x(n) = \cos\sqrt{2}\,\pi n$

   (ii) $x(n) = \cos \pi n / 3$

  (iii)   x(n) is periodic with period N = 4 and x(n) = {1, 1, 0, 0}

Solution:

  (i) For $\omega_0 = \sqrt{2}\pi \rightarrow f_0 = 1/\sqrt{2}$. Since f_0 is irrational number, the sequence is not periodic. Hence the sequence cannot be expanded in Discrete Fourier Series.

 (ii) In this case $\omega_0 = \pi/3 \rightarrow f_0 = 1/6$ and hence x(n) is periodic with fundamental period

    N = 6

$$X(k) = \frac{1}{N} \sum_{n=0}^{N-1} x(n) e^{-j2\pi kn/N}$$

$$= \frac{1}{6} \sum_{n=0}^{5} x(n) e^{-j2\pi kn/6} \qquad k = 0, 1,\dots 5 \qquad\qquad \dots\dots(2.3.1)$$

    x(n) can be expressed as

$$x(n) = \cos\frac{2\pi n}{6} = \frac{1}{2}e^{j2\pi n/6} + \frac{1}{2}e^{-j2\pi n/6} \qquad\qquad .....(2.3.2)$$

Synthesis equation $x(n) = \sum_{k=0}^{5} X(k)e^{j2\pi kn/6}$ $\qquad\qquad .....(2.3.3)$

Comparing (2.3.2) and (2.3.3)

$$X(1) = \frac{1}{2}$$

In 2nd term of eq. (2.3.2) $k = -1$

$\qquad\therefore\qquad\qquad e^{-j2\pi n/6}$ which can be written as

$$e^{-j2\pi n/6} = e^{j2\pi(5-6)/6} = e^{j2\pi 5n/6}$$

i.e., $\qquad\qquad k = 5$

$\therefore\qquad\qquad X(-1) = X(5)$

$$X(5) = 1/2$$

$$X(0) = X(2) = X(3) = X(4) = 0.$$

(iii) $\quad X(k) = \frac{1}{4}\sum_{n=0}^{3} x(n)e^{-j2\pi kn/6}, \qquad\qquad k = 0, 1, 2, 3$

$$X(k) = \frac{1}{4}\left(1 + e^{-j\pi k/2}\right)$$

$k = 0, 1, 2, 3$, we get

$$X(0) = \frac{1}{2}, \quad X(1) = \frac{1}{4}(1-j), \quad X(2) = 0, \quad X(3) = \frac{1}{4}(1+j)$$

The magnitude and phase spectra are

$$|X(0)| = \frac{1}{2}, \quad |X(1)| = \frac{\sqrt{2}}{4}, \quad |X(2)| = 0, \quad |X(3)| = \frac{\sqrt{2}}{4}$$

$$\angle X(0) = 0, \angle X(1) = -\frac{\pi}{4}, \quad \angle X(2) = \text{undefined}, \angle X(3) = \frac{\pi}{4}$$

2.4 Discrete-Fourier Transform

2.4.1 Development of Discrete Fourier Transform (DFT)

The DFT is the most important discrete transform, used to perform Fourier analysis in many practical applications. In digital signal processing, the function is any quantity or signal that varies over time, such as the pressure of a sound wave, a radio signal, or daily temperature readings, sampled over a finite time interval (often defined by a window function). In image processing, the samples can be the values of pixels along a row or column of a raster image. The DFT is also used to efficiently solve partial differential equations, and to perform other operations such as convolutions or multiplying large integers.

Since it deals with a finite amount of data, it can be implemented in computers by numerical algorithms or even dedicated hardware. These implementations usually employ efficient Fast Fourier Transform (FFT) algorithms; so much so that the terms "FFT" and "DFT" are often used interchangeably. The terminology is further blurred by the (now rare) synonym finite Fourier transform for the DFT, which apparently predates the term "fast Fourier transform" but has the same initialism.

We know Fourier Transform of aperiodic continuous-time signal x(t) as

$$X(f) = \int_{-\infty}^{\infty} x(t) e^{-j2\pi ft} dt \qquad .....(2.4.1)$$

This equation is also called analysis equation or direct transform

$$x(t) = \frac{1}{2\pi} \int_{-\infty}^{\infty} X(f) e^{j2\pi ft} d\omega \qquad .....(2.4.2)$$

(or)

$$= \int_{-\infty}^{\infty} X(f) e^{j2\pi ft} df$$

This equation is called synthesis equation or Inverse Transform.

(i)   If a time domain signal is continuous and aperiodic, it's frequency representation is also continuous and aperiodic as shown in Fig. 2.1.

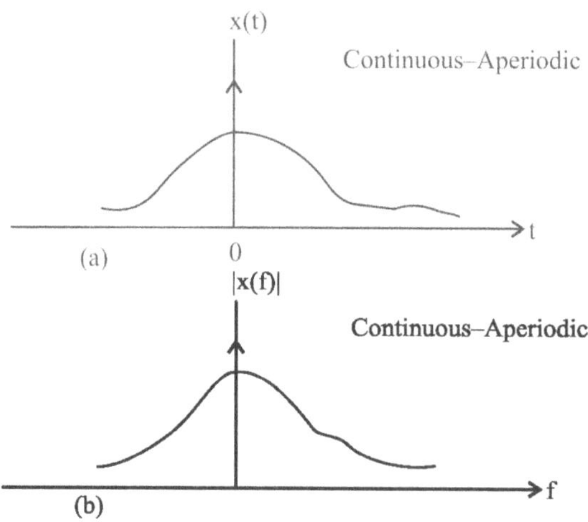

Fig. 2.1 (a) Time domain representation of continuous-a periodic
(b) Frequency domain representation of continuous-a periodic.

(ii)   If a time domain signal is continuous with period t_p, it's frequency representation will be discrete with spacing $f = \dfrac{1}{t_p}$ but Aperiodic in nature as shown in Fig. 2.2.

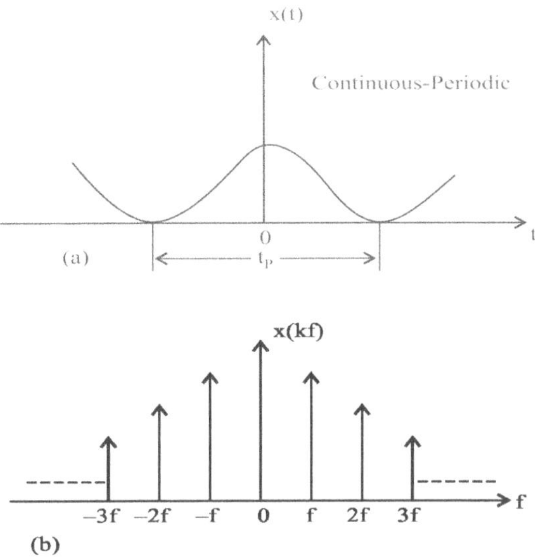

Fig.2.2 (a) Time domain representation of continuous-periodic (b) Frequency domain representation of discrete-a periodic.

Analysis equation (2.4.1) becomes

$$X(Kf) = \frac{1}{t_p} \int_{t_p} x(t) e^{-j2\pi kft} dt \qquad\qquad .....(2.4.3)$$

Synthesis equation (2.4.2) becomes

$$x(t) = \sum_{K=-\infty}^{\infty} X(kf) e^{j2\pi kft}$$

$$= \sum_{k=-\infty}^{\infty} X(k) e^{j2\pi kft} \qquad\qquad .....(2.4.4)$$

Which is exponential fourier series representation of continuous time signal x(t).

(iii) If a time domain signal is sampled (discrete) but Aperiodic, it's frequency representation will be continuous and it repeats with a period $f_s = \dfrac{1}{T}$. as shown in Fig. 2.3.

Equation (2.4.1) becomes

$$X(f) = \int\limits_{-\infty}^{\infty} x(nT)e^{-j2\pi fnT} d(nT)$$

Consider uniform sampling with T = 1 sec

$$= \int\limits_{-\infty}^{\infty} x(n)e^{-j2\pi fn} dn.$$

Discrete-sequence x(n) can be written as sum of impulse sequences.

$$\therefore \qquad X(f) = \int\limits_{-\infty}^{\infty} \sum_{n=-\infty}^{\infty} \left[x(n)\delta(t-n) \right] e^{-j2\pi fn} dn.$$

$$X(f) = \sum_{n=-\infty}^{\infty} x(n)e^{-j2\pi fn}$$

$$\because \qquad \delta(t-n) = 1 \text{ only at } t = n.$$

$$X(\omega) = \sum_{n=-\infty}^{\infty} x(n)e^{-j\omega n} \qquad\qquad .....(2.4.5)$$

This is a Discrete-time Fourier Transform (DTFT) of Aperiodic discrete sequence x(n). It is also called Analysis equation (or) direct Transform.

X(w) is a frequency response of aperiodic discrete sequence. Equation (2.4.2) becomes

$$x(t) = x(nT) = \frac{1}{f_s} \int\limits_{f_s} X(f)e^{j2\pi fnT} df$$

$$x(n) = \frac{1}{2\pi} \int\limits_{2\pi} X(\omega)e^{j\omega n} d\omega \qquad\qquad .....(2.4.6)$$

which is synthesis equation (or) Inverse Transform

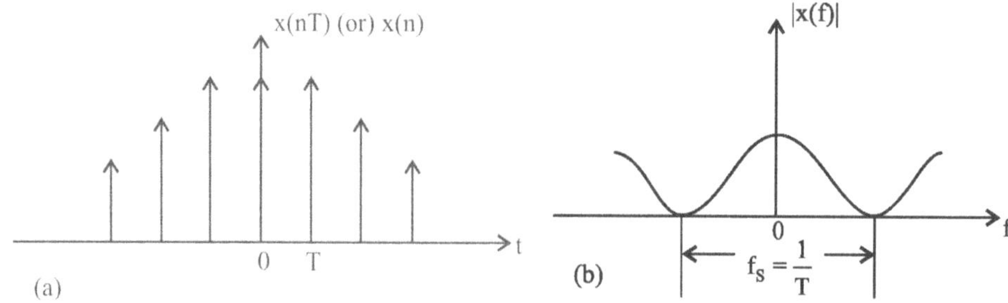

Fig. 2.3 (a) Represents aperiodic discrete time sequence.
(b) Frequency domain representation of periodic continuous signal.

(iv) If a time domain signal is sampled (discrete) and periodic with period t_p, it's frequency representation will be discrete and periodic with period $f_s = \dfrac{1}{T}$, as shown in Fig. 2.4.

Equation (2.4.1) becomes

$$X(kf) = \sum_n x(nT) e^{-j2\pi kf\,nT}$$

where $\qquad f = \dfrac{1}{t_p}, \qquad\qquad t_p = \text{period of } x(t) \text{ or } x(nT)$

$$T = \dfrac{1}{f_s}, \qquad\qquad f_s = \text{period of } X(f)$$

Consider

$$fT = \frac{1}{t_p}.T = \frac{1}{NT}.T = \frac{1}{N},$$

N is no. of samples in a period t_p i.e. $0 \le n \le N-1$

$$\therefore \qquad X(k) = \sum_{n=0}^{N-1} x(n) e^{-j2\pi kn/N} \qquad\qquad \dots\dots(2.4.7)$$

Which is a Discrete Fourier Transform (DFT)

Equation (2.4.2) becomes

$$x(t) = x(nT) = \frac{1}{N} \sum_{K=0}^{N-1} X(kf) e^{j2\pi kfnT}$$

$$x(n) = \frac{1}{N} \sum_{K=0}^{N-1} X(k) e^{j2\pi kn/N} \qquad\qquad \dots\dots(2.4.8)$$

Which is an Inverse Discrete Fourier Transform (IDFT)

Here N is number of samples in a period f_s. i.e., $0 \le k \le N-1$

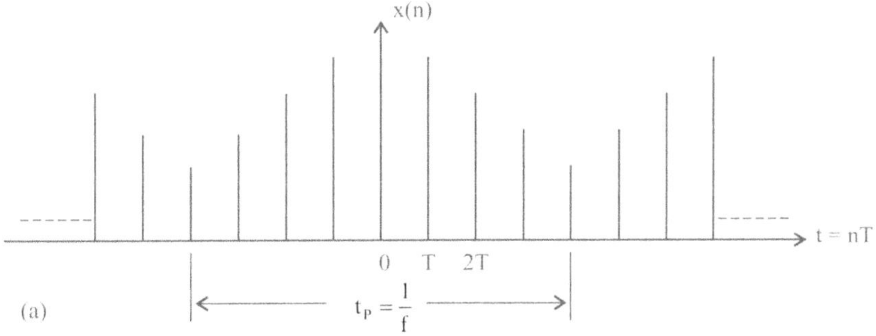

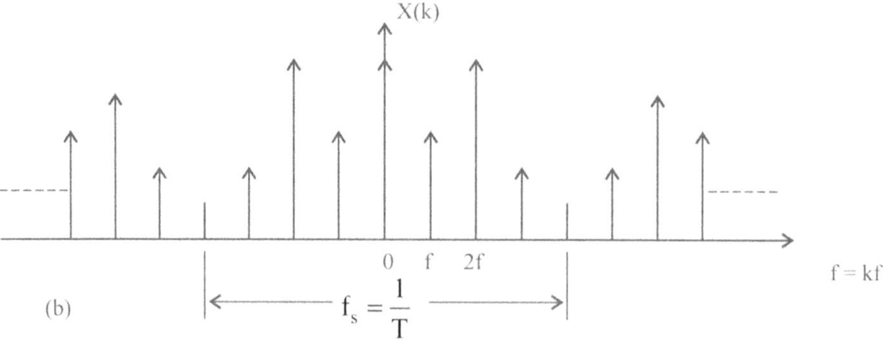

Fig. 2.4 (a) Represents periodic discrete sequence (b) Represents periodic discrete spectrum.

Example 2.2: Show that the frequency response of a discrete system is a periodic function of frequency.

Solution: equation (2.4.5) is a frequency response of a discrete sequence.

Similarly we can write frequency response of a discrete system using system function $(H(\omega))$ and impulse response h(n).

$$H(\omega) = \sum_{n=-\infty}^{\infty} h(n)e^{-j\omega n}$$

Consider

$$H(\omega + 2\pi k) = \sum_{n=-\infty}^{\infty} h(n)e^{-j(\omega + 2\pi k)n}$$

$$= \sum_{n=-\infty}^{\infty} h(n)e^{-j\omega n}.e^{-j2\pi kn}$$

$$= \sum_{n=-\infty}^{\infty} h(n)e^{-j\omega n}$$

$$\because \quad e^{-j2\pi kn} = 1$$

Hence frequency response of a discrete system is a periodic function of frequency.

Example 2.3: If x(n) and $X(e^{j\omega})$ constitute a Fourier transform pair, find the corresponding pairs of the following:

(i) $R_e\big[x(n)\big]$;

(ii) $I_m\big[X(e^{j\omega})\big]$

Solution:

(i) equation 2.4.6 is

$$x(n) = \frac{1}{2\pi} \int_{2\pi} X\left(e^{j\omega}\right) e^{j\omega n} d\omega$$

$$R_e\left[x(n)\right] = \frac{1}{2\pi} \int_{2\pi} R_e\left[X\left(e^{j\omega}\right)(\cos\omega n + j\sin\omega n)\right] d\omega$$

$$X\left(e^{j\omega}\right)(\cos\omega n + j\sin\omega n) = \left(X_R\left(e^{j\omega}\right) + jX_I\left(e^{j\omega}\right)\right)(\cos\omega n + j\sin\omega n)$$

$$X_R\left(e^{j\omega}\right)\cos\omega n + jX_R\left(e^{j\omega}\right)\sin\omega n + jX_I\left(e^{j\omega}\right)\cos\omega n - X_I\left(e^{j\omega}\right)\sin\omega n$$

$$R_e\left[X\left(e^{j\omega}\right)(\cos\omega n + j\sin\omega n)\right] = X_R\left(e^{j\omega}\right)\cos\omega n - X_I\left(e^{j\omega}\right)\sin\omega n$$

$$\therefore\ R_e\left[x(n)\right] = \frac{1}{2\pi} \int_{2\pi} (X_R\left(e^{j\omega}\right)\cos\omega n - X_I\left(e^{j\omega}\right)\sin\omega n)\, d\omega$$

(ii) equation (2.4.5) is

$$X(\omega) = \left[X\left(e^{j\omega}\right)\right] = \sum_{n=-\infty}^{\infty} x(n)e^{-j\omega n}$$

$$I_m\left[X\left(e^{j\omega}\right)\right] = \sum_{n=-\infty}^{\infty} I_m\left[x(n)(\cos\omega n - j\sin\omega n)\right]$$

$$= x(n)(\cos\omega n - j\sin\omega n) = \left(x_R(n) + jx_I(n)\right)(\cos\omega n - j\sin\omega n)$$

$$= x_R(n)\cos\omega n - jx_R(n)\sin\omega n + jx_I(n)\cos\omega n + x_I(n)\sin\omega n$$

$$\therefore\ I_m\left[x(n)(\cos\omega n - j\sin\omega n)\right] = x_I(n)\cos\omega n - x_R(n)\sin\omega n$$

$$\therefore\ I_m\left[X\left(e^{j\omega}\right)\right] = \sum_{n=-\infty}^{\infty} (x_I(n)\cos\omega n - x_R(n)\sin\omega n)$$

Example 2.4: Find the Fourier transform of the following signals

(i) $x(n) = (\cos\omega_0 n)u(n)$

(ii) $x(n) = a^n u(n)$ $-1 < a < 1$

Solution:

Since the given signals are discrete, we should use DTFT.

We know $X(\omega) = \displaystyle\sum_{n=-\infty}^{\infty} x(n) e^{-j\omega n}$

(i) $X(\omega) = \displaystyle\sum_{n=0}^{\infty} \cos\omega_0 n . e^{-j\omega n}$

$$= \sum_{n=0}^{\infty}\left(\frac{e^{j\omega_0 n} + e^{-j\omega_0 n}}{2}\right) e^{-j\omega n}$$

$$= \frac{1}{2}\left[\sum_{n=0}^{\infty} e^{jn(\omega_0 - \omega)} + \sum_{n=0}^{\infty} e^{-jn(\omega_0 + \omega)}\right]$$

$$= \frac{1}{2}\left[\frac{1}{1 - e^{j(\omega_0 - \omega)}} + \frac{1}{1 - e^{-j(\omega_0 + \omega)}}\right]$$

$$= \frac{1}{2}\left[\frac{1 - e^{-j\omega_0}.e^{-j\omega} + 1 - e^{j\omega_0}.e^{-j\omega}}{1 - e^{-j\omega_0}.e^{-j\omega} - e^{j\omega_0}.e^{-j\omega} + e^{-j2\omega}}\right]$$

$$= \frac{1}{2}\left[\frac{2 - e^{-j\omega}\left(e^{j\omega_0} + e^{-j\omega_0}\right)}{1 - e^{-j\omega}\left(e^{j\omega_0} + e^{-j\omega_0}\right) + e^{-j2\omega}}\right]$$

$$= \frac{1}{2}\left[\frac{2 - e^{-j\omega}\, 2\cos\omega_0}{1 - e^{-j\omega}\, 2\cos\omega_0 + e^{-j2\omega}}\right]$$

$$X(\omega) = \frac{1 - e^{-j\omega}\cos\omega_0}{1 - e^{-j\omega}\, 2\cos\omega_0 + e^{-j2\omega}}$$

(ii) $X(\omega) = \displaystyle\sum_{n=0}^{\infty} a^n\, e^{-j\omega n}$

$$= \sum_{n=0}^{\infty}\left(a\, e^{-j\omega}\right)^n$$

$$= \frac{1}{1 - a\, e^{-j\omega}} \qquad \text{if} \quad |a| < 1$$

2.5 Properties of DFT

1. *Periodicity*

If $x(n) \xleftrightarrow{\text{DFT}} X(k)$ is a DFT pair, then $\text{DFT}\{x(n)\}$ and $\text{I DFT}\{X(k)\}$ are periodic with period N

Proof: (i) $DFT\{x(n)\} = X(k) = \sum\limits_{n=0}^{N-1} x(n)\, e^{-j2\pi kn/N}$

consider $X(k+N) = \sum\limits_{n=0}^{N-1} x(n)\, e^{-j2\pi(k+N)n/N}$

$$= \sum_{n=0}^{N-1} x(n)\, e^{-j2\pi kn/N} \times e^{-j2\pi n}$$

$$= \sum_{n=0}^{N-1} x(n)\, e^{-j2\pi kn/N} \quad \because \quad e^{-j2\pi n} = 1$$

$$= X(k)$$

Hence $DFT\{x(n)\}$ is periodic with period N

(ii) $IDFT\{X(k)\} = x(n) = \dfrac{1}{N}\sum\limits_{k=0}^{N-1} X(k)\, e^{j2\pi kn/N}$

consider $x(n+N) = \dfrac{1}{N}\sum\limits_{k=0}^{N-1} X(k)\, e^{j2\pi k(n+N)/N}$

$$= \dfrac{1}{N}\sum_{k=0}^{N-1} X(k)\, e^{j2\pi kn/N}\, e^{j2\pi k}$$

$$= \dfrac{1}{N}\sum_{k=0}^{N-1} X(k)\, e^{j2\pi kn/N} \quad \because \quad e^{j2\pi k} = 1$$

$$= x(n)$$

Hence $IDFT\{X(k)\}$ is periodic with period N

2. Linearity

If $x_1(n) \xleftrightarrow{\text{DFT}} X_1(k)$

and $x_2(n) \xleftrightarrow{\text{DFT}} X_2(k)$ are DFT pairs, then

$$DFT\{a_1 x_1(n) + a_2 x_2(n)\} = a_1 X_1(k) + a_2 X_2(k)$$

3. Symmetry properties

If $x(n) \xleftrightarrow{\text{DFT}} X(k)$ is a pair

then, when $x(n)$ is complex

(i) $x^*(n) \xleftrightarrow{\text{DFT}} X^*(-k)$ is a pair

Proof: We know DFT equation

$$\text{DFT}\{x(n)\} = X(k) = \sum_{n=0}^{N-1} x(n)\, e^{-j2\pi kn/N}$$

Put
$$e^{-j2\pi/N} = W_N$$

$$X(k) = \sum_{n=0}^{N-1} x(n)\, W_N^{kn}$$

consider
$$X(-k) = \sum_{n=0}^{N-1} x(n)\, W_N^{-kn}$$

$$X^*(-k) = \sum_{n=0}^{N-1} x^*(n)\, W_N^{kn}$$

$\therefore$
$$x^*(n) \xleftrightarrow{\text{DFT}} X^*(-k) \text{ is a pair}$$

(ii) $x^*(-n) \xleftrightarrow{\text{DFT}} X^*(k)$ is a pair

Proof: We know IDFT equation

$$\text{IDFT}\{X(k)\} = x(n) = \frac{1}{N}\sum_{k=0}^{N-1} X(k)\, W_N^{-kn}$$

where
$$W_N = e^{-j2\pi/N}$$

consider
$$x(-n) = \frac{1}{N}\sum_{k=0}^{N-1} X(k)\, W_N^{kn}$$

$$x^*(-n) = \frac{1}{N}\sum_{k=0}^{N-1} X^*(k)\, W_N^{-kn}$$

$\therefore$
$$x^*(-n) \xleftrightarrow{\text{DFT}} X^*(k) \text{ is a pair}$$

(iii) $x_e(n) \xleftrightarrow{\text{DFT}} X_R(k)$ and $x_o(n) \xleftrightarrow{\text{DFT}} jX_I(k)$ are pairs

Proof: $x_e(n)$ is even part of $x(n)$

$X_R(k)$ is real part of $X(k)$

$x_o(n)$ is odd part of $x(n)$

$X_I(k)$ is imaginary part of $X(k)$

we know $x_e(n) = \dfrac{1}{2}\left[x(n) + x^*(-n)\right]$

IDFT equation is

$$x(n) = \frac{1}{N}\sum_{k=0}^{N-1} X(k)\, W_N^{-kn} = \frac{1}{N}\sum_{k-0}^{N-1}\left[X_R(k) + j\,X_I(k)\right] W_N^{-kn}$$

$$x(-n) = \frac{1}{N}\sum_{k=0}^{N-1}\left[X_R(k) + j\,X_I(k)\right] W_N^{kn}$$

$$x^*(-n) = \frac{1}{N}\sum_{k=0}^{N-1}\left[X_R(k) - j\,X_I(k)\right] W_N^{-kn}$$

$$x_e(n) = \frac{1}{2}\left[\frac{1}{N}\sum_{k=0}^{N-1}\left[X_R(k) + j\,X_I(k)\right] W_N^{-kn} + \frac{1}{N}\sum_{k=0}^{N-1}\left[X_R(k) - j\,X_I(k)\right] W_N^{-kn}\right]$$

$$x_e(n) = \frac{1}{N}\sum_{k=0}^{N-1} X_R(k)\, W_N^{-kn}$$

$$\therefore\ x_e(n) \xleftrightarrow{\;DFT\;} X_R(k)\ \text{is a pair}$$

We know $x_0(n) = \dfrac{1}{2}\left[x(n) - x^*(-n)\right]$

$$\therefore\qquad x_0(n) = \frac{1}{2}\left[\frac{1}{N}\sum_{k=0}^{N-1}\left[X_R(k) + jX_I(k)\right] W_N^{-kn} - \frac{1}{N}\sum_{k=0}^{N-1}\left[X_R(k) - jX_I(k)\right] W_N^{-kn}\right]$$

$$x_0(n) = \frac{1}{N}\sum_{k=0}^{N-1} j\,X_I(k)\, W_N^{-kn}$$

$$\therefore\qquad x_0(n) \xleftrightarrow{\;DFT\;} jX_I(k)\ \text{is a pair}$$

When x(n) is Real

(i) $X(k) = X^*(-k)$ DFT is conjugate - symmetric

Proof: $X(k) = \displaystyle\sum_{n=0}^{N-1} x(n)\, W_N^{kn}$ DFT equation

$$X(-k) = \sum_{n=0}^{N-1} x(n)\, W_N^{-kn}$$

$$X^*(-k) = \sum_{n=0}^{N-1} x(n)\, W_N^{kn}$$

$$\therefore\qquad X(k) = X^*(-k)$$

(ii) $X_R(k) = X_R(-k)$ Real part is even

and $X_I(k) = -X_I(-k)$ Imaginary part is odd

Proof: Let $X(k) = X_R(k) + j X_I(k)$

DFT equation is

$$X(k) = \sum_{n=0}^{N-1} x(n)\ W_N^{kn} = \sum_{n=0}^{N-1} x(n)\ e^{-j2\pi kn/N}$$

$\therefore$ $X_R(k) + jX_I(k) = \sum_{n=0}^{N-1} x(n)\left[\cos\ (2\pi kn/N) - j\sin\ (2\pi kn/N)\right]$

Equate Real and Imaginary parts

$$X_R(k) = \sum_{n=0}^{N-1} x(n)\ \cos\ (2\pi kn/N)$$

and $X_I(k) = -\sum_{n=0}^{N-1} x(n)\ \sin\ (2\pi kn/N)$

$$X_R(-k) = \sum_{n=0}^{N-1} x(n)\ \cos\ (-2\pi kn/N)$$

$$= \sum_{n=0}^{N-1} x(n)\ \cos\ (2\pi kn/N) = X_R(k)$$

$\therefore$ $X_R(-k) = X_R(k)$

Hence real part is even symmetry

$$X_I(-k) = -\sum_{n=0}^{N-1} x(n)\ \sin\ (-2\pi kn/N)$$

$$= \sum_{n=0}^{N-1} x(n)\ \sin\ (2\pi kn/N)$$

$\therefore$ $X_I(-k) = -X_I(k)$

Hence Imaginary part is odd symmetry

4. *DFT of delayed sequence (or Shift of a Sequence)*

If $x(n) \xleftarrow{\ \ DFT\ \ } X(k)$ is pair

then $x(n-m) \xleftarrow{\ \ DFT\ \ } e^{-j2\pi km/N}\ X(k)$ is a pair

Proof:
$$\text{IDFT}\{X(k)\} = x(n) = \frac{1}{N}\sum_{k=0}^{N-1} X(k)\, e^{j2\pi kn/N}$$

Consider

$$x(n-m) = \frac{1}{N}\sum_{k=0}^{N-1} X(k)\, e^{j2\pi k(n-m)/N}$$

$$x(n-m) = \frac{1}{N}\sum_{k=0}^{N-1} X(k)\, e^{-j2\pi km/N} \times e^{j2\pi kn/N}$$

$$\therefore \quad x(n-m) \xleftrightarrow{\text{DFT}} e^{-j2\pi\, km/N} X(k) \text{ is a pair}$$

Hence proved

5. *DFT of time reversed sequence*

If
$$x(n) \xleftrightarrow{\text{DFT}} X(k) \text{ is a pair}$$

then
$$x(N-n) \xleftrightarrow{\text{DFT}} X(N-k) \text{ is a pair}$$

Proof: We know

$$\text{DFT}\{x(n)\} = \sum_{n=0}^{N-1} x(n) e^{-j2\pi kn/N}$$

$$\text{DFT}\{x(N-n)\} = \sum_{n=0}^{N-1} x(N-n) e^{-j2\pi kn/N}$$

Let change index from n to m, where $m = N - n$ $\therefore$ $n = N - m$

$$\therefore \ \text{DFT}\{x(N-n)\} = \sum_{m=0}^{N-1} x(m) e^{-j2\pi k(N-m)/N}$$

$$= \sum_{m=0}^{N-1} x(m) e^{-j2\pi kN/N} \times e^{j2\pi km/N}$$

$$= \sum_{m=0}^{N-1} x(m) e^{j2\pi km/N} \times 1 \quad \because e^{-j2\pi kN/N} = e^{-j2\pi k} = 1; \text{ and } 1 = e^{-j2\pi m}$$

$$= \sum_{m=0}^{N-1} x(m) e^{j2\pi km/N} \times e^{-j2\pi m}$$

$$= \sum_{m=0}^{N-1} x(m) e^{-j2\pi m(N-k)/N}$$

$$= X(N-k)$$

Hence proved

6. *Multiplication of Two DFTs and Circular Convolution (or) Periodic Convolution*

If $x_1(n) \xleftrightarrow{\text{DFT}} X_1(k)$ and $x_2(n) \xleftrightarrow{\text{DFT}} X_2(k)$ are

DFT pairs, then

$$x_1(n) \otimes x_2(n) \xleftrightarrow{\text{DFT}} X_1(k) X_2(k)$$

Proof: Let $\quad X_1(k) = \displaystyle\sum_{n=0}^{N-1} x_1(n)\, e^{-j2\pi\, kn/N}, \qquad\qquad k = 0, 1, ..., N-1$

and $\quad X_2(k) = \displaystyle\sum_{n=0}^{N-1} x_2(n)\, e^{-j2\pi\, kn/N}, \qquad\qquad k = 0, 1, ..., N-1$

multiplication of two DFTs $X_1(k)$ and $X_2(k)$, results a DFT say $X_3(k)$

$$\therefore \qquad X_3(k) = X_1(k) X_2(k), \qquad\qquad k = 0, 1, 2\ ...., N-1$$

$$\text{IDFT}\{X_3(k)\} = x_3(m) = \frac{1}{N} \sum_{k=0}^{N-1} X_3(k)\, e^{j2\pi km/N}$$

$$= \frac{1}{N} \sum_{k=0}^{N-1} X_1(k)\, X_2(k)\, e^{j2\pi km/N}$$

$$x_3(m) = \frac{1}{N} \sum_{k=0}^{N-1} \left[\sum_{n=0}^{N-1} x_1(n)\, e^{-j2\pi kn/N} \right] \left[\sum_{\ell=0}^{N-1} x_2(\ell)\, e^{-j2\pi \ell/N} \right] e^{j2\pi km/N}$$

$$= \frac{1}{N} \sum_{n=0}^{N-1} x_1(n) \sum_{\ell=0}^{N-1} x_2(\ell) \left[\sum_{k=0}^{N-1} e^{j2\pi k(m-n-\ell)/N} \right]$$

Then inner sum in the brackets has the form

$$\sum_{k=0}^{N-1} a^k = \begin{cases} N, & a = 1 \\ \dfrac{1-a^N}{1-a}, & a \neq 1 \end{cases}$$

where $\qquad a = e^{j2\pi(m-n-\ell)/N}$

we observe that $a = 1$ when $m - n - \ell$ is a multiple of N, on the other hand, $a^N = 1$ for any value of $a \neq 0$ since $x_2(\ell \pm pN) = x_2(\ell)$

$$\therefore \qquad \sum_{k=0}^{N-1} a^k = \begin{cases} N & \ell = m-n+PN = ((m-n))_N, \quad \text{P an integer} \\ & \qquad\qquad\text{or} \\ & \qquad = ((m-n),\, \text{mod } N) \\ 0 & \quad\text{otherwise} \end{cases}$$

$$\therefore \qquad x_3(m) = \sum_{n=0}^{N-1} x_1(n) x_2\big((m-n)\big)_N, \qquad m = 0, 1, ..., N-1 \quad .....(2.5.1)$$

$$= \sum_{n=0}^{N-1} x_1(n) \; x_2\big((m-n), \text{mod} N\big), \quad m = 0, \; 1, ..., N-1$$

$$\therefore \qquad \text{IDFT} \; \{X_1(k) X_2(k)\} = x_3(m) = x_1(m) \otimes x_2(m)$$

Example 2.5: If x(n) and x(k) constitute a DFT pair find the corresponding pair of the following. [JNTU 2002]

   (i) $x^*(n)$ (ii) $X_{ep}(k)$

Solution:

  (i)   We know

$$x(n) = \frac{1}{N} \sum_{k=0}^{N-1} X(k) \; W_N^{-kn}, \qquad\qquad n = 0, 1, 2, ..., N-1$$

$$= \frac{1}{N} \sum_{k=0}^{N-1} X(k) \; e^{j2\pi kn/N} \qquad\qquad\qquad .....(2.5.2)$$

take $\qquad x^*(n) = \dfrac{1}{N} \displaystyle\sum_{k=0}^{N-1} X^*(k) \; e^{-j2\pi kn/N}$

put $\qquad k = -k$

$$x^*(n) = \frac{1}{N} \sum_{k=0}^{N-1} X^*(-k) \; e^{j2\pi kn/N}$$

which is similar to eqn.2.5.2

$\therefore \qquad x^*(n)$ and $X^*(-k)$ is DFT pair.

$\therefore \qquad x^*(n) \xleftarrow{\quad\text{DFT}\quad} X^*(-k)$

 (ii) $X_{ep}(k)$ is even part of X(k)

we know $X_{ep}(k) = \dfrac{X(k) + X^*(-k)}{2} \qquad\qquad .....(2.5.3)$

$$X(k) = \sum_{n=0}^{N-1} x(n) \; e^{-j2\pi kn/N}, \qquad\qquad k = 0, 1, 2, ..., N-1$$

take $\qquad X(-k) = \displaystyle\sum_{n=0}^{N-1} \big[x_R(n) + jx_I(n)\big] \; e^{j2\pi kn/N}$

where $\quad$ $x_R(n)$ is real part of $x(n)$

$\quad$ $x_I(n)$ is imaginary part of $x(n)$

$$X^*(-k) = \sum_{n=0}^{N-1} \left[x_R(n) - jx_I(n) \right] e^{-j2\pi kn/N}$$

$$X_{ep}(k) = \frac{X(k) + X^*(-k)}{2}$$

$$= \frac{1}{2} \left[\sum_{n=0}^{N-1} \left[x_R(n) + jx_I(n) \right] e^{-j2\pi kn/N} + \sum_{n=0}^{N-1} \left[x_R(n) - jx_I(n) \right] e^{-j2\pi kn/N} \right]$$

$$X_{ep}(k) = \sum_{n=0}^{N-1} x_R(n) \, e^{-j2\pi kn/N}$$

$\therefore \qquad X_{ep}(k) \xleftrightarrow{\text{DFT}} x_R(n)$ is a DFT pair

Example 2.6: Let $X(k)$ denote the N – pt DFT of the N – pt sequence $x(n)$. Show that with N solution and if $x(n) = x(N - 1 - n)$ then $X(N/2) = 0$. $\qquad$ [JNTU 2002]

Solution: We know N – pt DFT of $x(n)$

$$X(k) = \sum_{n=0}^{N-1} x(n) W_N^{kn}, \qquad k = 0, 1, 2 \dots N - 1$$

consider $N = 4$

$\therefore \qquad \displaystyle X(k) = \sum_{n=0}^{3} x(n) W_N^{kn}, \qquad k = 0, 1, 2, 3$

$$X(k) = x(0) W_4^0 + x(1) W_4^k + x(2) W_4^{2k} + x(3) W_4^{3k}$$

At $\qquad \displaystyle k = \frac{N}{2} = \frac{4}{2} = 2$

$\therefore \qquad X(2) = x(0) W_4^0 + x(1) W_4^2 + x(2) W_4^4 + x(3) W_4^6 \qquad\qquad \dots(2.5.4)$

by using periodicity properties of W_N

i. e., $\qquad W_N^{k+N} = W_N^k$

$\therefore \qquad W_4^{4+0} = W_4^0 \qquad\qquad$ and $\quad W_4^6 = W_4^{4+2} = W_4^2$

$\qquad W_4^0 = 1; \; W_4^2 = -1$

and given $\quad x(n) = x(N - 1 - n)$

$\qquad n = 0 \; \rightarrow \; x(0) = x(4 - 1 - 0)$

$$\therefore \qquad x(0) = x(3)$$

$$n = 1 \rightarrow x(1) = x(4-1-1)$$

$$\therefore \qquad x(1) = x(2)$$

Substitute in eq. (2.5.4)

$$\therefore \qquad X\left(\frac{N}{2}\right) = X(2) = x(0) - x(1) + x(1) - x(0)$$

$$= 0$$

Hence proved

Example 2.7: Sketch the magnitude and phase of the DFTs of the following sequence

$$x\,(n) = (0,\, 1,\, 2,\, -2,\, -1) \hspace{4cm} \text{[JNTU99/S]}$$

Solution: N = 5;

$$X(k) = \sum_{n=0}^{4} x(n)\,W_5^{xn}, \qquad x = 0,\, 1,\, 2,\, 3,\, 4$$

$$
\begin{bmatrix} X(0) \\ X(1) \\ X(2) \\ X(3) \\ X(4) \end{bmatrix}
=
\begin{bmatrix}
W_5^0 & W_5^0 & W_5^0 & W_5^0 & W_5^0 \\
W_5^0 & W_5^1 & W_5^2 & W_5^3 & W_5^4 \\
W_5^0 & W_5^2 & W_5^4 & W_5^6 & W_5^8 \\
W_5^0 & W_5^3 & W_5^6 & W_5^9 & W_5^{12} \\
W_5^0 & W_5^4 & W_5^8 & W_5^{12} & W_5^{16}
\end{bmatrix}
\begin{bmatrix} x(0) \\ x(1) \\ x(2) \\ x(3) \\ x(4) \end{bmatrix}
$$

$$W_5^0 = 1,\ W_5^1 = 0.3 - j\,0.9,\ W_5^2 = -\,0.8 - j\,0.6$$

$$W_5^3 = W_5^{-2+5} = W_5^{-2} = -\,0.8 + j\,0.6$$

$$W_5^4 = W_5^{-1+5} = W_5^{-1} = 0.3 + j\,0.9$$

$$W_5^6 = W_5^{1+5} = W_5^1 = 0.3 - j\,0.9$$

$$W_5^8 = W_5^{3+5} = W_5^3 = W_5^{-2} = -\,0.8 + j\,0.6$$

$$W_5^9 = W_5^{4+5} = W_5^4 = W_5^{-1} = 0.3 + j\,0.9$$

$$W_5^{12} = W_5^{7+5} = W_5^7 = W_5^{2+5} = W_5^2 = -\,0.8 - j\,0.6$$

$$W_5^{16} = W_5^{11+5} = W_5^{6+5} = W_5^{1+5} = W_5^1 = 0.3 - j\,0.9$$

$$\begin{bmatrix} X(0) \\ X(1) \\ X(2) \\ X(3) \\ X(4) \end{bmatrix} = \begin{bmatrix} 1 & 1 & 1 & 1 & 1 \\ 1 & 0.3-j\,0.9 & -0.8-j\,0.6 & -0.8+j\,0.6 & 0.3+j\,0.9 \\ 1 & -0.8-j\,0.6 & 0.3+j\,0.9 & 0.3-j\,0.9 & -0.8+j\,0.6 \\ 1 & -0.8+j\,0.6 & 0.3-j\,0.9 & 0.3+j\,0.9 & -0.8-j\,0.6 \\ 1 & 0.3+j\,0.9 & -0.8+j\,0.6 & -0.8-j\,0.6 & 0.3-j\,0.9 \end{bmatrix} \begin{bmatrix} 0 \\ 1 \\ 2 \\ -2 \\ -1 \end{bmatrix}$$

$$\begin{bmatrix} X(0) \\ X(1) \\ X(2) \\ X(3) \\ X(4) \end{bmatrix} = \begin{bmatrix} 0 \\ -4.2\,j \\ 2.4j \\ -2.4j \\ 4.2j \end{bmatrix}$$

Magnitude plot

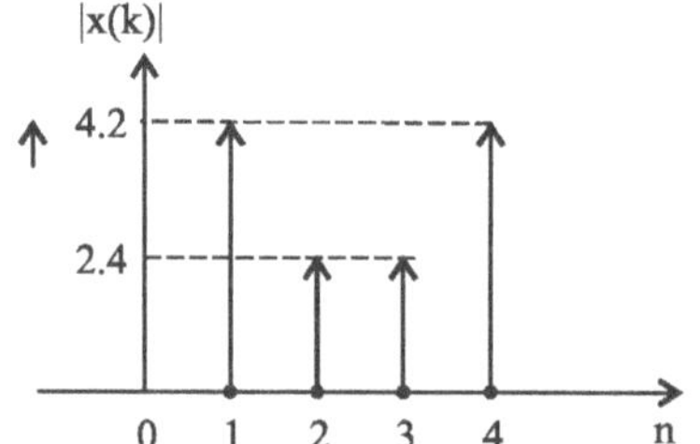

Phase plot

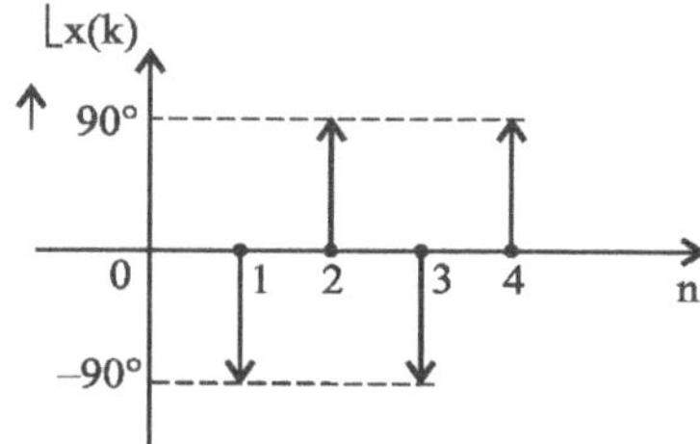

2.6 Linear Convolution of Sequences using DFT

2.6.1 Linear Convolution

Response $y(n)$ of any discrete system can be expressed as convolution sum of input sequence $x(n)$ and impulse response $h(n)$

$$\therefore \quad y(n) = \sum_{k=-\infty}^{\infty} x(k)h(n-k) \qquad \qquad .....(2.6.1)$$

which is also called linear convolution.

The linear convolution of any two sequences can be found by any one of the following methods.

Let us consider the sequences $x(n) = \{2, 3, 1, 2\}$ and $h(n) = \left\{1, \underset{\uparrow}{2}, 3, 4\right\}$

Graphical Method:

Procedure:

Step 1: Choose an initial index n of the output sequence y(n) as if x(n) starts at $n = n_1$ and h(n) starts at $n = n_2$, then $n = n_1 + n_2$.

Step 2: Calculate the length of output sequence $L = M + N - 1$, where M is length of x(n) and N is the length of h(n).

Step 3: Represent graphically both sequences in terms of k.

Step 4: Fold h(k) about $k = 0$ to obtain h(–k) and shift by 'n' to the right if 'n' is positive and left if 'n' is negative to obtain h(n–k).

Step 5: Multiply x(k) and h(n–k) element by element and sum the products to get y(n).

Step 6: Increment the index 'n', shift the sequence h(n–k) to right by one sample and do step 5.

Step 7: Repeat step 6 until the length of output sequence is L.

Solution:

x(n) starts at $n_1 = 0$ and h(n) starts at $n_2 = -1$

∴ Starting index of output sequence $n = n_1 + n_2 = 0 + (-1) = -1$

x(n) length $M = 4$ and h(n) length $N = 4$

∴ Output sequence length $L = 4 + 4 - 1 = 7$

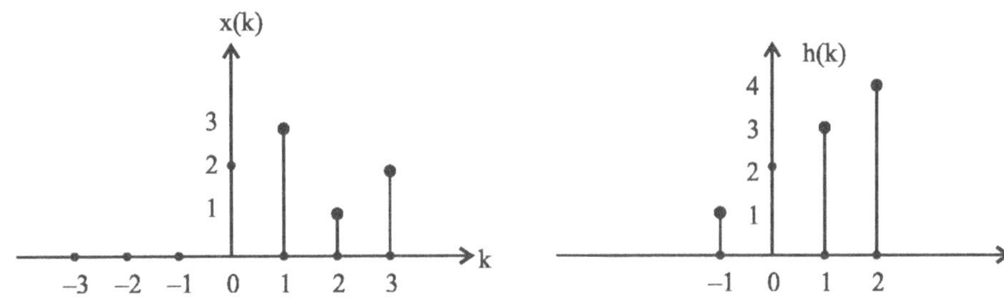

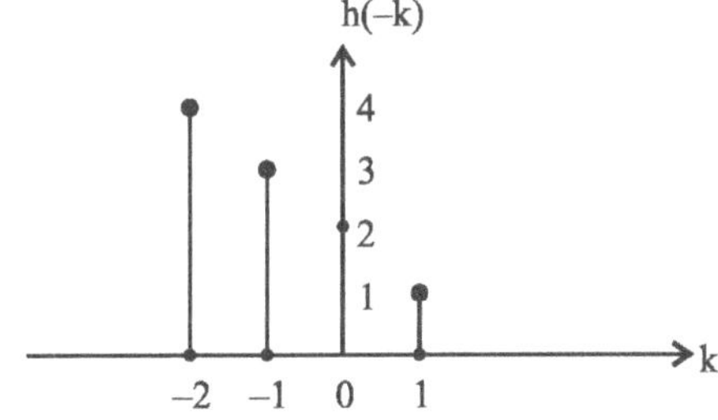

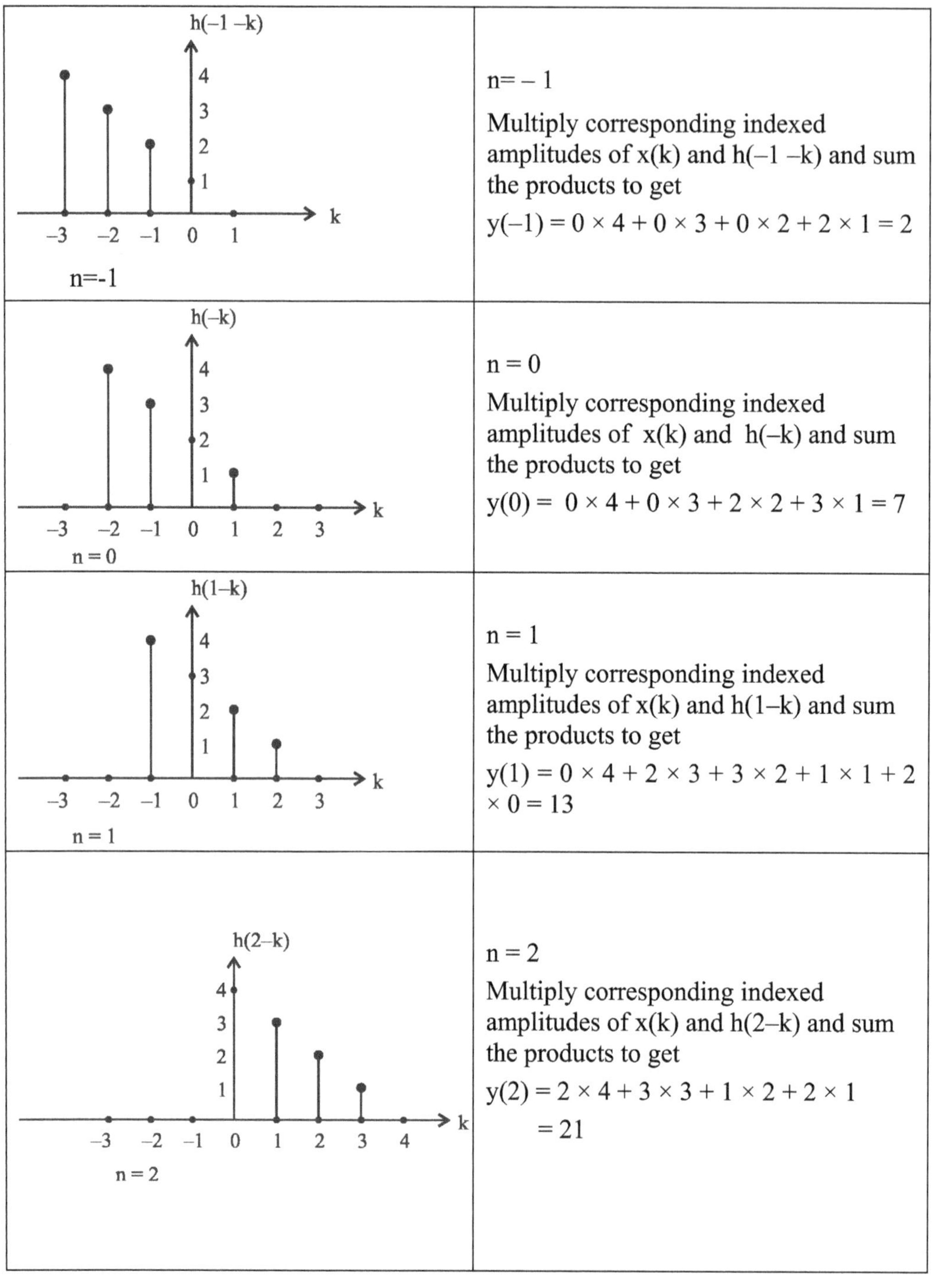

$n = -1$

Multiply corresponding indexed amplitudes of $x(k)$ and $h(-1-k)$ and sum the products to get

$$y(-1) = 0 \times 4 + 0 \times 3 + 0 \times 2 + 2 \times 1 = 2$$

$n = 0$

Multiply corresponding indexed amplitudes of $x(k)$ and $h(-k)$ and sum the products to get

$$y(0) = 0 \times 4 + 0 \times 3 + 2 \times 2 + 3 \times 1 = 7$$

$n = 1$

Multiply corresponding indexed amplitudes of $x(k)$ and $h(1-k)$ and sum the products to get

$$y(1) = 0 \times 4 + 2 \times 3 + 3 \times 2 + 1 \times 1 + 2 \times 0 = 13$$

$n = 2$

Multiply corresponding indexed amplitudes of $x(k)$ and $h(2-k)$ and sum the products to get

$$y(2) = 2 \times 4 + 3 \times 3 + 1 \times 2 + 2 \times 1 = 21$$

Contd...

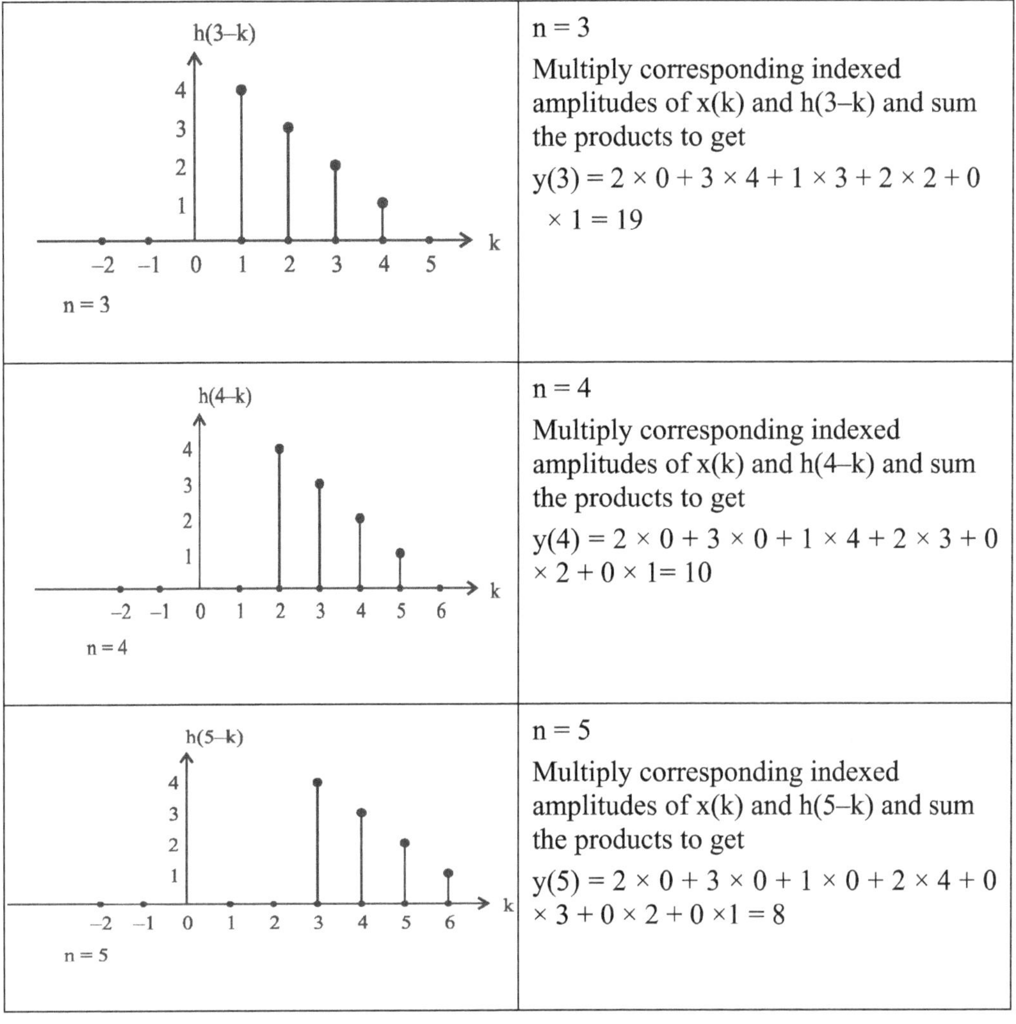

n = 3

Multiply corresponding indexed amplitudes of x(k) and h(3–k) and sum the products to get

$y(3) = 2 \times 0 + 3 \times 4 + 1 \times 3 + 2 \times 2 + 0 \times 1 = 19$

n = 4

Multiply corresponding indexed amplitudes of x(k) and h(4–k) and sum the products to get

$y(4) = 2 \times 0 + 3 \times 0 + 1 \times 4 + 2 \times 3 + 0 \times 2 + 0 \times 1 = 10$

n = 5

Multiply corresponding indexed amplitudes of x(k) and h(5–k) and sum the products to get

$y(5) = 2 \times 0 + 3 \times 0 + 1 \times 0 + 2 \times 4 + 0 \times 3 + 0 \times 2 + 0 \times 1 = 8$

Fig. 2.5 Operations on sequence to perform linear convolution.

Now output sequence length is 7.

$$\therefore \quad y(n) = \left\{ 2,\ \underset{\uparrow}{7},\ 13,\ 21, 19, 10, 8 \right\}$$

To check the answer for correctness, sum all the samples of x(n) and multiply with the sum of all the samples in h(n). This value must be equal to sum of all samples of y(n).

In the given problem $\sum_{n} x(n) = 8,\ \sum_{n} h(n) = 10,\ \sum_{n} y(n) = 80$

$$\therefore \quad \left[\sum_n x(n)\right] \cdot \left[\sum_n h(n)\right] = \sum_n y(n)$$

$8 . 10 = 80$ which is equal to $\sum_n y(n)$. Hence the result is correct.

Direct Method:

Procedure:

Step 1: Calculate the length of output sequence $L = M + N - 1$, where M is the length of $x(n)$ and N is the length of $h(n)$.

Step 2: Choose an initial index 'n' of the output sequence $y(n)$ as if $x(n)$ starts at $n = n_1$ and $h(n)$ starts at $n = n_2$, then $n = n_1 + n_2$.

Step 3: In the convolution summation formula take k values from 0 to M–1, if the formula is $\sum_k x(k) \, h(n{-}k)$ or take k values from 0 to N–1; if the formula is $\sum_k h(k) \, x(n{-}k)$ and calculate the value of $y(n)$.

Step 4: Increment 'n' and do step 3.

Step 5: Repeat step 4 until the length of output sequence becomes L.

Solution:

$$L = 4 + 4 - 1 = 7$$

$$n = n_1 + n_2 = 0 + (-1) = -1$$

$$\therefore \; y(-1) = \sum_{k=0}^{3} x(k) \, h(-1 - k)$$

$$= x(0) \, h(-1) + x(1) \, h(-2) + x(2) \, h(-3) + x(3) \, h(-4)$$

$$= 2 \times 1 + 3 \times 0 + 1 \times 0 + 2 \times 0$$

$$y(-1) = 2$$

$$y(0) = \sum_{k=0}^{3} x(k) \, h(-k)$$

$$= x(0) \, h(0) + x(1) \, h(-1) + x(2) \, h(-2) + x(3) \, h(-3)$$

$$= 2 \times 2 + 3 \times 1 + 1 \times 0 + 2 \times 0$$

$$y(0) = 7$$

$$y(1) = \sum_{k=0}^{3} x(k)\, h(1-k)$$

$$= x(0)\, h(1) + x(1)\, h(0) + x(2)\, h(-1) + x(3)\, h(-2)$$

$$= 2 \times 3 + 3 \times 2 + 1 \times 1 + 2 \times 0$$

$$y(1) = 13$$

$$y(2) = \sum_{k=0}^{3} x(k)\, h(2-k)$$

$$= x(0)\, h(2) + x(1)\, h(1) + x(2)\, h(0) + x(3)\, h(-1)$$

$$= 2 \times 4 + 3 \times 3 + 1 \times 2 + 2 \times 1$$

$$y(2) = 21$$

$$y(3) = \sum_{k=0}^{3} x(k)\, h(3-k)$$

$$= x(0)\, h(3) + x(1)\, h(2) + x(2)\, h(1) + x(3)\, h(0)$$

$$= 2 \times 0 + 3 \times 4 + 1 \times 3 + 2 \times 2$$

$$y(3) = 19$$

$$y(4) = \sum_{k=0}^{3} x(k)\, h(4-k)$$

$$= x(0)\, h(4) + x(1)\, h(3) + x(2)\, h(2) + x(3)\, h(1)$$

$$= 2 \times 0 + 3 \times 0 + 1 \times 4 + 2 \times 3$$

$$y(4) = 10$$

$$y(5) = \sum_{k=0}^{3} x(k)\, h(5-k)$$

$$= x(0)\, h(5) + x(1)\, h(4) + x(2)\, h(3) + x(3)\, h(2)$$

$$= 2 \times 0 + 3 \times 0 + 1 \times 0 + 2 \times 4$$

$$y(5) = 8$$

$$\therefore \quad y(n) = \left\{ 2,\ \underset{\uparrow}{7},\ 13,\ 21, 19, 10, 8 \right\}$$

Matrix method:

Procedure:

Step 1: Choose starting index for x(n) as n_1 and for h(n) as n_2, then the output sequence y(n) starting index will be $n_1 + n_2$.

Step 2: Calculate the length of output sequence $L = M + N - 1$, where M is the length of x(n) and N is the length of h(n).

Step 3: Form $L \times L$ matrix using either x(n) or h(n) as shown below

Consider x(n)

$$\begin{bmatrix} x(n_1) & 0 & 0 & 0 & \cdots & 0 \times (L-1) \\ x(n_1 + 1) & & & & & \\ x(n_1 + 2) & & & & & \\ \vdots & & & & & \\ x(n_1 + L - 1) & & & & & \end{bmatrix}$$

Remaining rows elements are obtained by shifting their preceding rows towards right by one as shown below:

$$\begin{bmatrix} x(n_1) & 0 & 0 & 0 & \cdots & 0 \times (L-1) \\ x(n_1 + 1) & x(n_1) & 0 & 0 & \cdots & 0 \\ x(n_1 + 2) & x(n_1 + 1) & x(n_1) & 0 & \cdots & 0 \\ \vdots & & & \ddots & & \\ \vdots & & & & \ddots & \\ x(n_1 + L - 1) & & & & \ddots & x(n_1) \end{bmatrix}$$

Finally upper half diagonal contains all zeros.

Step 4: Form $L \times 1$ matrix using either x(n) or h(n) as shown below

Consider h(n)

$$\begin{bmatrix} h(n_2) \\ h(n_2 + 1) \\ h(n_2 + 2) \\ \vdots \\ h(n_2 + L - 1) \end{bmatrix}$$

***Step* 5:** Multiply $L \times L$ matrix and $L \times 1$ matrix to get output sequence y(n), which will be $L \times 1$ matrix as shown below

$$\begin{bmatrix} y(n_1 + n_2) \\ y(n_1 + n_2 + 1) \\ y(n_1 + n_2 + 2) \\ \vdots \\ y(n_1 + n_2 + L - 1) \end{bmatrix}$$

Solution:

$n_1 = 0;\ n_2 = -1$

$n_1 + n_2 = -1$

$L = 4 + 4 - 1 = 7$

$$\begin{bmatrix} 2 & 0 & 0 & 0 & 0 & 0 & 0 \\ 3 & 2 & 0 & 0 & 0 & 0 & 0 \\ 1 & 3 & 2 & 0 & 0 & 0 & 0 \\ 2 & 1 & 3 & 2 & 0 & 0 & 0 \\ 0 & 2 & 1 & 3 & 2 & 0 & 0 \\ 0 & 0 & 2 & 1 & 3 & 2 & 0 \\ 0 & 0 & 0 & 2 & 1 & 3 & 2 \end{bmatrix} \begin{bmatrix} 1 \\ 2 \\ 3 \\ 4 \\ 0 \\ 0 \\ 0 \end{bmatrix} = \begin{bmatrix} 2 \\ 7 \\ 13 \\ 21 \\ 19 \\ 10 \\ 8 \end{bmatrix}$$

$\therefore\ y(n) = \left\{ 2,\ \underset{\uparrow}{7},\ 13,\ 21, 19, 10, 8 \right\}$

Diagonal method:

Procedure:

***Step* 1:** Write down the sequences x(n) and h(n) as shown below

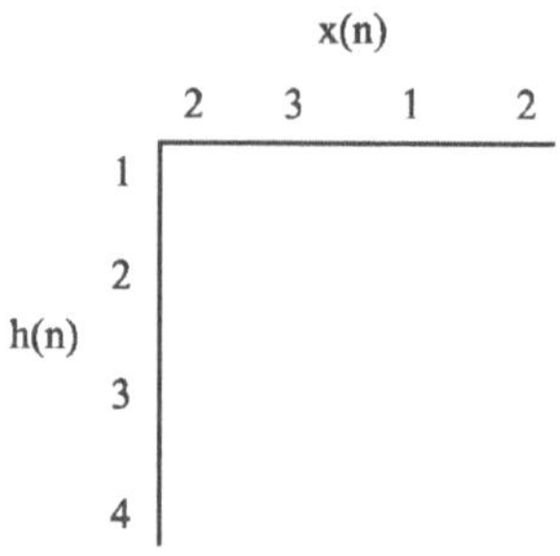

Step 2: Multiply each and every sample in x(n) with the samples of h(n) and tabulate the values

Step 3: Separate the elements in the table by drawing diagonal lines as shown below

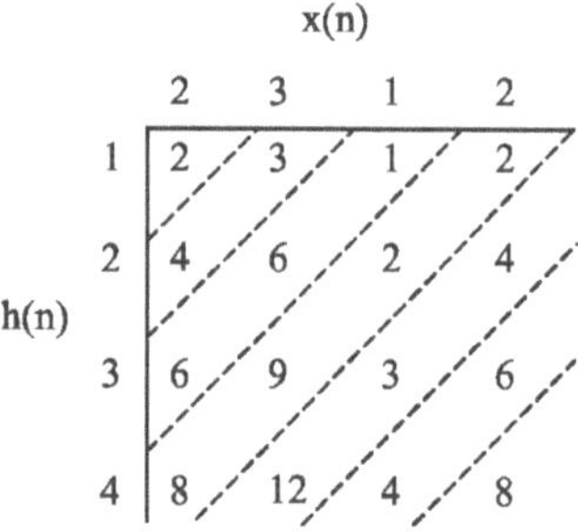

Step 4: Starting from the top sum all the elements in each diagonal and write down in the same order.

$$= 2, 4 + 3, 6 + 6 + 1, 8 + 9 + 2 + 2, 12 + 3 + 4, 4 + 6, 8$$

$$= 2, 7, 13, 21, 19, 10, 8$$

Step 5: The starting index of output sequence y(n) is $n_1 + n_2$, where n_1 is starting index of x(n) and n_2 is the starting index of h(n).

$$\therefore \qquad n_1 = 0, n_2 = -1$$

$$n_1 + n_2 = -1$$

$$\therefore \qquad y(n) = \left\{2, \underset{\uparrow}{7}, 13, 21, 19, 10, 8\right\}$$

Tabular method:

Tabulate the sequence x(k) and shifted versions of h(k) as shown below.

The starting index of output sequence y(n) is $n = n_1 + n_2 = 0 + (-1) = -1$

k	−3	−2	−1	0	1	2	3	4	5	6
x(k)				2	3	1	2			
h(k)			1	2	3	4				
h(−k)		4	3	2	1					
n = −1 h(−1 −k)	4	3	2	1						
n = 0 h(−k)		4	3	2	1					
n = 1 h(1−k)			4	3	2	1				
n = 2 h(2−k)				4	3	2	1			
n = 3 h(3−k)					4	3	2	1		
n = 4 h(4−k)						4	3	2	1	
n = 5 h(5−k)							4	3	2	1

$$y(-1) = \sum_{k=-3}^{6} x(k)\, h(-1-k) = 1 \times 2 = 2$$

$$y(0) = \sum_{k=-3}^{6} x(k)\, h(-k) = 2 \times 2 + 3 \times 1 = 7$$

$$y(1) = \sum_{k=-3}^{6} x(k)\, h(1-k) = 2 \times 3 + 3 \times 2 + 1 \times 1 = 13$$

$$y(2) = \sum_{k=-3}^{6} x(k)\, h(2-k) = 2 \times 4 + 3 \times 3 + 2 \times 1 + 1 \times 2 = 21$$

$$y(3) = \sum_{k=-3}^{6} x(k)\, h(3-k) = 3 \times 4 + 1 \times 3 + 2 \times 2 = 19$$

$$y(4) = \sum_{k=-3}^{6} x(k)\, h(4-k) = 4 \times 1 + 3 \times 2 = 10$$

$$y(5) = \sum_{k=-3}^{6} x(k)\, h(5-k) = 4 \times 2 = 8$$

$$\therefore \quad y(n) = \left\{ 2,\ \underset{\uparrow}{7},\ 13,\ 21, 19, 10, 8 \right\}$$

Linear Convolution using DFT:

If the length of sequence x(n) is M and the length of sequence h(n) is N, then the length of output sequence y(n) will be M + N − 1

So to perform linear convolution using DFT,

***Step* 1:** Make both the sequence i.e., x(n) and h(n) of length M + N − 1 by padding with zeros.

***Step* 2:** Find out DFT{x(n)} = X(k) and DFT{h(n)} = H(k).

***Step* 3:** Multiply X(k) and H(k) which gives Y(k).

***Step* 4:** Find inverse DFT{Y(k)} which gives the output sequence y(n).

Consider x(n) = {1, 0, 1} and h(n) = {1, 1}

$$M = 3, N = 2$$

∴　　　$$M + N - 1 = 4$$

∴　　pad x(n) and h(n) with zeros to get the lengths 4:

∴　　　$$x(n) = \{1, 0, 1, 0\}$$

and　　$$h(n) = \{1, 1, 0, 0\}$$

we know DFT$\{x(n)\}$ = X(k) = $\sum\limits_{n=0}^{N-1} x(n) W_N^{kn}$,　　k = 0, 1, 2.....N−1

where　　$W_N = e^{-j2\pi/N}$

This can also be represented in matrix form as

$$\begin{bmatrix} X(0) \\ X(1) \\ X(2) \\ \vdots \\ X(N-1) \end{bmatrix} = \begin{bmatrix} W_N^0 & W_N^0 & W_N^0 & \cdots & W_N^0 \\ W_N^0 & W_N^1 & W_N^2 & \cdots & W_N^{N-1} \\ W_N^0 & W_N^2 & W_N^4 & \cdots & W_N^{2(N-1)} \\ \vdots & \vdots & \vdots & & \vdots \\ W_N^0 & W_N^{N-1} & W_N^{2(N-1)} & \cdots & W_N^{(N-1)(N-1)} \end{bmatrix} \begin{bmatrix} x(0) \\ x(1) \\ x(2) \\ \vdots \\ x(N-1) \end{bmatrix}$$

Here N = 4

$$X(k) = \sum_{n=0}^{3} x(n) W_N^{kn}, \qquad x = 0, 1, 2, 3$$

$$\begin{bmatrix} X(0) \\ X(1) \\ X(2) \\ X(3) \end{bmatrix} = \begin{bmatrix} W_4^0 & W_4^0 & W_4^0 & W_4^0 \\ W_4^0 & W_4^1 & W_4^2 & W_4^3 \\ W_4^0 & W_4^2 & W_4^4 & W_4^6 \\ W_4^0 & W_4^3 & W_4^6 & W_4^9 \end{bmatrix} \begin{bmatrix} x(0) \\ x(1) \\ x(2) \\ x(3) \end{bmatrix}$$

By using periodicity property of W_N i.e., $W_N^{k+N} = W_N^k$

$\therefore \qquad W_4^{k+4} = W_4^k$

$W_4^3 = W_4^{-1+4} = W_4^{-1}$, which is conjugate of W_4^1

$W_4^4 = W_4^{0+4} = W_4^0$

$W_4^6 = W_4^{2+4} = W_4^2$

$W_4^9 = W_4^{5+4} = W_4^5 = W_4^{1+4} = W_4^1$

$$\begin{bmatrix} X(0) \\ X(1) \\ X(2) \\ X(3) \end{bmatrix} = \begin{bmatrix} W_4^0 & W_4^0 & W_4^0 & W_4^0 \\ W_4^0 & W_4^1 & W_4^2 & W_4^{-1} \\ W_4^0 & W_4^2 & W_4^0 & W_4^2 \\ W_4^0 & W_4^{-1} & W_4^2 & W_4^1 \end{bmatrix} \begin{bmatrix} x(0) \\ x(1) \\ x(2) \\ x(3) \end{bmatrix}$$

$W_4^0 = 1$

$W_4^1 = e^{-j\frac{2\pi}{4}} = \cos(90) - j\sin(90) = -j$

$W_4^2 = e^{-j\frac{2\pi}{4} \cdot 2} = \cos(\pi) - j\sin(\pi) = -1$

$W_4^{-1} = j$

$$\begin{bmatrix} X(0) \\ X(1) \\ X(2) \\ X(3) \end{bmatrix} = \begin{bmatrix} 1 & 1 & 1 & 1 \\ 1 & -j & -1 & j \\ 1 & -1 & 1 & -1 \\ 1 & j & -1 & -j \end{bmatrix} \begin{bmatrix} 1 \\ 0 \\ 1 \\ 0 \end{bmatrix}$$

$$\begin{bmatrix} X(0) \\ X(1) \\ X(2) \\ X(3) \end{bmatrix} = \begin{bmatrix} 1+0+1+0 \\ 1+0-1+0 \\ 1+0+1+0 \\ 1+0-1+0 \end{bmatrix} = \begin{bmatrix} 2 \\ 0 \\ 2 \\ 0 \end{bmatrix}$$

$\therefore X(k) = \{2, 0, 2, 0\}$

Similarly
$$\begin{bmatrix} H(0) \\ H(1) \\ H(2) \\ H(3) \end{bmatrix} = \begin{bmatrix} 1 & 1 & 1 & 1 \\ 1 & -j & -1 & j \\ 1 & -1 & 1 & -1 \\ 1 & j & -1 & -j \end{bmatrix} = \begin{bmatrix} 1 \\ 1 \\ 0 \\ 0 \end{bmatrix}$$

$$\begin{bmatrix} H(0) \\ H(1) \\ H(2) \\ H(3) \end{bmatrix} = \begin{bmatrix} 1+1+0+0 \\ 1-j+0+0 \\ 1-1+0+0 \\ 1+j+0+0 \end{bmatrix} = \begin{bmatrix} 2 \\ 1-j \\ 0 \\ 1+j \end{bmatrix}$$

$\therefore \qquad H(k) = \{2,\, 1-j,\, 0,\, 1+j\}$

$\therefore \qquad Y(k) = X(k)H(k),\ k = 0,\, 1,\, 2,\, 3$

$\qquad k = 0,\ Y(0) = X(0)H(0) = 4$

$\qquad k = 1,\ Y(1) = X(1)H(1) = 0$

$\qquad k = 2;\ Y(2) = X(2)H(2) = 0$

$\qquad k = 3;\ Y(3) = X(3)H(3) = 0$

$\therefore \qquad y(n) = \text{IDFT}\{Y(k)\} = \dfrac{1}{N} \sum_{k=0}^{N-1} Y(k) W_N^{-kn}, \qquad n = 0,\, 1,\, 2,\, 3 \ldots .N-1$

where $\qquad W_N^{-kn}$ is conjugate of W_N^{kn}

$$\therefore \qquad \begin{bmatrix} y(0) \\ y(1) \\ y(2) \\ y(3) \end{bmatrix} = \frac{1}{4} \underbrace{\begin{bmatrix} 1 & 1 & 1 & 1 \\ 1 & j & -1 & -j \\ 1 & -1 & 1 & -1 \\ 1 & -j & -1 & j \end{bmatrix}}_{W_4^*} \begin{bmatrix} 4 \\ 0 \\ 0 \\ 0 \end{bmatrix}$$

$$\begin{bmatrix} y(0) \\ y(1) \\ y(2) \\ y(3) \end{bmatrix} = \frac{1}{4} \begin{bmatrix} 4+0+0+0 \\ 4+0+0+0 \\ 4+0+0+0 \\ 4+0+0+0 \end{bmatrix} = \begin{bmatrix} 1 \\ 1 \\ 1 \\ 1 \end{bmatrix}$$

$\therefore \qquad y(n) = \{1,\, 1,\, 1,\, 1\}$

Check: $\quad \sum_{n} x(n) = 2;\ \sum_{n} h(n) = 2,\ \sum_{n} y(n) = 4.$

$\therefore \qquad \left[\sum_{n} x(n) \right]\left[\sum_{n} h(n) \right] = 2 \times 2 = 4,$ which is equal to $\sum_{n} y(n).$

Hence the obtained result is correct.

2.6.2 Circular Convolution (or) Periodic Convolution

Circular Convolution is one of the Property of DFT.

Circular convolution of sequences $x_1(n)$ and $x_2(n)$ can be performed by

$$y(n) = x_1(n) \otimes x_2(n) = \sum_{k=0}^{N-1} x_1(k) x_2\big((n-k)\big)_N$$

where $\otimes$ symbol for circular convolution.

This equation can also be written as

$$y(n) = \sum_{k=0}^{N-1} x_1(k) x_2(n-k)$$

$$= \sum_{k=0}^{N-1} x_1(k) x_2\big((n-k) \bmod N\big)$$

Circular convolution can be performed by using various methods as follows.

Consider $x_1(n) = \{1, 0, 1, 1\}$ and $x_2(n) = \{1, 1\}$

If the length of $x_1(n)$ be M and $x_2(n)$ be N, then the length of the output sequence $y(n)$ will be maximum of M and N.

To perform circular convolution lengths of two sequences must be made equal by padding with zeros as follows.

If $M > N$, then pad $x_2(n)$ with $(M - N)$ zeros.

If $N > M$, then pad $x_1(n)$ with $(N-M)$ zeros.

Graphical Method:

Step 1: Make lengths of two input sequences equal, if they are not equal, by padding with zeros.

$$\therefore \ x_1(n) = \{1, 0, 1, 1\} \ \text{and} \ \ x_2(n) = \{1, 1, 0, 0\}$$

Step 2: Represent $x_1(n)$ and $x_2(n)$ on the circle in the anticlockwise direction as shown below.

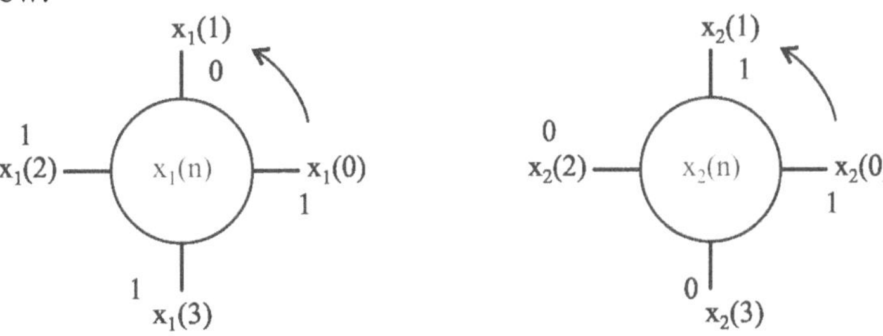

***Step* 3:** Fold either $x_1(n)$ or $x_2(n)$, say $x_2(n)$ is folded as shown below i.e., writing, $x_2(n)$ in clockwise direction.

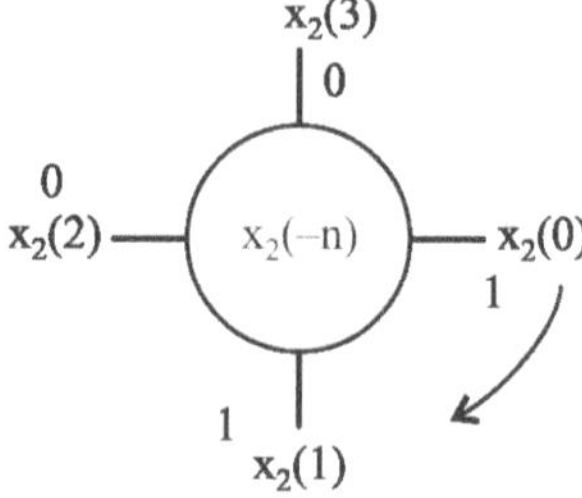

***Step* 4:** Multiply corresponding position values of $x_1(n)$ and $x_2(-n)$ to get $y(0)$.

$$y(0) = x_1(0)\, x_2(0) + x_1(1)\, x_2(3) + x_1(2)\, x_2(2) + x_1(3)\, x_2(1)$$

$$= 1 \times 1 + 0 \times 0 + 1 \times 0 + 1 \times 1 = 2$$

***Step* 5:** Rotate $x_2(-n)$ by one position in anticlockwise direction and multiply corresponding position values to get $y(1)$.

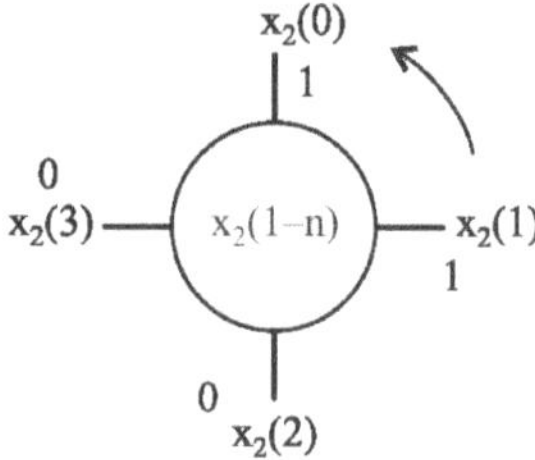

$$y(1) = x_1(0)\, x_2(1) + x_1(1)\, x_2(0) + x_1(2)\, x_2(3) + x_1(3)\, x_2(2)$$

$$= 1 \times 1 + 0 \times 1 + 1 \times 0 + 1 \times 0 = 1$$

***Step* 6:** similarly rotate and multiply corresponding position values to get $y(2)$ and $y(3)$.

to get $y(2)$:

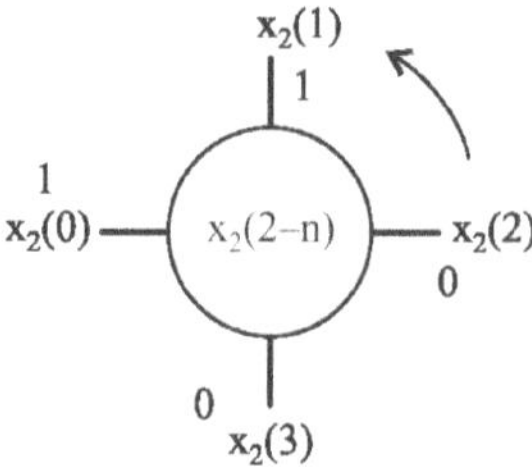

$$y(2) = x_1(0)\, x_2(2) + x_1(1)\, x_2(1) + x_1(2)\, x_2(0) + x_1(3)\, x_2(3)$$

$$= 1 \times 0 + 0 \times 1 + 1 + 1 \times 1 + 1 \times 0 = 1$$

to get y(3):

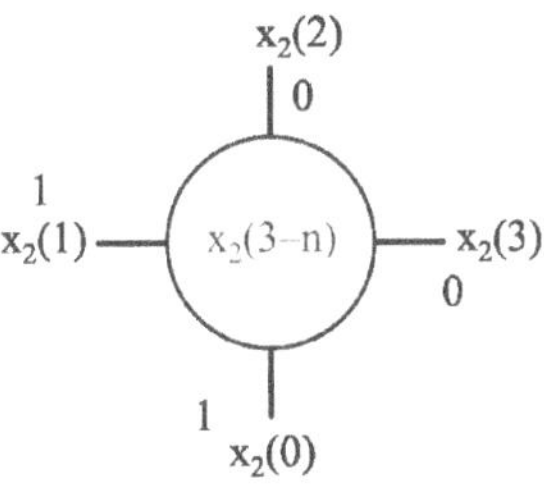

$$y\,(3) = x_1(0)\,x_2(3) + x_1(1)\,x_2(2) + x_1(2)x_2(1) + x_1(3)x_2(0)$$

$$= 1 \times 0 + 0 \times 0 + 1 \times 1 + 1 \times 1 = 2$$

$$\therefore\ y(n) = \{2,\, 1,\, 1,\, 2\}$$

Direct Method:

Step **1:** Make lengths of two input sequences equal, if they are not equal, by padding with zeros.

$$\therefore\ x_1(n) = \{1,\, 0,\, 1,\, 1\}\ \text{and}\ x_2(n) = \{1,\, 1,\, 0,\, 0\}$$

Step **2:** Perform circular convolution by using

$$y(n) = \sum_{k=0}^{N-1} x_1(k)\,x_2\big((n{-}k)\bmod N\big), \qquad n = 0,\, 1,\, 2,\, \ldots,\, N{-}1$$

where N is the length of input or output sequence(s).

Here N = 4.

$$y(n) = \sum_{k=0}^{3} x_1(k)\,x_2\big((n{-}k)\bmod N\big), \qquad n = 0,\, 1,\, 2,\, 3$$

$$y(n) = x_1(0)\,x_2\big((n)\bmod 4\big) + x_1(1)\,x_2\big((n{-}1)\bmod 4\big)$$
$$+\, x_1(2)\,x_2\big((n{-}2)\bmod 4\big) + x_1(3)\,x_2\big((n{-}3)\bmod 4\big)$$

$$y(o) = x_1(0)\,x_2\big((0)\bmod 4\big) + x_1(1)\,x_2\big((-1)\bmod 4\big)$$
$$+\, x_1(2)\,x_2\big((-2)\bmod 4\big) + x_1(3)\,x_2\big((-3)\bmod 4\big)$$

$$= x_1(0)\,x_2(0) + x_1(1)\,x_2(4-1) + x_1(2)\,x_2(4-2) + x_1(3)\,x_2(4-3)$$

$$= x_1(0)\,x_2(0) + x_1(1)\,x_2(3) + x_1(2)\,x_2(2) + x_1(3)\,x_2(1)$$

$$= 1\times1 + 0\times0 + 1\times0 + 1\times1 = 2$$

$$y(1) = x_1(0)\,x_2\big((1)\bmod 4\big) + x_1(1)\,x_2\big((0)\bmod 4\big)$$
$$+\, x_1(2)\,x_2\big((-1)\bmod 4\big) + x_1(3)\,x_2\big((-2)\bmod 4\big)$$

$$= x_1(0)\,x_2(1) + x_1(1)\,x_2(0) + x_1(2)\,x_2(4-1) + x_1(3)\,x_2(4-2)$$

$$= x_1(0)\, x_2(1) + x_1(1)\, x_2(0) + x_1(2)\, x_2(3) + x_1(3)\, x_2(2)$$

$$= 1 \times 1 + 0 \times 1 + 1 \times 0 + 1 \times 0 = 1$$

$$y(2) = x_1(0)\, x_2((2) \bmod 4) + x_1(1)\, x_2\,((1) \bmod 4)$$

$$+ x_1(2)\, x_2((0) \bmod 4) + x_1(3)\, x_2\,((-1) \bmod 4)$$

$$= x_1(0)\, x_2(2) + x_1(1)\, x_2(1) + x_1(2)\, x_2(0) + x_1(3)\, x_2(4-1)$$

$$= x_1(0)\, x_2(2) + x_1(1)\, x_2(1) + x_1(2)\, x_2(0) + x_1(3)\, x_2(3)$$

$$= 1 \times 0 + 0 \times 1 + 1 \times 1 + 1 \times 0 = 1$$

$$y(3) = x_1(0)\, x_2((3) \bmod 4) + x_1(1)\, x_2\,((2) \bmod 4)$$

$$+ x_1(2)\, x_2((1) \bmod 4) + x_1(3)\, x_2\,((0) \bmod 4)$$

$$= x_1(0)\, x_2(3) + x_1(1)\, x_2(2) + x_1(2)\, x_2(1) + x_1(3)\, x_2(0)$$

$$= 1 \times 0 + 0 \times 0 + 1 \times 1 + 1 \times 1 = 2$$

$$\therefore \quad y(n) = \{2, 1, 1, 2\}$$

Matrix Method:

***Step* 1:** Make lengths of two input sequences equal, if they are not equal, by padding with zeros.

$$\therefore \ x_1(n) = \{1, 0, 1, 1\} \text{ and } x_2(n) = \{1, 1, 0, 0\}$$

***Step* 2:** Form N×N matrix either by using $x_1(n)$ or $x_2(n)$ as follows.

Let us consider $x_1(n)$, N is the length of input or output sequence(s).

$$\begin{bmatrix}
x_1(0) & x_1(N-1) & x_1(N-2) & \cdots & x_1(2) & x_1(1) \\
x_1(1) & x_1(0) & x_1(N-1) & \cdots & x_1(3) & x_1(2) \\
x_1(2) & x_1(1) & x_1(0) & \cdots & x_1(4) & x_1(3) \\
\vdots & \vdots & \vdots & \vdots & \vdots & \vdots \\
x_1(N-2) & x_1(N-3) & x_1(N-4) & \cdots & x_1(0) & \cdots & x_1(N-1) \\
x_1(N-1) & x_1(N-2) & x_1(N-3) & \cdots & x_1(1) & x_1(0)
\end{bmatrix}$$

***Step* 3:** Form N×1 matrix either by using $x_2(n)$ or $x_1(n)$ as follows.

Let us consider $x_2(n)$.

$$\begin{bmatrix} x_2(0) \\ x_2(1) \\ x_2(2) \\ \vdots \\ x_2(N-2) \\ x_2(N-1) \end{bmatrix}$$

***Step* 4:** Multiply N×N matrix and N×1 matrix to get N×1 matrix which is output sequence $y(n)$ as shown below.

$$\begin{bmatrix} x_1(0) & x_1(N-1) & x_1(N-2) & \cdots & x_1(2) & x_1(1) \\ x_1(1) & x_1(0) & x_1(N-1) & \cdots & x_1(3) & x_1(2) \\ x_1(2) & x_1(1) & x_1(0) & \cdots & x_1(4) & x_1(3) \\ \vdots & \vdots & \vdots & \vdots & \vdots & \vdots \\ x_1(N-2) & x_1(N-3) & x_1(N-4) & \cdots & x_1(0) & x_1(N-1) \\ x_1(N-1) & x_1(N-2) & x_1(N-3) & \cdots & x_1(1) & x_1(0) \end{bmatrix} \begin{bmatrix} x_2(0) \\ x_2(1) \\ x_2(2) \\ \vdots \\ x_2(N-2) \\ x_2(N-1) \end{bmatrix} = \begin{bmatrix} y(0) \\ y(1) \\ y(2) \\ \vdots \\ y(N-2) \\ y(N-1) \end{bmatrix}$$

$$\begin{bmatrix} 1 & 1 & 1 & 0 \\ 0 & 1 & 1 & 1 \\ 1 & 0 & 1 & 1 \\ 1 & 1 & 0 & 1 \end{bmatrix} \begin{bmatrix} 1 \\ 1 \\ 0 \\ 0 \end{bmatrix} = \begin{bmatrix} 1+1+0+0 \\ 0+1+0+0 \\ 1+0+0+0 \\ 1+1+0+0 \end{bmatrix} = \begin{bmatrix} 2 \\ 1 \\ 1 \\ 2 \end{bmatrix}$$

$\therefore \quad y(n) = \{2, 1, 1, 2\}$

Circular Convolution using DFT:

***Step* 1:** Make lengths of two input sequences equal, if they are not equal, by padding with zeros.

$\therefore\ x_1(n) = \{1, 0, 1, 1\}$ and $x_2(n) = \{1, 1, 0, 0\}$

***Step* 2:** Find DFT$\{x_1(n)\}$ and DFT$\{x_2(n)\}$, which are $X_1(k)$ and $X_2(k)$ respectively, by using

$$X(k) = \sum_{n=0}^{N-1} x(n) W_N^{kn}, \qquad k = 0, 1, 2, \ldots N-1$$

***Step* 3:** Multiply $X_1(k)$ and $X_2(k)$ to get $Y(k)$.

***Step* 4:** Find IDFT$\{Y(k)\}$, which is $y(n)$, by using

$$y(n) = \frac{1}{N} \sum_{k=0}^{N-1} Y(k)\, W_N^{-kn}, \qquad n = 0, 1, 2, \ldots, N-1$$

Here $N = 4$

$$\therefore \qquad X_1(k) = \sum_{n=0}^{3} x_1(n)\, W_4^{kn}, \qquad k = 0, 1, 2, 3$$

$$\begin{bmatrix} X_1(0) \\ X_1(1) \\ X_1(2) \\ X_1(3) \end{bmatrix} = \begin{bmatrix} w_4^0 & w_4^0 & w_4^0 & w_4^0 \\ w_4^0 & w_4^1 & w_4^2 & w_4^{-1} \\ w_4^0 & w_4^2 & w_4^0 & w_4^2 \\ w_4^0 & w_4^{-1} & w_4^2 & w_4^1 \end{bmatrix} \begin{bmatrix} x_1(0) \\ x_1(1) \\ x_1(2) \\ x_1(3) \end{bmatrix}$$

$$= \begin{bmatrix} 1 & 1 & 1 & 1 \\ 1 & -j & -1 & j \\ 1 & -1 & 1 & -1 \\ 1 & j & -1 & -j \end{bmatrix} \begin{bmatrix} 1 \\ 0 \\ 1 \\ 1 \end{bmatrix}$$

$$= \begin{bmatrix} 1+0+1+1 \\ 1+0-1+j \\ 1+0+1-1 \\ 1+0-1-j \end{bmatrix} = \begin{bmatrix} 3 \\ j \\ 1 \\ -j \end{bmatrix}$$

$$\therefore \qquad X_1(k) = \{3,\, j,\, 1,\, -j\}$$

$$\begin{bmatrix} X_2(0) \\ X_2(1) \\ X_2(2) \\ X_2(3) \end{bmatrix} = \begin{bmatrix} 1 & 1 & 1 & 1 \\ 1 & -j & -1 & j \\ 1 & -1 & 1 & -1 \\ 1 & j & -1 & -j \end{bmatrix} \begin{bmatrix} 1 \\ 1 \\ 0 \\ 0 \end{bmatrix} = \begin{bmatrix} 1+1+0+0 \\ 1-j+0+0 \\ 1-1+0+0 \\ 1+j+0+0 \end{bmatrix} = \begin{bmatrix} 2 \\ 1-j \\ 0 \\ 1+j \end{bmatrix}$$

$$\therefore \qquad X_2(k) = \{2,\, 1-j,\, 0,\, 1+j\}$$

$$Y(k) = X_1(k)\, X_2(k)$$

$$Y(0) = X_1(0)\, X_2(0) = 6$$

$$Y(1) = X_1(1)\, X_2(1) = 1+j$$

$$Y(2) = X_1(2)\, X_2(2) = 0$$

$$Y(3) = X_1(3)\, X_2(3) = 1-j$$

$$\therefore \quad \begin{bmatrix} y(0) \\ y(1) \\ y(2) \\ y(3) \end{bmatrix} = \frac{1}{4} \begin{bmatrix} 1 & 1 & 1 & 1 \\ 1 & j & -1 & -j \\ 1 & -1 & 1 & -1 \\ 1 & -j & -1 & j \end{bmatrix} \begin{bmatrix} 6 \\ 1+j \\ 0 \\ 1-j \end{bmatrix}$$

$$= \frac{1}{4} \begin{bmatrix} 6+1+j+0+1-j \\ 6-1+j+0-1-j \\ 6-1-j+0-1+j \\ 6-j+1+0+1+j \end{bmatrix}$$

$$= \frac{1}{4} \begin{bmatrix} 8 \\ 4 \\ 4 \\ 8 \end{bmatrix} = \begin{bmatrix} 2 \\ 1 \\ 1 \\ 2 \end{bmatrix}$$

$$\therefore \quad y(n) = \{2, 1, 1, 2\}$$

Diagonal Method:

Step **1:** Make the lengths of two input sequences equal, if they are not equal, by padding with zeros.

$$\therefore \quad x_1(n) = \{1, 0, 1, 1\} \text{ and } x_2(n) = \{1, 1, 0, 0\}$$

Step **2:** Write down the sequences $x_1(n)$ and $x_2(n)$ as shown below.

$$\begin{array}{c c c c c}
 & & & x_1(n) & \\
 & 1 & 0 & 1 & 1 \\
\hline
1 & & & & \\
1 & & & & \\
x_2(n) \quad 0 & & & & \\
0 & & & &
\end{array}$$

Step **3:** Multiply each and every sample in $x_1(n)$ with the samples of $x_2(n)$ and tabulate the values.

Step **4:** Separate the elements in the table by drawing diagonal lines as shown below.

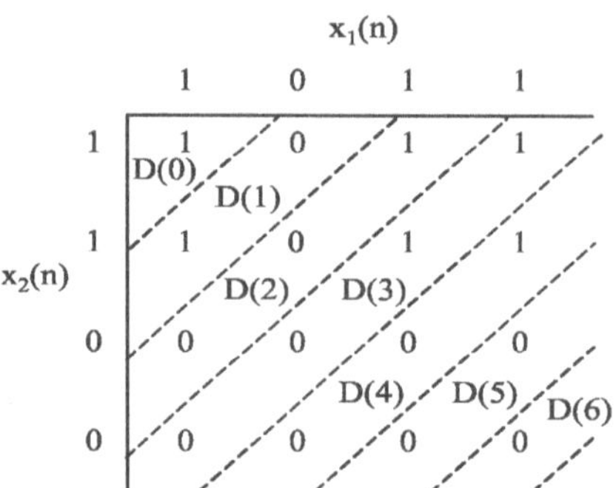

***Step* 5:** Denote the diagonals from top to bottom with D(0), D(1), …, D(N – 2), D(N – 1), D(N), D(N + 1), D(N + 2), …, D(2N – 1).

where N is the length of input sequence (either $x_1(n)$ or $x_2(n)$ after making their lengths equal).

***Step* 6:** Add the elements of diagonal D(0) with D(N) to get y(0)

Add the elements of diagonal D(1) with D(N + 1) to get y(1)

Add the elements of diagonal D(2) with D(N + 2) to get y(2)

$$\vdots \qquad\qquad \vdots \qquad\qquad \vdots$$

Add the elements of diagonal D(N – 2) with D(2N – 2) to get y(N – 2)

Add the elements of diagonal D(N – 1) itself to get y(N – 1)

where y(0), y(1), …,   y(N – 1) is the result of circular convolution.

Here N = 4,

We have diagonals D(0), D(1), D(2), D(3), D(4), D(5), D(6)

∴   To get y(0) = Add the elements of D(0) with D(4) = 2

y(1) = Add the elements of D(1) with D(5) = 1

y(2) = Add the elements of D(2) with D(6) = 1

y(3) = Add the elements of D(3) itself = 2

∴   y(n) = {2, 1, 1, 2}

Example 2.8: Given the two sequences $x_1(n) = a^n$ and $x_2(n) = b^n$ of length 4.

Find $x_1(n) \otimes x_2(n)$ (Circular convolution)   [JNTU 2000/S]

(i)   Direct form

(ii) Using DFT

(iii) Graphical method

Solution: $N = 4$

$$\therefore \quad x_1(n) = \{1,\ a,\ a^2,\ a^3\}, \quad x_2(n) = \{1,\ b,\ b^2,\ b^3\}$$

(i)   Direct form

$$y(n) = \sum_{k=0}^{N-1} x_1(k)\ x_2\big((n-k)\ \text{mod}\ N\big)$$

$$y(n) = x_1(0)\ x_2\big((n)\ \text{mod}\ 4\big) + x_1(1)x_2\big((n-1)\ \text{mod}\ 4\big) + x_1(2)x_2\big((n-2)\ \text{mod}\ 4\big)$$
$$+ x_1(3)\ x_2\big((n-3)\ \text{mod}\ 4\big)$$

$$y(0) = x_1(0)\ x_2\big((0)\ \text{mod}\ 4\big) + x_1(1)x_2\big((-1)\ \text{mod}\ 4\big) + x_1(2)x_2\big((-2)\ \text{mod}\ 4\big)$$
$$+ x_1(3)\ x_2\big((-3)\ \text{mod}\ 4\big)$$

$$= x_1(0)x_2(0) + x_1(1)x_2(3) + x_1(2)x_2(2) + x_1(3)x_2(1)$$

$$= 1 \times 1 + a \times b^3 + a^2 b^2 + a^3 b$$

$$y(0) = 1 + ab^3 + a^2 b^2 + a^3 b$$

$$y(1) = x_1(0)x_2(1) + x_1(1)x_2(0) + x_1(2)x_2(3) + x_1(3)x_2(2)$$

$$= 1 \times b + a \times 1 + a^2 b^3 + a^3 b^2$$

$$y(1) = b + a + a^2 b^3 + a^3 b^2$$

$$y(2) = x_1(0)x_2(2) + x_1(1)x_2(1) + x_1(2)x_2(0) + x_1(3)x_2(3)$$

$$= 1 \times b^2 + a \times b + a^2 \times 1 + a^3 b^3$$

$$y(2) = b^2 + ab + a^2 + a^3 b^3$$

$$y(3) = x_1(0)x_2(3) + x_1(1) + x_2(2) + x_1(2)x_2(1) + x_1(3)x_2(0)$$

$$= 1 \times b^3 + a \times b^2 + a^2 \times b + a^3 \times 1$$

$$y(3) = b^3 + ab^2 + a^2 b + a^3$$

$$\therefore \quad y(n) = \{1 + ab^3 + a^2 b^2 + a^3 b,\ b + a + a^2 b^3 + a^3 b^2,\ b^2 + ab + a^2 + a^3 b^3,$$
$$b^3 + ab^2 + a^2 b + a^3\}$$

(ii)   Using DFT

$$X_1(k) = \sum_{n=0}^{3} x_1(n) W_4^{kn}, \qquad k = 0, 1, 2, 3$$

$$\begin{bmatrix} X_1(0) \\ X_1(1) \\ X_1(2) \\ X_1(3) \end{bmatrix} = \begin{bmatrix} 1 & 1 & 1 & 1 \\ 1 & -j & -1 & j \\ 1 & -1 & 1 & -1 \\ 1 & j & -1 & -j \end{bmatrix} \begin{bmatrix} 1 \\ a \\ a^2 \\ a^3 \end{bmatrix}$$

$$\begin{bmatrix} X_1(0) \\ X_1(1) \\ X_1(2) \\ X_1(3) \end{bmatrix} = \begin{bmatrix} 1 + a + a^2 + a^3 \\ 1 - aj - a^2 + ja^3 \\ 1 - a + a^2 - a^3 \\ 1 + aj - a^2 - ja^3 \end{bmatrix}$$

Similarly
$$\begin{bmatrix} X_2(0) \\ X_2(1) \\ X_2(2) \\ X_2(3) \end{bmatrix} = \begin{bmatrix} 1 & 1 & 1 & 1 \\ 1 & -j & -1 & j \\ 1 & -1 & 1 & -1 \\ 1 & j & -1 & -j \end{bmatrix} \begin{bmatrix} 1 \\ b \\ b^2 \\ b^3 \end{bmatrix}$$

$$\begin{bmatrix} X_2(0) \\ X_2(1) \\ X_2(2) \\ X_2(3) \end{bmatrix} = \begin{bmatrix} 1 + b + b^2 + b^3 \\ 1 - jb - b^2 + jb^3 \\ 1 - b + b^2 - b^3 \\ 1 + jb - b^2 - jb^3 \end{bmatrix}$$

$$Y(0) = X_1(0) X_2(0) = 1 + a + b + ab + a^2 + b^2 + a^3 + b^3 + ba^2 + ba^3 + ab^2$$
$$+ ab^3 + b^2 a^2 + b^2 a^3 + a^2 b^3 + b^3 a^3$$

$$Y(1) = X_1(1) X_2(1) = 1 - aj - a^2 + a^3 j - bj - ab + a^2 bj + a^3 b - b^2$$
$$+ b^2 aj + a^2 b^2 - b^2 a^3 j + b^3 j + ab^3 - a^2 b^3 j - a^3 b^3$$

$$Y(2) = X_1(2) X_2(2) = 1 - a + a^2 - a^3 - b + ab - ba^2 + ba^3 + b^2 - ab^2 + a^2 b^2$$
$$- a^3 b^2 - b^3 + ab^3 - a^2 b^3 + a^3 b^3$$

$$Y(3) = X_1(3)X_2(3) = 1 + aj - a^2 - a^3 j + bj - ab - a^2 bj + a^3 b - b^2 - b^2 aj$$
$$+ a^2 b^2 + a^3 b^2 j - b^3 j + ab^3 + a^2 b^3 j - a^3 b^3$$

$$y(n) = \frac{1}{4}\begin{bmatrix} 1 & 1 & 1 & 1 \\ 1 & +j & -1 & -j \\ 1 & -1 & 1 & -1 \\ 1 & -j & -1 & j \end{bmatrix}\begin{bmatrix} Y(0) \\ Y(1) \\ Y(2) \\ Y(3) \end{bmatrix}$$

$$y(n) = \left\{ 1 + ab^3 + a^2 b^2 + a^3 b,\ a + b + a^2 b^3 + a^3 b^2,\ b^2 + ab + a^2 + a^3 b^3, b^3 + ab^2 + a^2 b + a^3 \right\}$$

(iii)   Graphical method

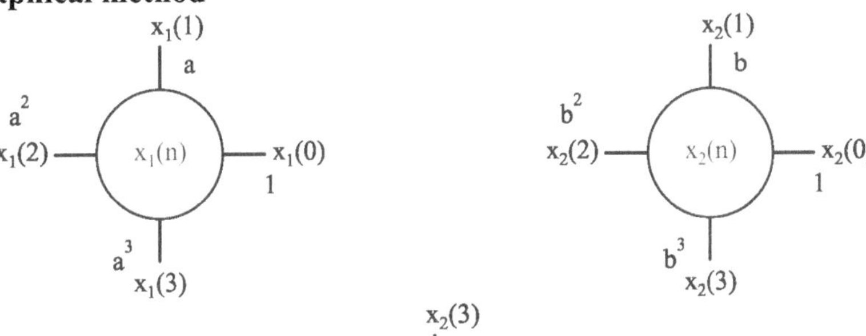

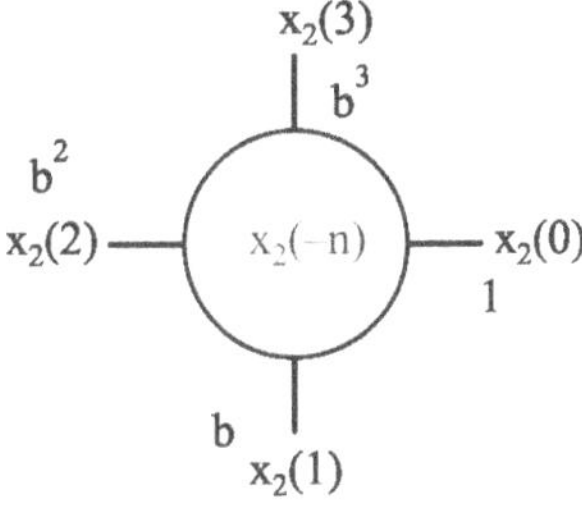

$$y(0) = x_1(n)\, x_2(-n) = 1 + ab^3 + a^2 b^2 + ba^3$$

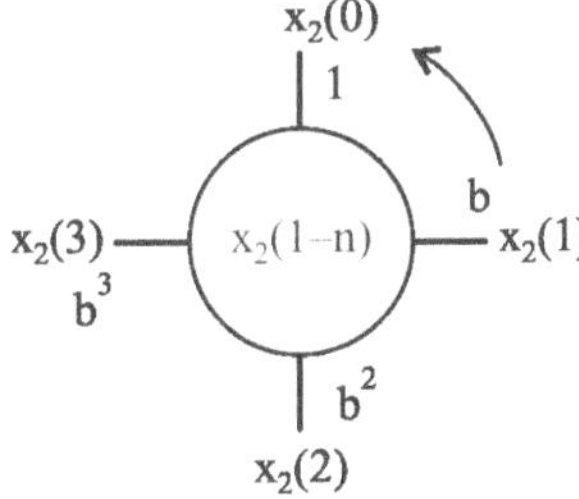

$$y(1) = x_1(n)\, x_2(1-n) = b + a + a^2 b^3 + a^3 b^2$$

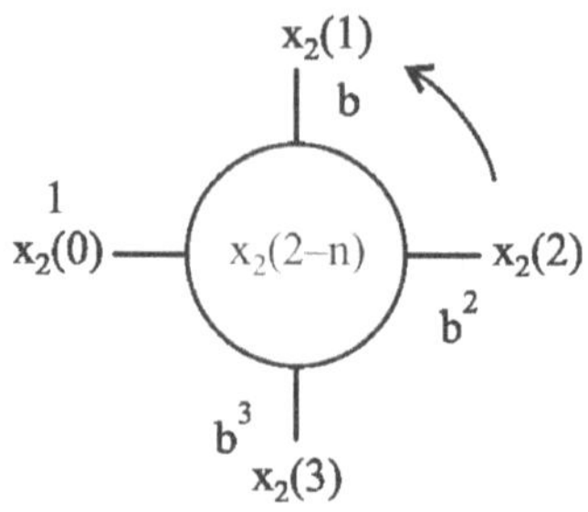

$$y(2) = x_1(n)\, x_2(2-n) = b^2 + ab + a^2 + a^3 b^3$$

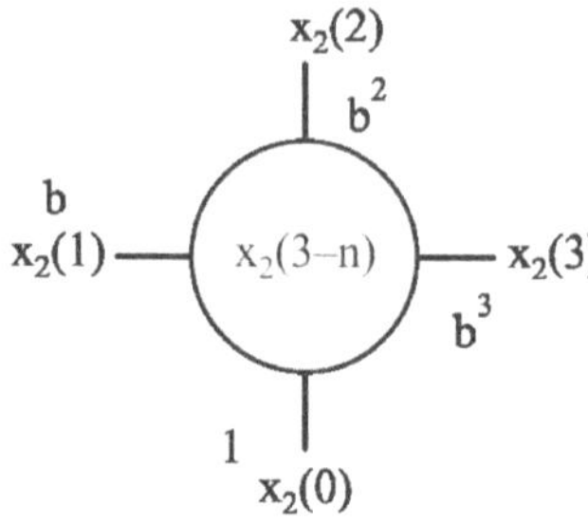

$$y(3) = x_1(n)\, x_2(3-n) = b^3 + ab^2 + a^2 b + a^3$$

$$y(n) = \left\{ 1 + ab^3 + a^2 b^2 + ba^3,\ b + a + a^2 b^3 + a^3 b^2,\ b^2 + ab + a^2 + a^3 b^3, b^3 + ab^2 + a^2 b + a^3 \right\}$$

Example 2.9: Compute $x_1(n) \otimes x_2(n)$ if

$$x_1(n) = \delta(n) + \delta(n-1) - \delta(n-2) - \delta(n-3)$$

and $\qquad x_2(n) = \delta(n) - \delta(n-2) + \delta(n-4)$

given $n = 5$ [JNTU 2000]

Solution: Let $y(n) = x_1(n) \otimes x_2(n)$

$$x_1(0) = \delta(0) + \delta(-1) - \delta(-2) - \delta(-3) = 1$$

$$x_1(1) = \delta(1) + \delta(0) - \delta(-1) - \delta(-2) = 1$$

$$x_1(2) = \delta(2) + \delta(1) - \delta(0) - \delta(-1) = -1$$

$$x_1(3) = \delta(3) + \delta(2) - \delta(1) - \delta(0) = -1$$

$$x_1(4) = \delta(4) + \delta(3) - \delta(2) - \delta(1) = 0$$

$$x_1(n) = \{1,\ 1,\ -1,\ -1,\ 0\}$$

$$x_2(0) = \delta(0) - \delta(-2) + \delta(-4) = 1$$

$$x_2(1) = \delta(1) - \delta(-1) + \delta(-3) = 0$$

$$x_2(2) = \delta(2) - \delta(0) + \delta(-2) = -1$$

$$x_2(3) = \delta(3) - \delta(1) + \delta(-1) = 0$$

$$x_2(4) = \delta(4) - \delta(2) + \delta(0) = 1$$

$$x_2(n) = \{1,\ 0,\ -1,\ 0,\ 1\}$$

By using Matrix method

$$\begin{bmatrix} y(0) \\ y(1) \\ y(2) \\ y(3) \\ y(4) \end{bmatrix} = \begin{bmatrix} 1 & 0 & -1 & -1 & 1 \\ 1 & 1 & 0 & -1 & -1 \\ -1 & 1 & 1 & 0 & -1 \\ -1 & -1 & 1 & 1 & 0 \\ 0 & -1 & -1 & 1 & 1 \end{bmatrix} \begin{bmatrix} 1 \\ 0 \\ -1 \\ 0 \\ 1 \end{bmatrix}$$

$$= \begin{bmatrix} 1+1+1 \\ 1-1 \\ -1-1-1 \\ -1-1 \\ 1+1 \end{bmatrix} = \begin{bmatrix} 3 \\ 0 \\ -3 \\ -2 \\ 2 \end{bmatrix}$$

$$y(n) = \{3,\ 0,\ -3,\ -2,\ 2\}$$

2.6.3   Linear Convolution using Circular Convolution

Let $x_1(n) = \{1,2,3\}$ and $x_2(n) = \{1,0,1,2\}$

The output sequence of linear convolution will have length $M + N - 1$

where M is the length of input sequence $x_1(n)$ and

   N is the length of input sequence $x_2(n)$

   To perform linear convolution using circular convolution, make the input sequences of length $M + N - 1$ by padding with zeros.

here $M = 3$ and $N = 4$

∴ $M + N - 1 = 3 + 4 - 1 = 6$

∴ $x_1(n) = \{1, 2, 3, 0, 0, 0\}$ and $x_2(n) = \{1, 0, 1, 2, 0, 0\}$

Now perform circular convolution on these two sequences which gives the output of Linear convolution.

$$
\begin{bmatrix} y(0) \\ y(1) \\ y(2) \\ y(3) \\ y(4) \\ y(5) \end{bmatrix} = \begin{bmatrix} 1 & 0 & 0 & 0 & 3 & 2 \\ 2 & 1 & 0 & 0 & 0 & 3 \\ 3 & 2 & 1 & 0 & 0 & 0 \\ 0 & 3 & 2 & 1 & 0 & 0 \\ 0 & 0 & 3 & 2 & 1 & 0 \\ 0 & 0 & 0 & 3 & 2 & 1 \end{bmatrix} \begin{bmatrix} 1 \\ 0 \\ 1 \\ 2 \\ 0 \\ 0 \end{bmatrix}
$$

$$
\begin{bmatrix} y(0) \\ y(1) \\ y(2) \\ y(3) \\ y(4) \\ y(5) \end{bmatrix} = \begin{bmatrix} 1 \\ 2 \\ 3+1 \\ 2+2 \\ 3+4 \\ 6 \end{bmatrix} \begin{bmatrix} 1 \\ 2 \\ 4 \\ 4 \\ 7 \\ 6 \end{bmatrix}
$$

$$y(n) = \{1, 2, 4, 4, 7, 6\}$$

2.6.4   Circular Convolution using Linear Convolution

Let $x_1(n) = \{1, 2, 3\}$ and $x_2(n) = \{1, 0, 1, 2\}$

To perform circular convolution, first perform Linear convolution on the given sequences as shown below:

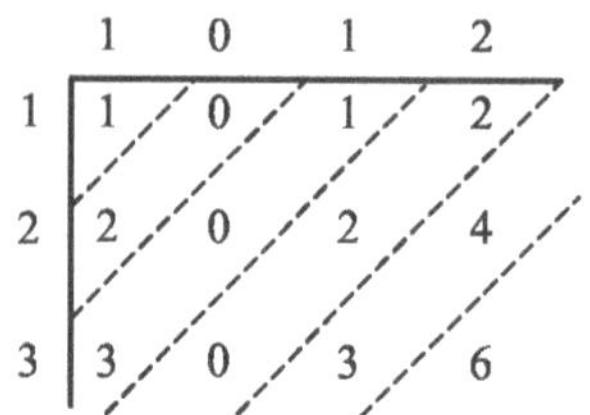

$$y(0) = 1$$
$$y(1) = 2 + 0 = 2$$
$$y(2) = 3 + 0 + 1 = 4$$
$$y(3) = 0 + 2 + 2 = 4$$
$$y(4) = 3 + 4 = 7$$
$$y(5) = 6$$
$$\therefore \quad y(n) = \{1, 2, 4, 4, 7, 6\}$$

Here the length of $y(n)$ is $M + N - 1 = 6$

but the length of output sequence of circular convolution will be maximum of M and N. Here it is 4.

So we need only $y(0)$, $y(1)$, $y(2)$ and $y(3)$

To get circular convolution output from the Linear convolution.

Observe in the above table $y(2)$ and $y(3)$ diagonals are having 3 elements each. To get 3 elements in $y(0)$ and $y(1)$, add $y(4)$ elements to $y(0)$ and $y(5)$ elements to $y(1)$.

$$\therefore \quad y(0) = 1 + 3 + 4 = 8$$

$$y(1) = 2 + 0 + 6 = 8$$

$$y(2) = 3 + 0 + 1 = 4$$

$$y(3) = 0 + 2 + 2 = 4$$

$\therefore$ Output of circular convolution is $y(n) = \{8, 8, 4, 4\}$

2.6.5 Differences between Linear and Circular Convolutions

Linear Convolution	Circular Convolution
1. The output sequence of Linear Convolution of $x_1(n)$ and $x_2(n)$ is $$y(n) = x_1(n) \times x_2(n)$$ $$= \sum_{k=-\infty}^{\infty} x_1(k)\, x_2(n-k)$$	1. The output sequence of Circular Convolution of $x_1(n)$ and $x_2(n)$ is $$y(n) = x_1(n) \otimes x_2(n)$$ $$= \sum_{k=0}^{N-1} x_1(k)\, x_2((n-k) \bmod N)$$ where N is the length of input sequences (after making of equal lengths) or output sequence.
2. Linear Convolution is performed by folding (time reversing) one sequence, shifting the folded sequence, multiplying the two sequences to obtain a product sequence and finally summing the values of the product sequence.	2. Circular Convolution is performed by folding and shifting (rotating) one of the sequence in circular fashion by computing the index of the sequence modulo N. In Linear Convolution there is no modulo N operation.
3. If lengths of input sequences are M and N, then the output sequence length will be $M + N - 1$.	
4. Output sequence is not a periodic.	3. If lengths of input sequences are M and N, then a) If M = N, output length will be M or N. b) If M > N, output

	length will be M.
	c) If $M < N$, output length will be N.
	4. Output sequence is periodic.

2.7 Computation of DFT

The DFT of a sequence x(n) can be found by using the formula

$$X(k) = \sum_{n=0}^{N-1} x(n) \, w_N^{kn}, \qquad k = 0, 1, 2, \ldots, N-1 \qquad \ldots\ldots(2.7.1)$$

where $\quad W_N = e^{-j2\pi/N}$, which is known as phase factor or twiddle factor.

Let us represent x(n) in complex form

i.e., $\quad x(n) = x_R(n) + jx_I(n)$

$$X(k) = \sum_{n=0}^{N-1} \left[x_R(n) + jx_I(n) \right] \left[\cos\left(\frac{2\pi kn}{N}\right) - j \sin\left(\frac{2\pi kn}{N}\right) \right] \qquad \ldots\ldots (2.7.2)$$

$$= \sum_{n=0}^{N-1} x_R(n)\cos\left(\frac{2\pi kn}{N}\right) + \sum_{n=0}^{N-1} x_I(n)\sin\left(\frac{2\pi kn}{N}\right) + j\sum_{n=0}^{N-1} x_I(n)\cos\left(\frac{2\pi kn}{N}\right)$$

$$-j\sum_{n=0}^{N-1} x_R(n)\sin\left(\frac{2\pi kn}{N}\right) \qquad \ldots\ldots (2.7.3)$$

From equation (2.7.2), it is evident that the direct computation of X(k), for each value of k, requires N complex multiplications and (N–1) complex additions. So to evaluate X(k) for all values of k, it requires N^2 complex multiplications and N(N–1) complete additions.

Similarly from equation (2.7.3), we can observe that the computation of X(k) for each k requires 4N real multiplications. Therefore to evaluate X(k) for all values of k requires $4N^2$ real multiplications.

In equation (2.7.3), each of the four sums of N terms requires N–1 real additions, and to combine the sum to get the real part and imaginary part requires two more additions.

Therefore to evaluate X(k) for each k requires 4(N–1) + 2 real additions. For all values of k a total number of real additions N(4N–2) required for direct evaluation of DFT.

Twiddle factor exhibits the following properties.

Symmetry property $\quad W_N^{K+\frac{N}{2}} = -W_N^K$

$$\text{Periodicity property} \quad W_N^{K+N} = W_N^K \qquad \qquad .....(2.7.4)$$

By exploiting the above properties of phase factor, number of computations can be reduced i.e., transformation becomes more faster. The algorithm that exploits the above properties of twiddle factor and makes the transformation faster is called Fast Fourier Transform (FFT).

2.7.1 Fast Fourier Transform (FFT)

As we discussed in the previous section, the Fast Fourier Transform algorithms exploit the two basic properties of phase factor given in equation (2.7.4) and reduces the number of complex multiplications and additions required to perform DFT from N^2 to $\dfrac{N}{2}\log_2 N$

and $N(N-1)$ to $N\log_2 N$ respectively.

FFT algorithms are based on the fundamental principle of decomposing the computation of discrete Fourier Transform of a sequence of length N into successively smaller discrete Fourier transforms. There are basically two classes of FFT algorithms. They are decimation-in-time and decimation-in-frequency.

2.7.2 Radix–2 FFT

By adopting Divide and Conquer method Cooley, Trikey have developed FFT algorithms. According to this algorithm, number of computations required to compute DFT can be reduced by breaking N–pt DFT, where N must be composite number (it should not be prime number), into even numbered samples DFT and odd numbered samples DFT. This process repeats till we get smaller size DFTs.

If N can be written as $N = r_1, r_2, r_3, ..., r_m$, where $r_1 = r_2 = r_3 = ... = r_m = r$, then $N = r^m$. So we can decimate N–pt DFT up to r–pt DFT. From the results of r–pt DFT, r^2 DFTs, From the results of r^2–pt DFT, r^3 DFTs..... From the results of r^{m-1} DFTs ,r^m DFT or N–pt DFTs can be calculated where r is the radix or base. Generally we use r = 2, hence the name Radix–2 FFT algorithm.

2.7.3 Radix–2 Decimation in Time (DIT) Algorithm

In decimation in time algorithm, time domain sequence x(n) in N–pt DFT is decimated into even numbered samples N/2–pt DFT and odd numbered samples N/2–pt DFT and this process repeats till we get 2–pt DFTs which is as discussed below:

$$\text{The N–pt DFT of sequence x(n) is } X(k) = \sum_{n=0}^{N-1} x(n)\, W_N^{kn}, \qquad k = 0, 1, ..., N-1$$

decimating x(n) into even and odd numbered samples we get

$$X(k) = \sum_{n \text{ even}} x(n) \, W_N^{kn} + \sum_{n \text{ odd}} x(n) \, W_N^{kn} \qquad\qquad \text{..... (2.7.5)}$$

Substitute $n = 2n$ for n even and $n = 2n + 1$ for n odd, then

$$X(k) = \sum_{n=0}^{\frac{N}{2}-1} x(2n) \, W_N^{2nk} + \sum_{n=0}^{\frac{N}{2}-1} x(2n+1) \, W_N^{(2n+1)k} \,, \ k = 0, 1, 2, ..., N-1 \qquad \text{..... (2.7.6)}$$

Put $x(2n) = g_1(n)$ and $x(2n + 1) = g_2(n)$

we know $W_N = e^{-j2\pi/N}$

$$\therefore \qquad W_N^{2nk} = e^{(-j2\pi/N)^{2nk}} = e^{(-j2\pi/N/2)^{nk}} = W_{N/2}^{nk}$$

$$X(k) = \sum_{n=0}^{\frac{N}{2}-1} g_1(n) \, W_{N/2}^{nk} + W_N^{K} \sum_{n=0}^{\frac{N}{2}-1} g_2(n) W_{N/2}^{nk} \,, \ k = 0, 1, 2, ..., N-1$$

$$X(k) = G_1(k) + W_N^{K} \, G_2(k), \qquad\qquad k = 0, 1, 2, ..., N-1 \ \text{.....(2.7.7)}$$

where $G_1(k)$ and $G_2(k)$ are N/2–pt DFTs of even numbered samples and odd numbered samples respectively. It can be shown that $G_1(k)$ and $G_2(k)$ are periodic with period N/2, hence k ranges from 0 to N/2–1.

For $N = 8$

From equation (2.7.7)

$$X(k) = G_1(k) + W_8^{K} \, G_2(k), \qquad k = 0, 1, 2, 3, 4, 5, 6, 7$$

$$X(0) = G_1(0) + W_8^{0} \, G_2(0)$$

$$X(1) = G_1(1) + W_8^{1} \, G_2(1)$$

$$X(2) = G_1(2) + W_8^{2} \, G_2(2)$$

$$X(3) = G_1(3) + W_8^{3} \, G_2(3) \qquad\qquad \text{.....(2.7.8)}$$

$$X(4) = G_1(4) + W_8^{4} \, G_2(4)$$

$$X(5) = G_1(5) + W_8^{5} \, G_2(5)$$

$$X(6) = G_1(6) + W_8^{6} \, G_2(6)$$

$$X(7) = G_1(7) + W_8^{7} \, G_2(7) \qquad\qquad \text{.....(2.7.9)}$$

Symmetry property of twiddle factor is $W_N^{K+N/2} = -W_N^K$

$\therefore$ by using symmetry property of twiddle factor

$$W_8^4 = W_8^{0+4} = -W_8^0$$
$$W_8^5 = W_8^{1+4} = -W_8^1$$
$$W_8^6 = W_8^{2+4} = -W_8^2$$
$$W_8^7 = W_8^{3+4} = -W_8^3$$

Since $G_1(k)$ and $G_2(k)$ are periodic with period N/2.

$$G_1(4) = G_1(0+4) = G_1(0) \text{ and } G_2(4) = G_2(0+4) = G_2(0)$$
$$G_1(5) = G_1(1) \text{ and } G_2(5) = G_2(1)$$
$$G_1(6) = G_1(2) \text{ and } G_2(6) = G_2(2)$$
$$G_1(7) = G_1(3) \text{ and } G_2(7) = G_2(3)$$

equation: (2.7.9) becomes

$\therefore$

$$X(4) = G_1(0) - W_8^0 G_2(0)$$
$$X(5) = G_1(1) - W_8^1 G_2(1)$$
$$X(6) = G_1(2) - W_8^2 G_2(2)$$
$$X(7) = G_1(3) - W_8^3 G_2(3)$$

$$.....(2.7.10)$$

Equations (2.7.8) & (2.7.10) can be drawn as shown in Fig.2.6.

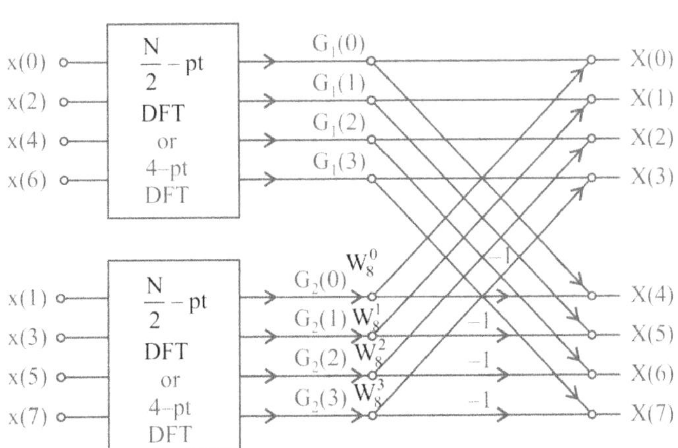

Fig. 2.6 Flow graph of the first stage decimation in time FFT algorithm for N = 8.

If we compute directly N-pt DFT, it requires N^2 complex multiplications and additions. This N-pt DFT is decimated into two $\dfrac{N}{2}-$ pt of DFTs as in eq. (2.7.7) in which each require $\left(\dfrac{N}{2}\right)^2$ complex multiplications and additions and W_N^k is multiplied with $G_2(k)$, which requires N computations. Hence total computations required are $N + 2\left(\dfrac{N}{2}\right)^2$.

For $N = 8$ Direct computation requires $N^2 = 8^2 = 64$

To compute N-pt DFT with two $\dfrac{N}{2}-$pt DFTs, computations required are

$$N + 2\left(\dfrac{N}{2}\right)^2 = 8 + 2(4)^2 = 40$$

Which is almost half of the computations required for direct computations.

Similarly $\dfrac{N}{2}-$pt DFTs $G_1(k)$ and $G_2(k)$ are decimated to get two $\dfrac{N}{4}-$pt DFTs for $G_1(k)$ and two $\dfrac{N}{4}-$pt DFTs for $G_2(k)$ as discussed below.

$$G_1(k) = \sum_{n=0}^{\frac{N}{2}-1} g_1(n) w_{N/2}^{nk}, \quad k = 0, 1, 2\ldots\ldots\dfrac{N}{2}-1 \qquad \ldots\ldots(2.7.11)$$

Decimating $g_1(n)$ into even and add numbered samples, we get

$$G_1(k) = \sum_{n\,\text{even}} g_1(n) W_{N/2}^{kn} + \sum_{n\,\text{odd}} g_1(n) W_{N/2}^{kn}$$

Substitute $n = 2n$ for n even and $n = 2n + 1$ for n odd, then

$$G_1(k) = \sum_{n=0}^{\frac{N}{4}-1} g_1(2n) W_{N/2}^{k2n} + \sum_{n=0}^{\frac{N}{4}-1} (2n+1) W_{N/2}^{k(2n+1)}, \quad k = 0, 1, 2\ldots\ldots\dfrac{N}{2}-1 \qquad \ldots\ldots(2.7.12)$$

Put $\quad g_1(2n) = a_1(n)$ and $g_1(2n+1) = a_2(n)$

$W_{N/2}^{2kn}$ can be written as $W_{N/4}^{kn}$

$$\therefore \ G_1(k) = \sum_{n=0}^{\frac{N}{4}-1} a_1(n) W_{N/2}^{kn} + W_{N/2}^{k} \sum_{n=0}^{\frac{N}{4}-1} a_2(n) W_{N/4}^{kn}, \ k = 0, 1 \dots \frac{N}{2}-1 \qquad \dots(2.7.13)$$

$$G_1(k) = A_1(k) + W_{N/2}^{k} A_2(k), \qquad k = 0, 1, \dots \frac{N}{2} - 1 \qquad \dots(2.7.14)$$

$$\text{Similarly} \ G_2(k) = \sum_{n=0}^{\frac{N}{4}-1} b_1(n) W_{N/4}^{Kn} + W_{N/2}^{k}, \sum_{n=0}^{\frac{N}{4}-1} b_2(n) W_{N/4}^{Kn}$$

$$k = 0, 1 \dots \frac{N}{2} - 1 \qquad \dots(2.7.15)$$

$$G_2(k) = B_1(k) + W_{N/2}^{k} B_2(k), \ k = 0, 1, \dots \frac{N}{2} - 1 \qquad \dots(2.7.16)$$

For $N = 8$

From Eq. (2.7.15)

$$G_1(k) = A_1(k) + W_{8/2}^{k} A_2(k), \ \ k = 0, 1, 2, 3$$

$$G_1(0) = A_1(0) + W_{8/2}^{0} A_2(0) \ \Big\}$$

$$G_1(1) = A_1(1) + W_{8/2}^{1} A_2(1) \qquad \qquad \dots(2.7.17)$$

$$G_1(2) = A_1(2) + W_{8/2}^{2} A_2(2) \ \Big\}$$

$$G_1(3) = A_1(3) + W_{8/2}^{3} A_2(3) \qquad \qquad \dots(2.7.18)$$

By using symmetry property of twiddle factor

$$W_{8/2}^{2} = W_4^{2} = W_4^{D+2} = -W_4^{0} = -W_{8/2}^{0} = -W_8^{0}$$

$$W_{8/2}^{3} = W_4^{3} = W_4^{1+2} = -W_4^{1} = -W_{8/2}^{1} = -W_8^{2}$$

$A_1(k)$ and $A_2(k)$ are periodic with period $\dfrac{N}{4}$.

$$A_1(0) = A_1(2) \ \text{and} \ A_2(0) = A_2(2)$$

$$A_1(1) = A_1(3) \ \text{and} \ A_2(1) = A_2(3)$$

$\therefore$ Equation (2.7.18) becomes

$$\left.\begin{array}{l} G_1(2) = A_1(0) - W_{8/2}^0 A_2(0) \\[2mm] G_1(3) = A_1(1) - W_{8/2}^1 A_2(1) \end{array}\right\} \qquad \ldots\ldots(2.7.19)$$

For $N = 8$

From eq. (2.7.16)

$$G_2(k) = B_1(k) + W_{8/2}^k B_2(k)$$

$$\left.\begin{array}{l} G_2(0) = B_1(0) + W_{8/2}^0 B_2(0) \\[2mm] G_2(1) = B_1(1) + W_{8/2}^1 B_2(1) \end{array}\right\} \qquad \ldots\ldots(2.7.20)$$

$$\left.\begin{array}{l} G_2(2) = B_1(2) + W_{8/2}^2 B_2(2) \\[2mm] G_2(3) = B_1(3) + W_{8/2}^3 B_2(3) \end{array}\right\} \qquad \ldots\ldots(2.7.21)$$

$$W_{8/2}^2 = -W_{8/2}^0 = -W_8^0 \quad \text{and} \quad W_{8/2}^3 = -W_{8/2}^1 = -W_8^2$$

$B_1(k)$ and $B_2(k)$ are periodic with period $\dfrac{N}{4}$

$\therefore \qquad B_1(0) = B_1(2)$ and $B_2(0) = B_2(2)$

$\qquad\qquad B_1(1) = B_1(3)$ and $B_2(1) = B_2(3)$

eqn. (2.7.22) becomes

$$\left.\begin{array}{l} G_2(2) = B_1(0) - W_{8/2}^0 B_2(0) \\[2mm] G_2(3) = B_1(1) - W_{8/2}^1 B_2(1) \end{array}\right\} \qquad \ldots\ldots(2.7.22)$$

By clubbing (2.7.17) and (2.7.19) upper block of Fig. 2.6 can be drawn as shown in Fig. 2.7 and by clubbing Eq. (2.7.20) and Eq. (2.7.22) lower block of Fig. 2.6 can be drawn as shown in Fig. 2.7.

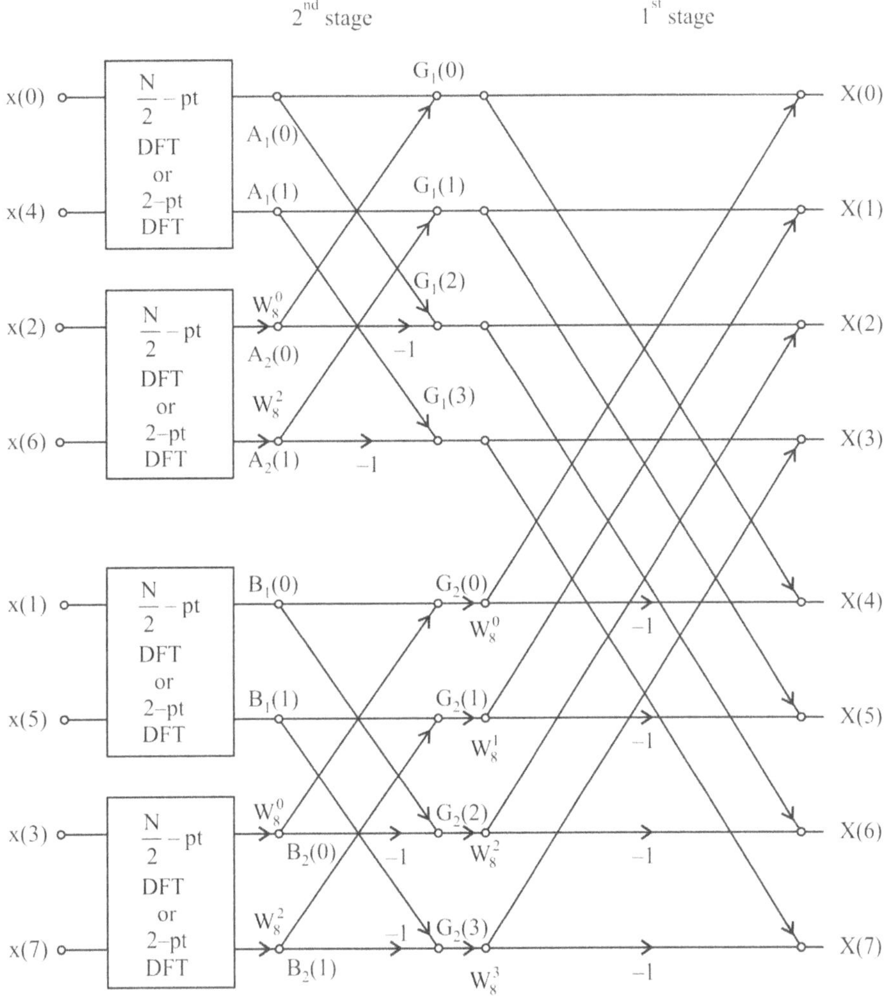

Fig. 2.7 Flow graph of the first and second stages decimation in time FFT algorithm for N = 8.

Direct computation requires N^2 computations $N^2 = 8^2 = 64$ $\qquad\qquad$ …..(2.7.23)

With first stage decimation (or) computing N–pt DFT with two $\dfrac{N}{2}$–Pt DFT^s, computations required are

$$N + 2\left(\frac{N}{2}\right)^2 = 40 \qquad\qquad …..(2.7.24)$$

With first and second stage decimation (or) computing N-pt PFT with two $\dfrac{N}{2}$–Pt

DFTs and computing these two $\dfrac{N}{2}$ DFTs with four $\dfrac{N}{4}$ – pt DFTs, computations required are

$$2\left[\frac{N}{2} + 2\left(\frac{N}{4}\right)^2\right] = 24 \qquad \qquad .....(2.7.25)$$

If we observe Eqs. (2.7.23), (2.7.24) and (2.7.25), No. of computations are reduced to almost half of the computations required for their proceeding stage.

i.e., Eq. (2.7.24) computations are almost $\dfrac{1}{2}$ of the computation of Eq. (2.7.23) and Eq. (2.7.25) computations are almost $\dfrac{1}{2}$ of the computations of Eq. (2.7.24).

From eq. (2.7.13) and (2.7.14); $A_1(k)$ and $A_2(k)$ are

$$\frac{N}{4} - \text{pt (or)} \ (\text{for } N = 8) \ 2 - \text{pt DFTs}$$

and

$$\left.\begin{aligned}
A_1(k) &= \sum_{n=0}^{\frac{N}{4}-1} a_1(n) W_{N/4}^{kn}, \quad k = 0, 1, 2.....\frac{N}{4} - 1 \\
A_2(k) &= \sum_{n=0}^{\frac{N}{4}-1} a_2(n) W_{N/4}^{kn}, \quad k = 0, 1, 2.....\frac{N}{4} - 1
\end{aligned}\right\} \qquad .....(2.7.26)$$

From eq. (2.7.15) and (2.7.16); $B_1(k)$ and $B_2(k)$ are

$$\frac{N}{4} - \text{pt (or)} \ (\text{for } N = 8) \ 2 - \text{pt DFTs}$$

$$\left.\begin{aligned}
B_1(k) &= \sum_{n=0}^{\frac{N}{4}-1} b_1(n) W_{N/4}^{kn}, \quad k = 0, 1.....\frac{N}{4} - 1 \\
B_2(k) &= \sum_{n=0}^{\frac{N}{4}-1} b_2(n) W_{N/4}^{kn}, \quad k = 0, 1.....\frac{N}{4} - 1
\end{aligned}\right\} \qquad .....(2.7.27)$$

for $N = 8$

from eq.s (2.7.26)

$$A_1(k) = \sum_{n=0}^{1} a_1(n) W_{8/4}^{kn}, \qquad \qquad k = 0, 1$$

$$= a_1(0) W_{8/4}^{0} + a_1(1) W_{8/4}^{k}, \qquad \qquad k = 0, 1$$

$$A_1(0) = a_1(0)\, W_{8/4}^0 + a_1(1)\, W_{8/4}^0$$

$$A_1(1) = a_1(0)\, W_{8/4}^0 + a_1(1)\, W_{8/4}^1$$

$$W_{8/4}^0 = W_8^0 = 1$$

$$W_{8/4}^1 = W_8^4 = W_8^{0+4} = -W_8^0$$

$$\therefore \quad A_1(0) = a_1(0) + a_1(1)\, W_8^0$$

$$A_1(1) = a_1(0) - a_1(1)\, W_8^0$$

From Fig. 2.7, the 1st block outputs are A$_1$(0) and A$_1$(1), whose inputs are x(0) and x(4) respectively.

$$\therefore \quad a_1(0) = x(0) \quad \text{and} \quad a_1(1) = x(4)$$

$$\therefore \qquad \left. \begin{aligned} A_1(0) &= x(0) + x(4)\, W_8^0 \\ A_1(1) &= x(0) - x(4)\, W_8^0 \end{aligned} \right\} \qquad \qquad \dots\dots(2.7.28)$$

Eq.s (2.7.28) can be drawn as below

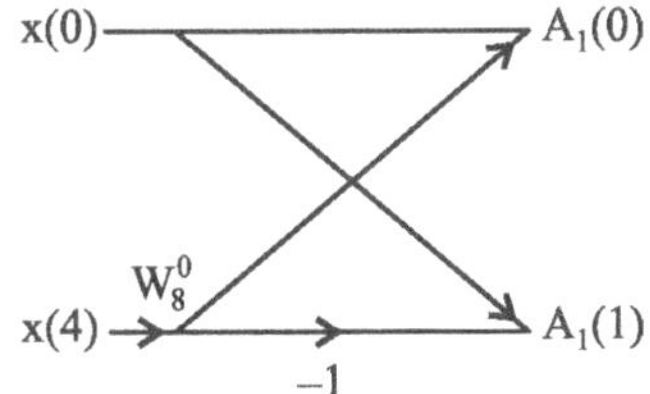

$$\text{Similarly} \qquad \left. \begin{aligned} A_2(0) &= x(2) + x(6)\, W_8^0 \\ A_2(1) &= x(2) - x(6)\, W_8^0 \end{aligned} \right\} \qquad \qquad \dots\dots(2.7.29)$$

and from eq. (2.7.27)

$$\left. \begin{aligned} B_1(0) &= x(1) + x(5)\, W_8^0 \\ B_1(1) &= x(1) - x(5)\, W_8^0 \end{aligned} \right\} \qquad \qquad \dots\dots(2.7.30)$$

$$\left. \begin{aligned} B_2(0) &= x(3) + x(7)\, W_8^0 \\ B_2(1) &= x(3) - x(7)\, W_8^0 \end{aligned} \right\} \qquad \qquad \dots\dots(2.7.31)$$

From eq.s (2.7.28), (2.7.29), (2.7.30) and (2.7.31), complete flow graph for N = 8, can be drawn as shown in Fig. 2.8.

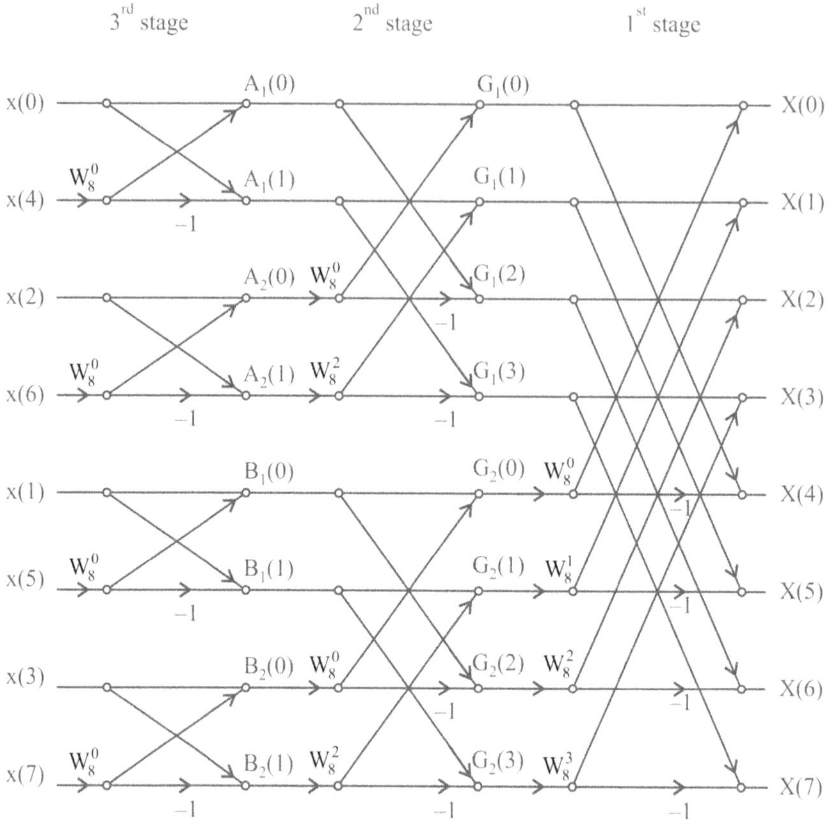

Fig. 2.8 Complete flow graph of decimation – in time FFT algorithm for N = 8.

From eqns (2.7.25), which is almost equal to $N \log_2^N$

$\therefore$ To compute N – pt DFT using FFT algorithm no. of computations required are $N \log_2^N$

If we observe Fig. 2.8, input samples are in Bit-Reversal order and output samples are in order.

2.7.4   Bit Reversal Operation (or) Scrambling Operation

Let $\hat{m}$ be the scrambled value of a given integer m. If m can be represented as a binary number as

$$m = m_{N-1} \, m_{N-2} \dots m_1 m_0$$

then $\qquad \hat{m} = m_0\ m_1 \ldots m_{N-2}\ m_{N-1}$

Example: Let $N = 16$: $\quad m = 1110 = (14)_{10}$

$$\hat{m} = 0111 = (7)_{10}$$

Let $N = 32$: $\quad m = 01110 = (14)_{10}$

$$\hat{m} = 01110 = (14)_{10}$$

Scrambled value changes with N

For $N = 8$

m	**m̂**
000 $(0)_{10}$	000 $(0)_{10}$
001 $(1)_{10}$	100 $(4)_{10}$
010 $(2)_{10}$	010 $(2)_{10}$
011 $(3)_{10}$	110 $(6)_{10}$
100 $(4)_{10}$	001 $(1)_{10}$
101 $(5)_{10}$	101 $(5)_{10}$
110 $(6)_{10}$	011 $(3)_{10}$
111 $(7)_{10}$	111 $(7)_{10}$

Example 2.10: Use eight point DIT FFT to find the DFT of the following sequence

$$x(n) = \left\{ \frac{1}{\sqrt{2}}, 1, \frac{1}{\sqrt{2}}, 0, -\frac{1}{\sqrt{2}}, -1, -\frac{1}{\sqrt{2}}, 0 \right\} \qquad \text{[JNTU 2000]}$$

Solution: We know $W_N = e^{-j2\pi/N}$, given $N = 8$

$$\therefore\ W_8^0 = \left(e^{-j2\pi/8}\right)^0 = 1$$

$$W_8^1 = \left(e^{-j2\pi/8}\right)^1 = \cos\frac{\pi}{4} - j\ \sin\frac{\pi}{4} = 0.707 - j\ 0.707$$

$$W_8^2 = \left(e^{-j2\pi/8}\right)^2 = \cos\frac{\pi}{2} - j\ \sin\frac{\pi}{2} = -j$$

$$W_8^3 = \left(e^{-j2\pi/8}\right)^3 = \cos\frac{3\pi}{4} - j\ \sin\frac{3\pi}{4} = -0.707 - j\ 0.707$$

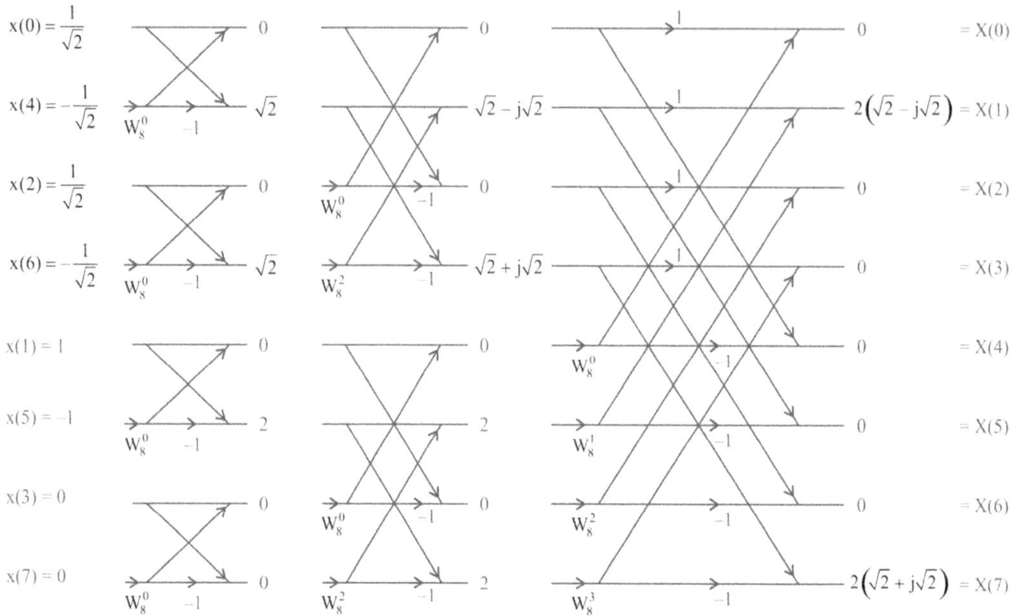

Fig. 2.9

Example 2.11: Find X(k) using DIT FFT algorithm of the following sequence x (n) = {0, 1, 1, 1}

Solution: Given N = 4

$$\therefore \qquad W_4^0 = 1, \qquad \text{and} \qquad W_4^1 = e^{-j\frac{2\pi}{4}} = -j$$

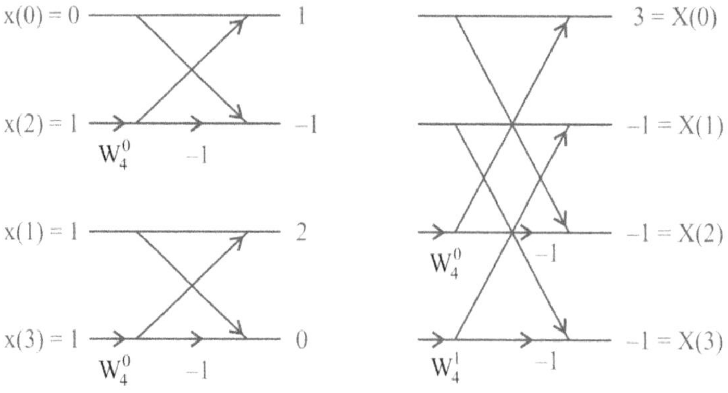

Fig. 2.10

Example 2.12: Find X(k) using DIT FFT algorithm of the following sequence x (n) = {1, 2, 3}

Solution: Here length of the sequence is 3, which cannot be represented as integer power of 2. So pad the sequence with 0

$$\therefore \quad x(n) = \{1, 2, 3, 0\}$$

Now length $N = 4 = 2^2$

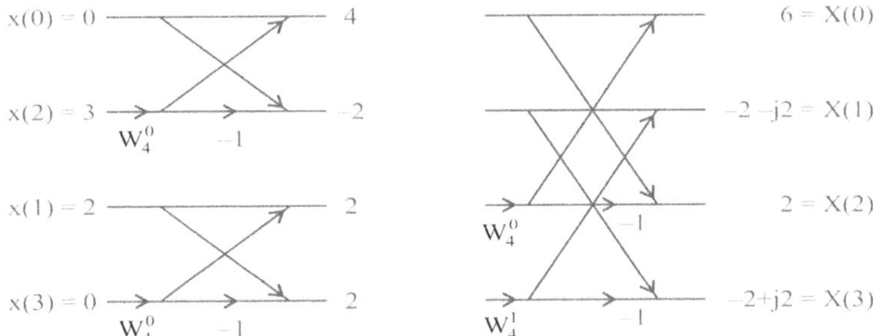

Fig. 2.11

2.7.5 Decimation-in-Frequency (DIF) FFT Algorithm

Decimation-in-Frequency FFT decomposes the DFT by splitting the sequence elements X(k) in the frequency domain into sets of smaller and smaller subsequences. The algorithm is as discussed below

We know N – pt DFT equation. $X(k) = \sum\limits_{n=0}^{N-1} x(n)\, W_N^{kn}, \quad k = 0, 1, 2...N-1$

where $\quad W_N = e^{-j2\pi/N}$

The summation is divided into two sums as shown in eqs (2.7.32).

$$X(k) = \sum\limits_{n=0}^{\frac{N}{2}-1} x(n)\, W_N^{kn} + \sum\limits_{n=\frac{N}{2}}^{N-1} x(n)\, W_N^{nk} \qquad .....(2.7.32)$$

Put $n = n + \dfrac{N}{2}$ in the second sum

$$\therefore \quad X(k) = \sum\limits_{n=0}^{\frac{N}{2}-1} x(n)\, W_N^{kn} + \sum\limits_{n=0}^{\frac{N}{2}-1} x\left(n + \frac{N}{2}\right) W_N^{k\left(n+\frac{N}{2}\right)}$$

$$= \sum\limits_{n=0}^{\frac{N}{2}-1} x(n)\, W_N^{kn} + W_N^{\frac{kN}{2}} \sum\limits_{n=0}^{\frac{N}{2}-1} x\left(n + \frac{N}{2}\right) W_N^{kn}$$

Since $\quad W_N^{\frac{kN}{2}} = e^{(-j2\pi/N)kN/2} = \left(e^{-j\pi}\right)^k = \left(\cos\pi - j\sin\pi\right)^k = (-1)^k$, we get

$$X(k) = \sum_{n=0}^{\frac{N}{2}-1} x(n)\, W_N^{kn} + (-1)^k \sum_{n=0}^{\frac{N}{2}-1} x\left(n+\frac{N}{2}\right) W_N^{kn}$$

$$X(k) = \sum_{n=0}^{\frac{N}{2}-1}\left[x(n)+(-1)^k x\left(n+\frac{N}{2}\right)\right] W_N^{kn} \qquad\qquad .....(2.7.33)$$

Decompose the sequence X(k) into even numbered and odd numbered samples by substituting $2k = 2k$ for even and $k = 2k + 1$ for odd in the eq (2.7.33)

$$X(2k) = \sum_{n=0}^{\frac{N}{2}-1}\left[x(n)+(-1)^{2k} x\left(n+\frac{N}{2}\right)\right] W_N^{2kn}, \quad k = 0,1,2,...\frac{N}{2}-1$$

Since $\quad (-1)^{2k} = 1 \quad$ and $\quad W_N^{2kn} = W_{N/2}^{kn}$

$$X(2k) = \sum_{n=0}^{\frac{N}{2}-1}\left[x(n)+x\left(n+\frac{N}{2}\right)\right] W_{N/2}^{kn}, \quad k = 0,1,2...,\frac{N}{2}-1 \qquad .....(2.7.34)$$

$$X(2k+1) = \sum_{n=0}^{\frac{N}{2}-1}\left[x(n)+(-1)^{2k+1} x\left(n+\frac{N}{2}\right)\right] W_N^{(2k+1)n}, \quad k = 0,1,2...,\frac{N}{2}-1$$

Since $\quad (-1)^{2k+1} = -1 \quad$ and $W_N^{(2k+1)n} = W_N^{2kn}\cdot W_N^n = W_{N/2}^{kn}\cdot W_N^n$

$$\therefore\quad X(2k+1) = \sum_{n=0}^{\frac{N}{2}-1}\left[x(n)-x\left(n+\frac{N}{2}\right)\right] W_N^n\cdot W_{N/2}^{kn}, \quad k = 0,1,2...,\frac{N}{2}-1 \quad .....(2.7.35)$$

eqs. (2.7.34) and (2.7.35) represent the $\dfrac{N}{2}-\text{pt}$ DFTs

Put
$$\left.\begin{aligned}
X(2k) &= G_1(k) \quad\text{and}\quad x(n)+x\left(n+\frac{N}{2}\right) = g_1(n) \\[2ex]
X(2k+1) &= G_2(k) \quad\text{and}\quad \left[x(n)-x\left(n+\frac{N}{2}\right)\right] W_N^n = g_2(n)
\end{aligned}\right\} \qquad .....(2.7.36)$$

For an $8 - \text{pt DFT}$, $N = 8$

$$g_1(n) = x(n) + x\left(n + \frac{8}{2}\right)$$

$$g_2(n) = \left[x(n) - x\left(n + \frac{8}{2}\right)\right]W_8^n, \quad n = 0, 1, 2, 3$$

$$
\begin{aligned}
g_1(0) &= x(0) + x(4) & g_2(0) &= \left[x(0) - x(4)\right]W_8^0 \\
g_1(1) &= x(1) + x(5) & g_2(1) &= \left[x(1) - x(5)\right]W_8^1 \\
g_1(2) &= x(2) + x(6) & g_2(2) &= \left[x(2) - x(6)\right]W_8^2 \\
g_1(3) &= x(3) + x(7) & g_2(3) &= \left[x(3) - x(7)\right]W_8^3
\end{aligned}
\qquad \dots\dots(2.7.37)
$$

From Eqs. (2.7.37), flow graph can be drawn as shown in Fig. 2.12.

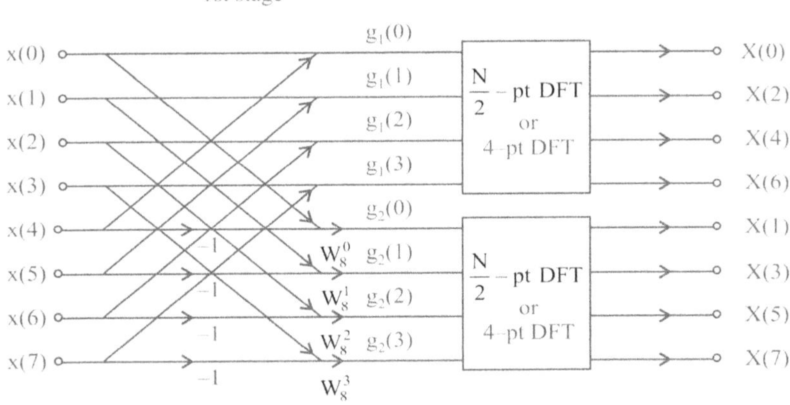

Fig. 2.12 Flow graph of 1st stage of DIF FFT for $N = 8$.

Eqs. (2.7.34) and (2.7.35) after substituting Eqs. (2.7.36), becomes

$$G_1(k) = \sum_{n=0}^{\frac{N}{2}-1} g_1(n)W_{N/2}^{kn} \quad \text{and} \quad G_2(k) = \sum_{n=0}^{\frac{N}{2}-1} g_2(n)W_{N/2}^{kn}, \qquad k = 0, 1 \dots \frac{N}{2} - 1.$$

Decimate $G_1(k)$ and $G_2(k)$ into even numbered and odd numbered sequences to get $\frac{N}{4}-\text{pt DFTs}$.

$$\therefore \quad G_1(k) = \sum_{n=0}^{\frac{N}{4}-1} g_1(n) W_{N/2}^{kn} + \sum_{n=N/4}^{\frac{N}{2}-1} g_1(n) W_{N/2}^{kn}$$

$$G_1(2k) = \sum_{n=0}^{\frac{N}{4}-1} g_1(n) W_{N/2}^{2kn} + \sum_{n=0}^{\frac{N}{4}-1} g_1\left(n + \frac{N}{4}\right) W_{N/2}^{2k\left(n+\frac{N}{4}\right)}, \quad k = 0, 1 \dots \frac{N}{4} - 1$$

$$= \sum_{n=0}^{\frac{N}{4}-1} g_1(n) W_{N/4}^{kn} + \sum_{n=0}^{\frac{N}{4}-1} g_1\left(n + \frac{N}{4}\right) W_{N/4}^{kn}$$

$$G_1(2k) = \sum_{n=0}^{\frac{N}{4}-1}\left[g_1(n) + g_1\left(n + \frac{N}{4}\right)\right] \cdot W_{N/4}^{kn}, \; k = 0, 1, \dots \frac{N}{4} - 1$$

Similarly $G_1(2k+1) = \sum_{n=0}^{\frac{N}{4}-1}\left[g_1(n) - g_1\left(n + \frac{N}{4}\right)\right] W_{N/4}^{n} W_{N/4}^{kn}, \; k = 0, 1, \dots \frac{N}{4} - 1 \quad \dots (2.7.38)$

$$G_2(2k) = \sum_{n=0}^{\frac{N}{4}-1}\left[g_2(n) + g_2\left(n + \frac{N}{4}\right)\right] W_{N/4}^{kn}, \; k = 0, 1, \dots \frac{N}{4} - 1$$

$$G_2(2k+1) = \sum_{n=0}^{\frac{N}{4}-1}\left[g_2(n) - g_2\left(n + \frac{N}{4}\right)\right] W_{N/2}^{n} W_{N/4}^{kn} \qquad \dots (2.7.39)$$

put $\quad G_1(2k) = A_1(k), \; G_1(2k+1) = A_2(k)$

$$g_1(n) + g_1\left(n + \frac{N}{4}\right) = a_1(n), \left[g_1(n) - g_1\left(n + \frac{N}{4}\right)\right] W_{N/2}^{n} = a_2(n)$$

and $\quad G_2(2k) = B_1(k), \; G_2(2k+1) = B_2(k)$ $\qquad \dots (2.7.40)$

$$g_2(n) + g_2\left(n + \frac{N}{4}\right) = b_1(n), \left[g_2(n) - g_2\left(n + \frac{N}{4}\right)\right] W_{N/2}^{n} = b_2(n)$$

In equation (2.7.38) and (2.7.39)

For 8–pt DFT, N = 8

$$g_1(n) + g_1\left(n + \frac{8}{4}\right) = a_1(n), \; , \left[g_1(n) - g_1\left(n + \frac{8}{4}\right)\right] W_{8/2}^{n} = a_2(n), n = 0, 1$$

$$g_1(0) + g_1(2) = a_1(0), \qquad \left[g_1(0) - g_1(2)\right] W_{8/2}^0 = a_2(0)$$

$$g_1(1) + g_1(3) = a_1(1), \qquad \left[g_1(1) - g_1(3)\right] W_{8/2}^1 = a_2(1) \qquad \Bigg\} \quad \ldots\ldots (2.7.41)$$

$$g_2(n) + g_2\left(n + \frac{8}{4}\right) = b_1(n), \quad , \qquad \left[g_2(n) - g_2\left(n + \frac{8}{4}\right)\right] W_{8/2}^n = b_2(n), n = 0,\ 1$$

$$g_2(0) + g_2(2) = b_1(0), \qquad \left[g_2(0) - g_2(2)\right] W_{8/2}^0 = b_2(0)$$

$$g_2(1) + g_2(3) = b_1(1), \qquad \left[g_2(1) - g_2(3)\right] W_{8/2}^1 = b_2(1) \qquad \Bigg\} \quad \ldots\ldots (2.7.42)$$

From equations (2.7.41) and (2.7.42), 2nd stage of fig 2.12 can be drawn as shown in Fig 2.13.

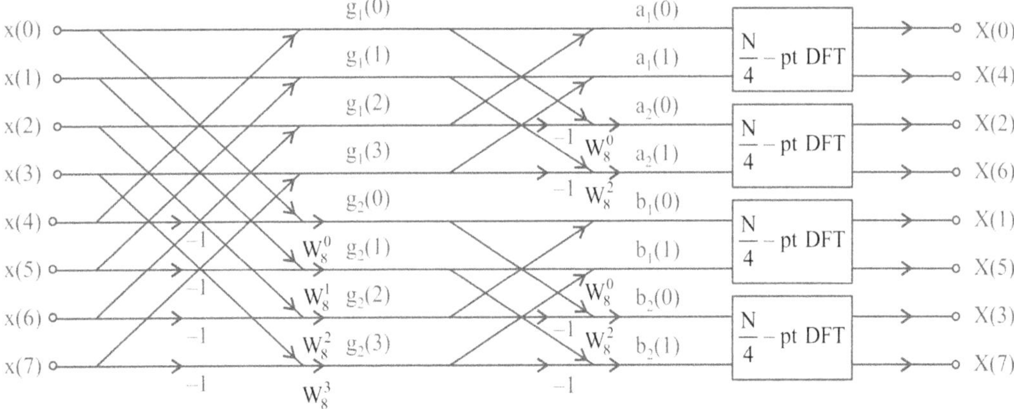

Fig 2.13 Flow graph of 1st and 2nd stages of DIF FFT Algorithm for $N = 8$.

Eqs. (2.7.38) and (2.7.39) after substituting equation (2.7.40), becomes

$$A_1(k) = \sum_{n=0}^{\frac{N}{4}-1} a_1(n)\, W_{N/4}^{kn}, \quad A_2(k) = \sum_{n=0}^{\frac{N}{4}-1} a_2(n)\, W_{N/4}^{kn}$$

$$k = 0,\ 1\ \ldots\ \frac{N}{4} - 1 \quad \Bigg\} \quad \ldots\ldots (2.7.43)$$

$$B_1(k) = \sum_{n=0}^{\frac{N}{4}-1} b_1(n)\, W_{N/4}^{kn}, \quad B_2(k) = \sum_{n=0}^{\frac{N}{4}-1} b_2(n)\, W_{N/4}^{kn}$$

For N = 8

$$A_1(k) = \sum_{n=0}^{1} a_1(n)\, W_{8/4}^{kn}, \qquad\qquad A_2(k) = \sum_{n=0}^{1} a_2(n)\, W_{8/4}^{kn} \qquad k = 0, 1$$

$$B_1(k) = \sum_{n=0}^{1} b_1(n)\, W_{8/4}^{kn}, \qquad\qquad B_2(k) = \sum_{n=0}^{1} b_2(n)\, W_{8/4}^{kn}$$

$$A_1(k) = a_1(0)\, W_{8/4}^{0} + a_1(1)\, W_{8/4}^{k}, \qquad A_2(0) = a_2(0)\, W_{8/4}^{0} + a_2(1)\, W_{8/4}^{k}$$

$$A_1(0) = a_1(0) + a_1(1) \qquad\qquad A_2(0) = a_2(0) + a_2(1)$$

$$A_1(1) = \left(a_1(0) - a_1(1)\right) W_8^{0} \qquad A_2(1) = \left(a_2(0) - a_2(1)\right) W_8^{0}$$

$$B_1(k) = b_1(0) W_{8/4}^{0} + b_1(1) W_{8/4}^{k}, \qquad B_2(k) = b_2(0) W_{8/4}^{0} + b_2(1) W_{8/4}^{k}$$

$$B_1(0) = b_1(0) + b_1(1) \qquad\qquad B_2(0) = b_2(0) + b_2(1)$$

$$B_1(1) = \left(b_1(0) - b_1(1)\right) W_8^{0} \qquad B_2(1) = \left(b_2(0) - b_2(1)\right) W_8^{0}$$

3rd stage of Fig. 2.13 can be drawn as shown in Fig. 2.14.

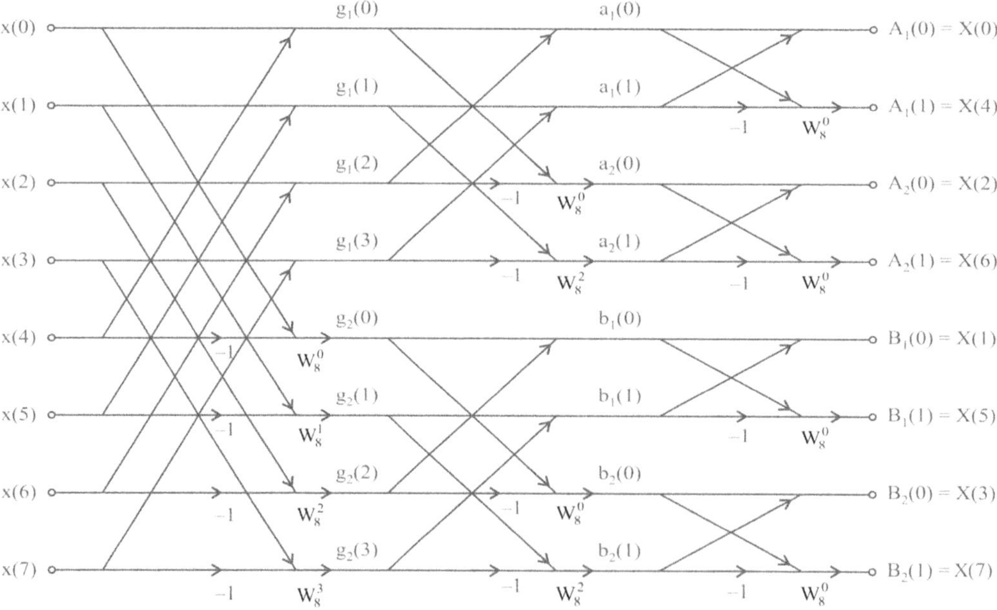

Fig. 2.14 Complete flow graph of DIF FFT algorithm for N = 8.

Note: Observe input is in normal order and output is in Bit reversal order.

2.7.5.1 Differences between Radix–2 DIT FFT and DIF FFT Flow Graphs

Radix–2 DIT FFT	Radix–2 DIF FFT
1. Decimation in time domain	1. Decimation in frequency domain
2. Input side samples will be in bit-reversal order	2. Input side samples will be in normal order
3. Output side samples will be in normal order	3. Output side samples will be in bit-reversal order
4. Basic butterfly is	4. Basic butterfly is
5. Number of multiplications required is $(N/2) \log_2 N$	5. Number of multiplications required is $(N/2) \log_2 N$
6. Number of additions required is $N\log_2 N$	6. Number of additions required is $N\log_2 N$

Example 2.13: Show that to any DIT FFT algorithm there corresponds to an inverse frequency algorithm that is DIF FFT. [JNTUH, 2001]

Solution:

Basic butterfly of DIT FFT for m^{th} stage is

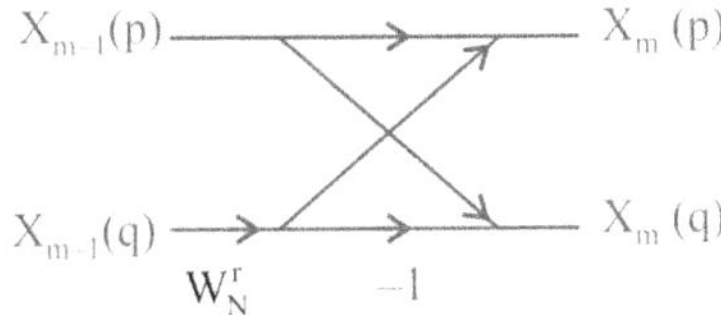

By transposing the above flow graph, we will get basic butterfly of DIF FFT for m^{th} stage.

i.e., by interchanging the inputs and outputs and reversing the directions.

$\therefore$ Reverse the directions in the above Fig.

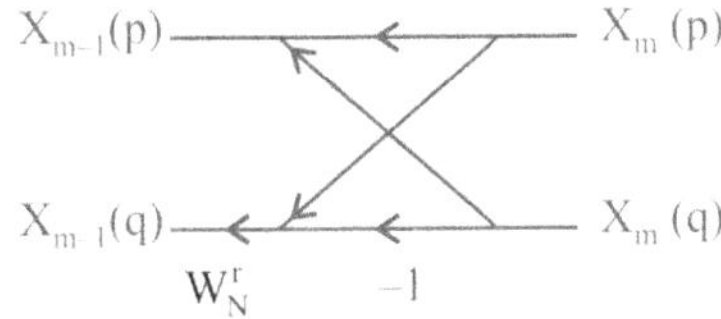

Now interchange the inputs and outputs.

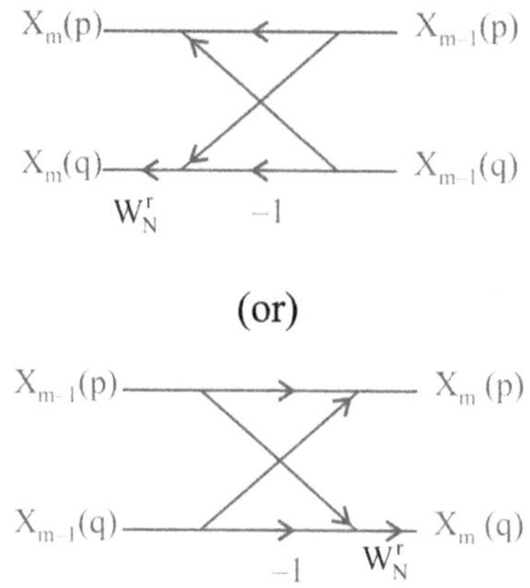

(or)

which is the basic butterfly of DIF FFT for mth stage

Example 2.14: Obtain the DFT of the following sequence using 8–points Radix–2, DIF FFT flow graph. [JNTU: 2001]

$$x(n) = \{0, 1, 2, 3, -3, -2, -1, -0\}$$

Solution:

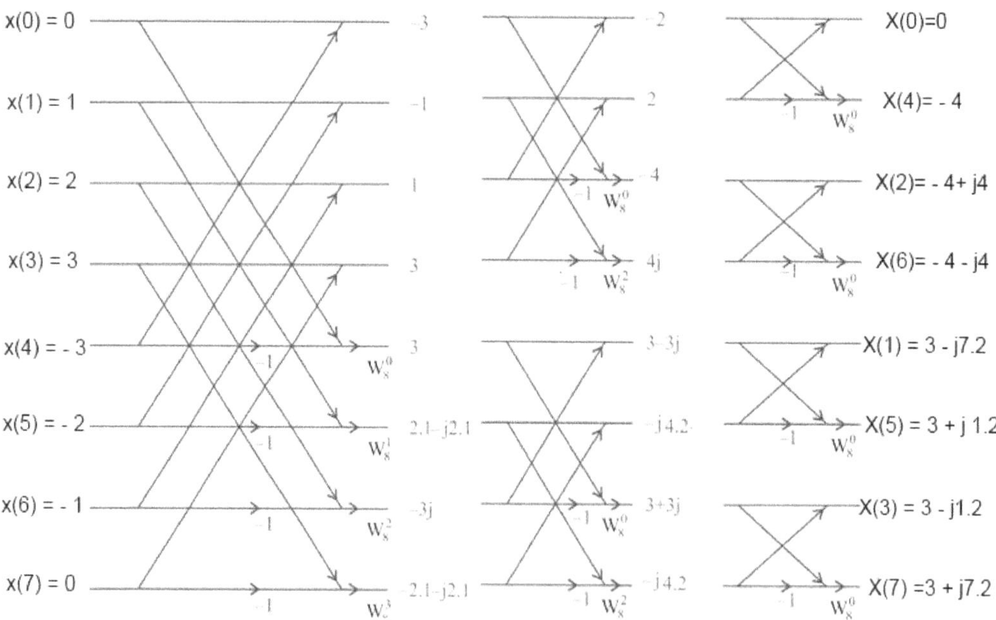

Fig. 2.15

Example 2.15: Given x(n) = {0, 1, 2, 3}, find X(k) using DIF FFT algorithm

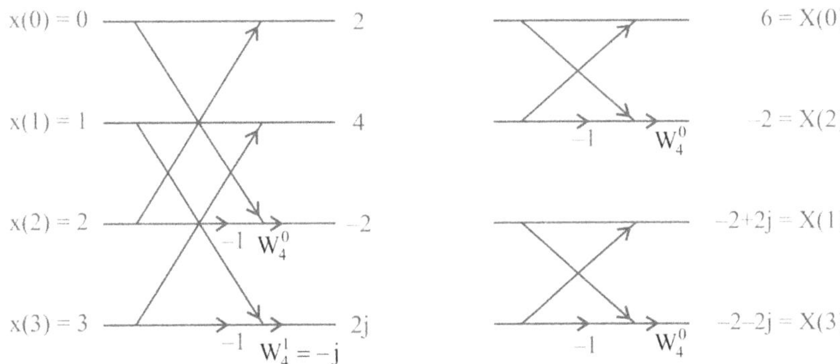

Fig. 2.16

2.7.6 Computing an Inverse DFT by using FFT

An FFT algorithm can also be used to compute efficiently the inverse DFT (IDFT).

The inverse DFT is given by

$$x(n) = \frac{1}{N}\sum_{k=0}^{N-1} X(k)\, W_N^{-nk}\, , n = 0, 1, \dots, N-1 \qquad \dots\dots(2.7.44)$$

Taking complex conjugate of the above equation, we get

$$N\, x^*(n) = \sum_{k=0}^{N-1} X^*(k)\, W_N^{nk} \qquad \dots\dots (2.7.45)$$

The right hand side is the DFT of the sequence $X^*(k)$

$$\therefore \qquad x^*(n) = \frac{1}{N}\mathrm{DFT}\left[X^*(k)\right] \qquad \dots\dots (2.7.46)$$

Taking complex conjugate of both sides, we get desired output sequence x(n).

$$x(n) = \frac{1}{N}\left[\sum_{k=0}^{N-1} X^*(k)\, W_N^{nk}\right]^*$$

$$\therefore \qquad x(n) = \frac{1}{N}\left[\mathrm{FFT}\left[X^*(k)\right]\right]^* \qquad \dots\dots (2.7.47)$$

Equation (2.7.45) is similar to an N-pt DFT with inputs $X^*(k)$ and outputs $N\, x^*(n)$.

For N = 8, DIT FFT algorithm flow graph is

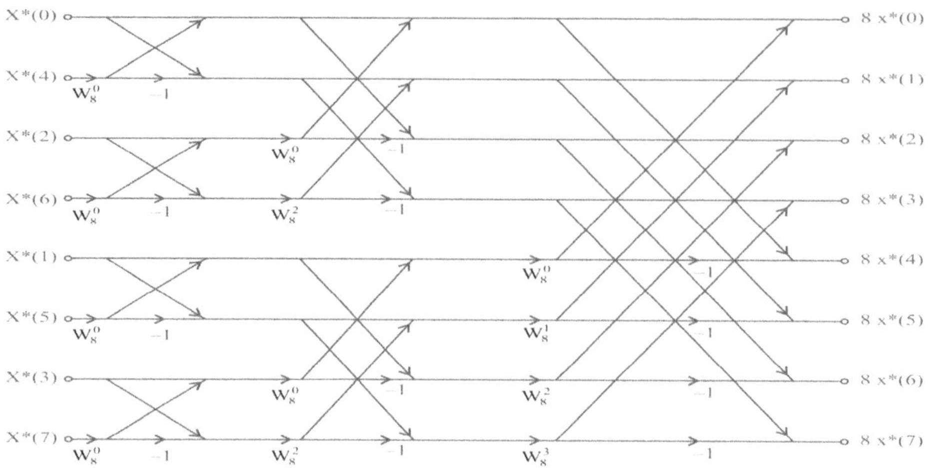

Fig. 2.17 Complete flow graph for computing IDFT using DIT FFT algorithm for $N = 8$.

Output $\quad x(n) = \dfrac{1}{8}\left[8x^*(0),\ 8x^*(1),\ 8x^*(2),\ 8x^*(3),\ 8x^*(4),\ 8x^*(5),\ 8x^*(6),\ 8x^*(7)\right]^*$

DIF FFT algorithm flow graph is

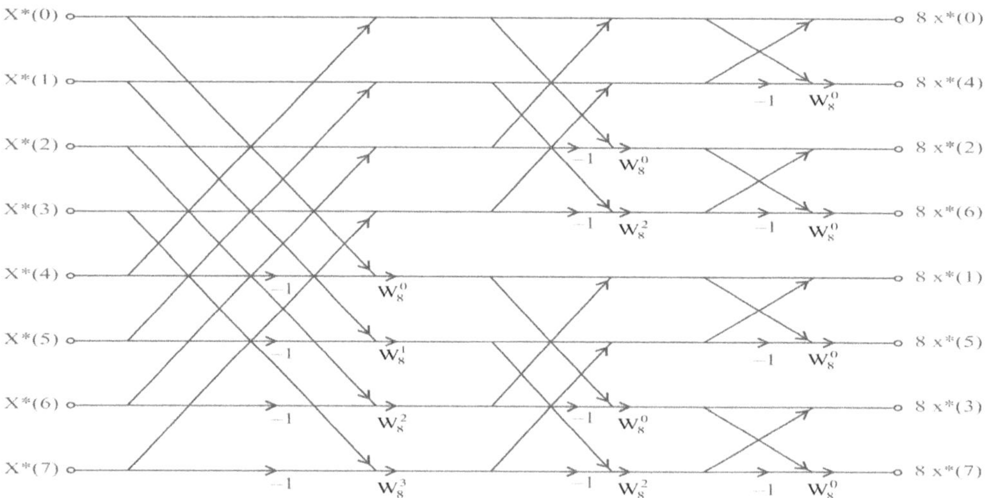

Fig. 2.18 Complete flow graph for computing IDFT using DIF FFT algorithm for $N = 8$.

Output $\quad x(n) = \dfrac{1}{8}\left[8x^*(0),\ 8x^*(4),\ 8x^*(2),\ 8x^*(6),\ 8x^*(1),\ 8x^*(5),\ 8x^*(3),\ 8x^*(7)\right]^*$

Example 2.16: Compute IDFT of the sequence

X(k) = [7, − 0.707 − j0.707, − j, 0.707 − j 0.707, 1, 0.707 + j 0.707, j, − 0.707 + j 0.707}

using DIT FFT algorithm

Solution:

Take complex conjugate of X(k) and apply bit reversed index inputs to the DIT FFT flow graph.

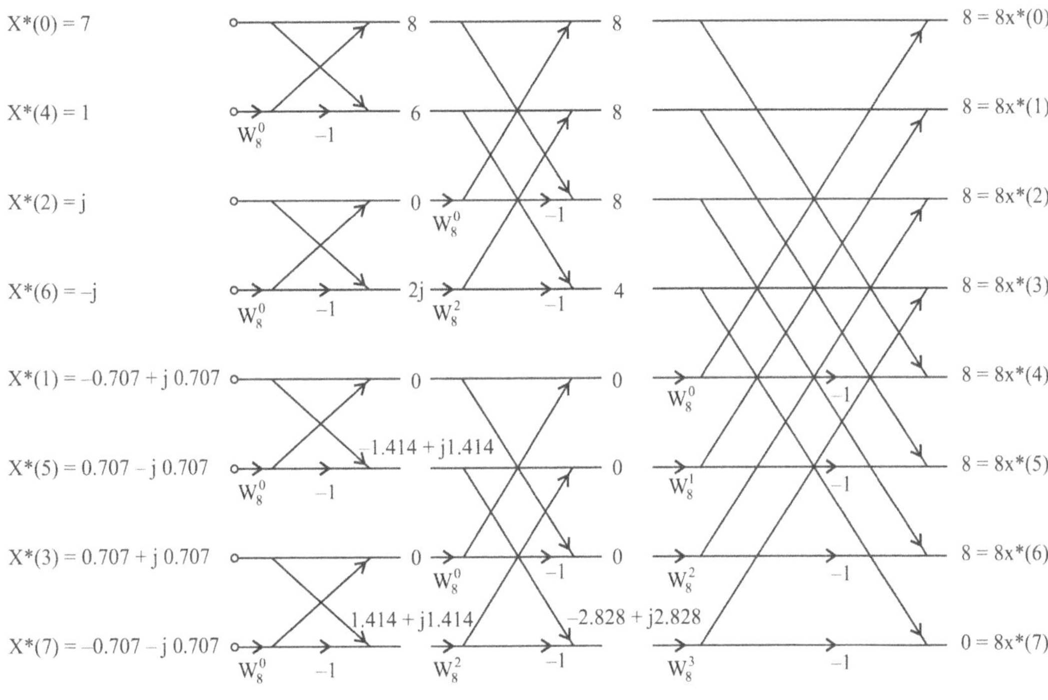

$$\therefore \quad \text{Output } N x^*(n) = \{8,8,8,8,8,8,0\}$$

$$\therefore \quad x(n) = \{1, 1, 1, 1, 1, 1, 1, 0\}$$

Example 2.17: Compute IDFT of the sequence

X (k) = {6, −2 + j2, −2, −2 − j2} using DIF FFT algorithm.

Solution:

Take complex conjugate of X(k) and apply in normal order as inputs to the DIF FFT algorithm.

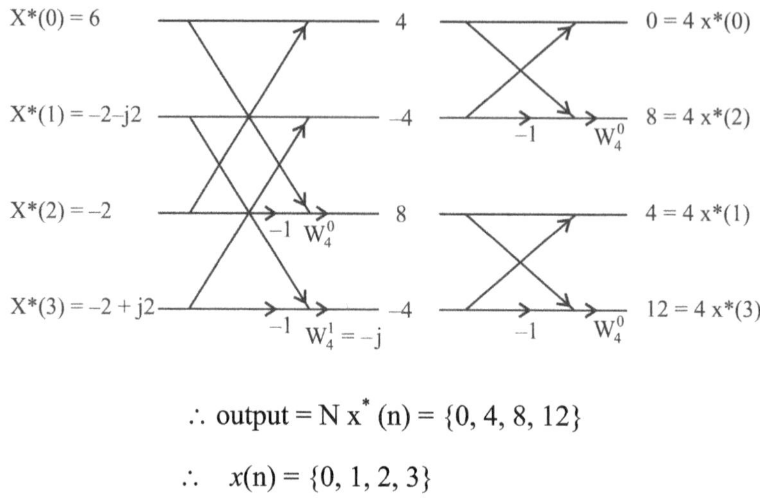

$$\therefore \text{ output} = N\, x^{*}(n) = \{0,\ 4,\ 8,\ 12\}$$

$$\therefore \quad x(n) = \{0,\ 1,\ 2,\ 3\}$$

Example 2.18: Show that DFT of a finite duration sequence corresponds to equally spaced samples on the unit circle of its z-transform.   [JNTU 2000]

Solution:

Consider Z-transform of finite duration sequence x(n)

$$X(z) = \sum_{n=0}^{N-1} x(n)z^{-n}$$

with the ROC that includes unit circle

If X(z) is sampled at N equally spaced points on the unit circle.

i.e.,
$$Z_k = e^{j\frac{2\pi k}{N}}, \quad k = 0,\ 1 \ldots (N-1)$$

we get
$$X\left(e^{j\frac{2\pi k}{N}}\right) = \sum_{n=0}^{N-1} x(n)e^{-j\frac{2\pi n k}{N}}$$

or
$$X(k) = \sum_{n=0}^{N-1} x(n)e^{-j2\pi kn/N}$$

Which is an expression for N – pt DFT.

Hence, Z-transform of finite duration sequence x(n) is determined by its N – pt DFT.

Review Questions

1. If a time domain sequence $x(n)$ is delayed by m, its corresponding DFS pair, when $x(n)$ and $X(k)$ is DFS pair, is

   (a) $e^{j2\pi km/N} \cdot x(k)$ 　　　　(b) $e^{-j2\pi km/N} \cdot x(k)$

   (c) $\dfrac{e^{-j2\pi km/N}}{x(k)}$ 　　　　(d) $e^{-j2\pi km/N} - x(k)$ 　　　*Ans :*[b]

2. If a time domain sequence $x(n)$ is advanced by m, its corresponding DFS pair, when $x(n)$ and $Xx(k)$ is DFS pair, is

   (a) $e^{j2\pi km/N} \cdot x(k)$ 　　　　(b) $e^{-j2\pi km/N} \cdot x(k)$

   (c) $\dfrac{e^{-j2\pi km/N}}{x(k)}$ 　　　　(d) $e^{-j2\pi km} - x(k)$ 　　　*Ans :*[a]

3. If $X(k)$ is delayed by l, its corresponding DFS pair, when $x(n)$ and $X(k)$ is DFS pair, is

   (a) $e^{j2\pi nl/N} \cdot x(n)$ 　　　　(b) $e^{-j2\pi nl/N} \cdot x(n)$

   (c) $\dfrac{e^{-j2\pi nl/N}}{x(n)}$ 　　　　(d) $e^{j2\pi nl/N} - x(n)$ 　　　*Ans :*[a]

4. If $X(k)$ is advanced by l, its corresponding DFS pair, when $x(n)$ and $X(k)$ is DFS pair, is

   (a) $e^{j2\pi nl/N} \cdot x(n)$ 　　　　(b) $e^{-j2\pi nl/N} \cdot x(n)$

   (c) $\dfrac{e^{-j2\pi nl/N}}{x(n)}$ 　　　　(d) $e^{j2\pi nl/N} - x(n)$ 　　　*Ans :*[b]

5. If $x(n)$ and $X(k)$ is a DFS pair, then $x(-k)$ corresponding DFS pair is
   (a) $N\,X(n)$ 　　　　(b) $X(-n)$
   (c) $x(n)$ 　　　　(d) $x(-n)$ 　　　*Ans:* [a]

6. If $x(n)$ (complex) and $X(k)$ is a DFS pair, then even part of $x(n)$ DFS pair is
   (a) even part of $X(k)$ 　　　　(b) odd part of $X(k)$
   (c) imaginary part of $X(k)$ 　　　　(d) real part of $X(k)$ 　　　*Ans:* [d]

7. If $x(n)$ (real) and $X(k)$ is a DFS pair, then real part of $X(k)$ is
   (a) odd symmetry 　　　　(b) even symmetry
   (c) zero 　　　　(d) none 　　　*Ans:* [b]

8. If $x(n)$ (real) and $X(k)$ is a DFS pair, then imaginary part of $X(k)$ is
   (a) odd symmetry 　　　　(b) even symmetry
   (c) zero 　　　　(d) none 　　　*Ans:* [a]

9.   The other name(s) of circular convolution is

   (a)   Linear convolution   (b)   Periodic convolution

   (c)   both (a) and (b)   (d)   none   *Ans:* [b]

10. The DFS coefficients cannot be found for the signal $x(n) = \cos\sqrt{3}\pi\, n$ because

   (a)   Period is rational number   (b)   Period is irrational

   (c)   Period is zero   (d)   none   *Ans:* [b]

11. Period of the signal $x(n) = \cos(\pi\, n / 2)$ is

   (a)   zero   (b)   4

   (c)   6   (d)   none   *Ans:* [b]

12. If a signal is continuous and non periodic its spectrum will be

   (a)   discrete and periodic   (b)   continuous and periodic

   (c)   continuous and aperiodic   (d)   discrete and aperiodic   *Ans:* [c]

13. If a signal is continuous and non periodic its spectrum can be found by

   (a)   fourier transform   (b)   fourier series

   (c)   discrete fourier series   (d)   discrete fourier transform   *Ans:* [a]

14. If a signal is continuous and periodic its spectrum will be

   (a)   discrete and periodic   (b)   continuous and periodic

   (c)   continuous and aperiodic   (d)   discrete and aperiodic   *Ans:* [d]

15. If a signal is continuous and periodic its spectrum can be found by

   (a)   fourier transform   (b)   fourier series
   (c)   discrete fourier series   (d)   discrete fourier transform   *Ans:* [b]

16. If a signal is discrete and aperiodic its spectrum can be found by

   (a)   fourier transform   (b)   DTFT

   (c)   fourier series   (d)   DFT   *Ans:* [b]

17. If a signal is discrete and aperiodic its spectrum will be

   (a)   discrete and aperiodic   (b)   continuous and periodic

   (c)   continuous and aperiodic   (d)   discrete and aperiodic   *Ans:* [b]

18. If a signal is discrete and periodic its spectrum will be

   (a)   discrete and periodic   (b)   continuous and periodic
   (c)   continuous and aperiodic   (d)   discrete and aperiodic   *Ans:* [a]

19. If x(n) (complex) and X(k) is a DFT pair, then $x^*(n)$ corresponding DFT pair is
    (a) X(K) (b) $X^*(K)$
    (c) $X^*(-K)$ (d) zero *Ans:* [c]

20. If x(n) and X(k) is a DFT pair, then x(N–n) corresponding DFT pair is
    (a) X(K) (b) X(K–N)
    (c) X(–K) (d) X(N–K) *Ans:* [d]

21. The following are the properties of DFT
    (a) Linear convolution (b) periodic convolution
    (c) circular convolution (d) both b and c *Ans:* [d]

22. The other name of twiddle factor is
    (a) phase factor (b) imaginary factor
    (c) real factor (d) none *Ans:* [a]

23. The symmetry properties of phase factor is
    (a) $W_N^{K+N} = W_N^K$ (b) $W_N^{K/N} = W_N^K$
    (c) $W_N^{K+\frac{N}{2}} = -W_N^K$ (d) none *Ans:* [c]

24. The periodicity property of phase factor is
    (a) $W_N^{K+N} = W_N^K$ (b) $W_N^{K/N} = W_N^K$
    (c) $W_N^{K+\frac{N}{2}} = -W_N^K$ (d) none *Ans:* [a]

25. W_N (phase factor) is equal to
    (a) $e^{j2\pi/N}$ (b) $e^{j2\pi N}$
    (c) $e^{-j2\pi N}$ (d) $e^{-j2\pi/N}$ *Ans:* [d]

26. Linear convolution can be obtained by using
    (a) DFT (b) circular convolution
    (c) both a and b (d) none *Ans:* [c]

27. Circular convolution can be obtained by using
    (a) DFT (b) Linear convolution
    (c) both a and b (d) none *Ans:* [c]

28. The response of a system whose impulse response h(n) = {1, 1, 1}, for the input
    x(n) = {1, 1, 1} is
    (a) {1, 3, 2, 1, 0} (b) {1, 2, 3, 2, 1}
    (c) {3, 2, 1, 2, 3} (d) {1, 2, 3, 1, 2} *Ans:* [b]

29. The length of output sequence of linear convolution of two sequences with lengths M and N is

    (a) $M + N$
    (b)   max $(M \& N)$
    (c) $M + N - 1$
    (d) min $(M \& N)$	*Ans:* [c]

30. The length of output sequence of circular convolution of two sequences with lengths M and N is

    (a) $M + N$
    (b)   max $(M \& N)$
    (c) $M + N - 1$
    (d) min $(M \& N)$	*Ans:* [b]

31. The purpose of FFT is

    (a)   to increase number of computations
    (b)   to reduce number of computations
    (c)   to increase speed
    (d) both b and c	*Ans:* [d]

32. The N-pt DFT, N must be ___________ to use Radix-2 FFT algorithms

    (a)   an integer power of 2
    (b)   odd number
    (c)   irrational number
    (d) none	*Ans:* [a]

33. The number of computations required to compute N-pt DFT directly

    (a)   N
    (b)   N–1
    (c)   N/2
    (d) N^2	*Ans:* [d]

34. The number of multiplications required to compute N-pt DFT using Radix-2 DIT FFT algorithm

    (a) $\dfrac{N}{2}\log_2 N$
    (b) $N \log_2 N$
    (c)   N
    (d) N^2	*Ans:* [a]

35. The number of additions required to compute N-pt DFT using Radix-2 DIF FFT algorithm

    (a) $\dfrac{N}{2}\log_2 N$
    (b) $N \log_2 N$
    (c)   N
    (d) N^2	*Ans:* [b]

36. In DIT FFT algorithm

    (a)   x(n) is decimated into even and odd numbered samples
    (b)   X(k) is decimated into even and odd numbered samples
    (c)   both (a) and (b)
    (d) None	*Ans:* [a]

37. In DIF FFT algorithm
    (a)   x(n) is decimated into even and odd numbered samaples
    (b)   x(k) is decimated into even and odd numbered samples
    (c)   both (a) and (b)
    (d)   None   *Ans:* [b]

38. The DIT FFT algorithm input samples will be in
    (a)   scrambled order   (b)   normal order
    (c)   bitreversal order   (d)   both a and c   *Ans:* [d]

39. The DIF FFT algorithm input samples will be in
    (a)   scrambled order   (b)   normal order
    (c)   bitreversal order   (d)   both a and c   *Ans:* [b]

40. The DIT FFT algorithm output samples will be in
    (a)   scrambled order   (b)   normal order
    (c)   bitreversal order   (d)   both a and c   *Ans:* [b]

41. The DIF FFT algorithm output samples will be in
    (a)   scrambled order   (b)   normal order
    (c)   bitreversal order   (d)   both a and c   *Ans:* [d]

42. Number of stages required to comput 16-pt DFT using Radix-2 DIT or DIF FFT algorithm
    (a)   3   (b)   4
    (c)   6   (d)   1   *Ans:* [b]

43. Since shape of the 2-pt DFT flow graph is butterfly, FFT algorithms are also called butterfly algorithms.

44. Number of butterflies required for each stage of N-pt DFT using FFT algorithms are
    (a)   N   (b) N^2
    (c) $\dfrac{N}{2}$ (d)   N–1   *Ans :* [c]

45. Inverse DFT can be computed efficiently by using FFT algorithms
    (a)   True   (b)   False
    (c)   cannot say   *Ans:* [a]

CHAPTER 3

Z-transform

3.0 Introduction

The Z-transform is a convenient tool for representing, analyzing and designing discrete-time signals and systems. It plays a similar role in discrete-time systems to that which the laplace transform plays in continuous-time systems.

In this chapter, we introduce the Z-transform and discuss its applications.

3.1 The Z-transform

The Z-transform of a sequence x(n), which is valid for all n is defined as

$$X(z) = \sum_{n=-\infty}^{\infty} x(n) z^{-n} \qquad \qquad .....(3.1.1)$$

Where z is a complex variable

In causal systems, x(n) may be non zero only in the interval $0 \le n \le \infty$ and Eq. (3.1.1) reduces to the so-called one sided Z-transform.

$$X^{+}(z) = \sum_{n=0}^{\infty} x(n) z^{-n} \qquad \qquad .....(3.1.2)$$

Clearly, the Z-transform is a power series with an infinite number of terms and so may not converge for all values of Z. The region where the Z-transform converges is known as the region of convergence (ROC), and in this region the values of X(z) are finite.

Let us discuss some examples

Example 3.1 Find the Z-transform and the region of convergence for each of the discrete-time sequences given in Fig. 3.1.

Solution:

(a) The sequence of Fig. 3.1(a) is non causal, since x(n) is not zero for n < 0, but it is of a finite duration.

From Eq. (3.1.1) the Z-transform is given by

$$X(z) = \sum_{n=-\infty}^{\infty} x(n)z^{-n}$$

$$= \sum_{n=-6}^{0} x(n)z^{-n} = z^5 + 3z^4 + 5z^3 + 3z^2 + z$$

It is verified that the value of X(z) becomes infinite, when $z = \infty$. Thus the ROC is everywhere in the Z-plane except at $z = \infty$.

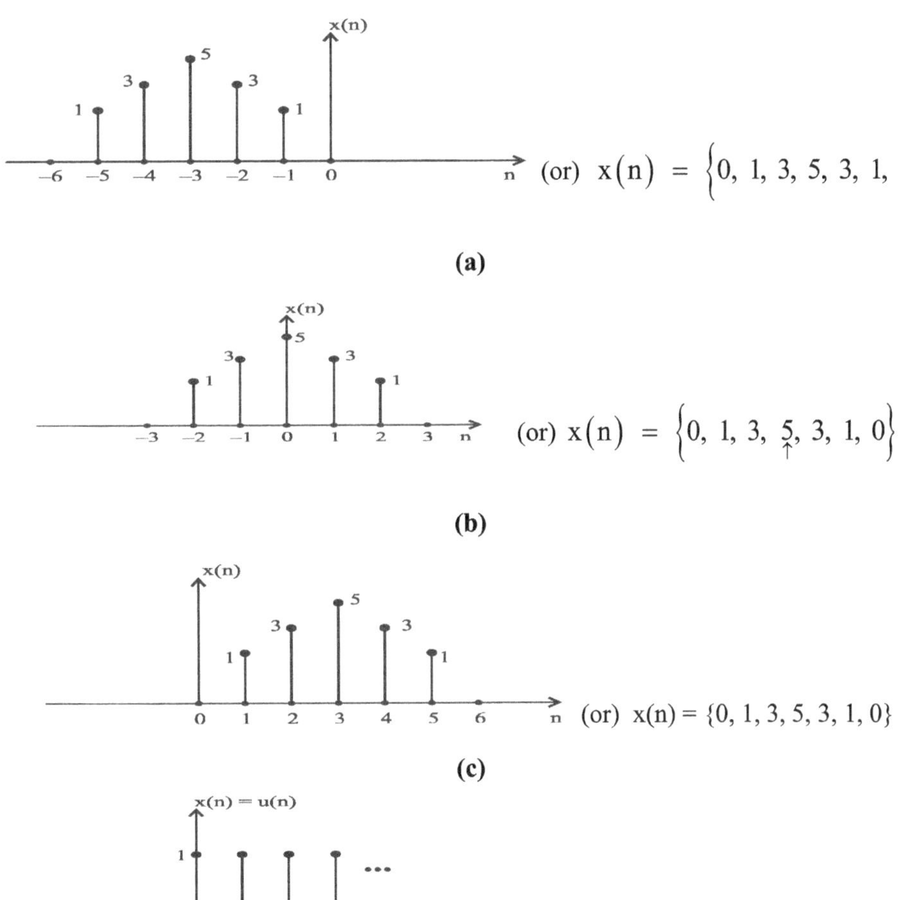

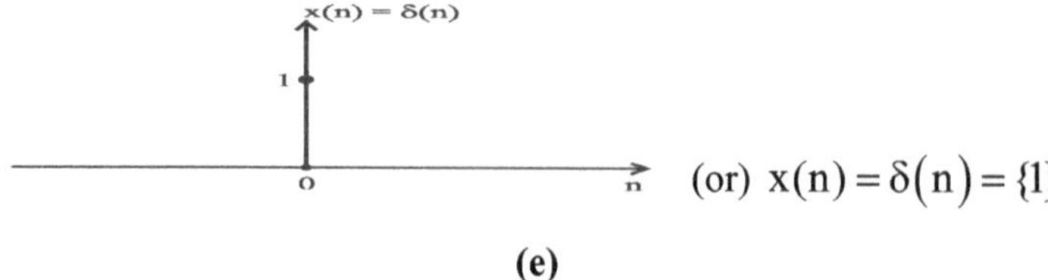

(or) $x(n) = \delta(n) = \{1\}$

(e)

Fig. 3.1 Causal and non causal discrete-time sequences.

(b)   Again, the sequence in Fig. 3.1(b) is not causal. It is of a finite duration, and double sided.

From Eq. (3.1), the Z-transform is given by

$$X(z) = \sum_{n=-\infty}^{\infty} x(n)z^{-n} = \sum_{n=-3}^{3} x(n)z^{-n} = z^2 + 3z + 5 + 3z^{-1} + z^{-2}$$

It is evident that the value of X(z) is infinite if z = 0 or if z = 0 or if z = ∞.

Therefore the ROC is everywhere except at z = 0 and z = ∞.

(c)   Fig. 3.1(c) represents a causal, finite duration sequence. The Z-transform is given by

$$X(z) = \sum_{n=-\infty}^{\infty} x(n)z^{-n} = \sum_{n=0}^{6} x(n)z^{-n} = z^{-1} + 3z^{-2} + 5z^{-3} + 3z^{-4} + z^{-5}$$

In this case, $X(z) = \infty$ for z = 0. Thus the ROC is everywhere except at z = 0.

(d)   The discrete-time sequence in Fig. 3.1(d) is unit step sequence, which is defined as

$$u(n) = 1 \qquad n \geq 1$$
$$= 0 \qquad n < 0$$

Clearly it is a causal sequence of infinite duration. The Z-transform is given by

$$X(z) = \sum_{n=-\infty}^{\infty} x(n) = \sum_{n=0}^{\infty} z^{-n}$$

$$= 1 + z^{-1} + z^{-2} + \ldots\ldots$$

This is a geometric series with a common ratio of z^{-1}. The series converges if $|z^{-1}| < 1$ or equivalently if $|z| > 1$. Thus we may express $X(z)$ in closed form provided that $|z| > 1$.

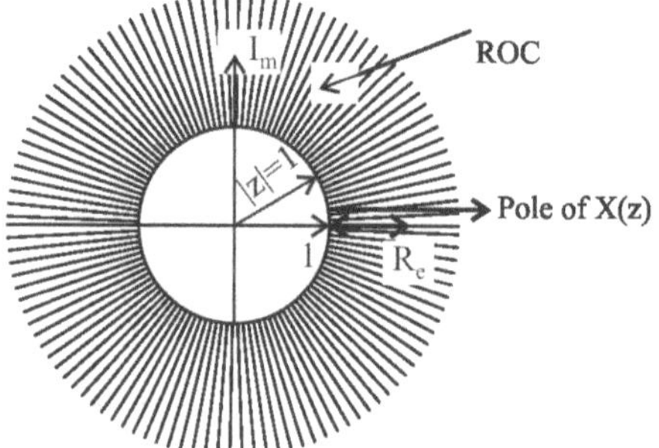

Fig. 3.2 Roc for example 3.1(a).

$$X(z) = 1 + z^{-1} + z^{-2} + \ldots$$

$$= \frac{1}{1 - z^{-1}} = \frac{z}{z - 1}$$

In this case, the Z-transform is valid everywhere outside a circle of unit radius whose center is at the origin. The exterior of the circle is the region of convergence (Fig. 3.2). We can readily verify that when $|z| > 1$ $X(z)$ converges where as when $|z| < 1$ $X(z)$ diverges.

For example, if we let $z = 2$(outside the unit circle),

$$X(z) = 1 + \frac{1}{2} + \left(\frac{1}{2}\right)^2 + \ldots = \frac{2}{2-1} = 2,$$

which is finite

On the other hand if $z = 1/2$ (inside the unit circle),

$$X(z) = 1 + \frac{1}{0.5} + \left(\frac{1}{0.5}\right)^2 + \ldots = 1 + 2 + 4 + 8 + \ldots$$

which is observed to be diverging.

(e)   In this case, it is an impulse sequence, which is defined as

$$\delta(n) = 1 \qquad n = 0$$
$$= 0, \qquad n \neq 0$$

The Z-transform is given by

$$X(z) = \sum_{n=-\infty}^{\infty} x(n)z^{-n} = 1,$$

$X(z)$ is finite for all values of z.

$\therefore$ ROC is entire z-plane.

From above examples we can infer that:

(i)   For causal sequences of finite duration the Z-transform converges everywhere except at $z = 0$.

(ii)   For causal sequences of infinite duration the Z-transform converges everywhere outside a circle bounded by the radius of the pole with the largest radius.

(iii)   For stable causal systems the ROC always encloses the circle of unit radius.

Example 3.2: Determine the Z-transform of the signal $x(n) = a^n u(n)$

Solution: $x(n)$ can be written as

$$x(n) = a^n, n \geq 0$$

$$= 0, \ n < 0$$

The Z-transform is given by

$$X(z) = \sum_{n=0}^{\infty} a^n z^{-n}$$

$$= \sum_{n=0}^{\infty} \left(az^{-1}\right)^n$$

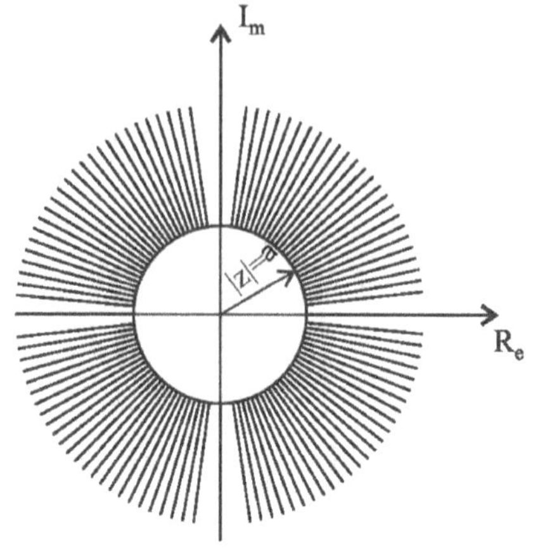

Fig. 3.3

If $\left|az^{-1}\right| < 1$ or equivalently, $|z| > |a|$, this power series converges to

$$\frac{1}{1 - az^{-1}} = \frac{z}{z - a}.$$

$\therefore$ Z transform pair

$$a^n u(n) \leftrightarrow \frac{1}{1 - az^{-1}}, \ \text{ROC: } |z| > |a|$$

The ROC is the exterior of a circle having radius $|a|$, which is shown in Fig. 3.3.

Example 3.3: Determine the Z-transform and ROC of the signal

$$x(n) = -b^n u(-n-1)$$

Solution: The sequence can be written as

$$x(n) = 0, \qquad n \geq 0$$

$$= -b^n, \qquad n \leq 1$$

The Z-transform is given by

$$X(z) = -\sum_{n=-\infty}^{-1} \left(bz^{-1}\right)^n$$

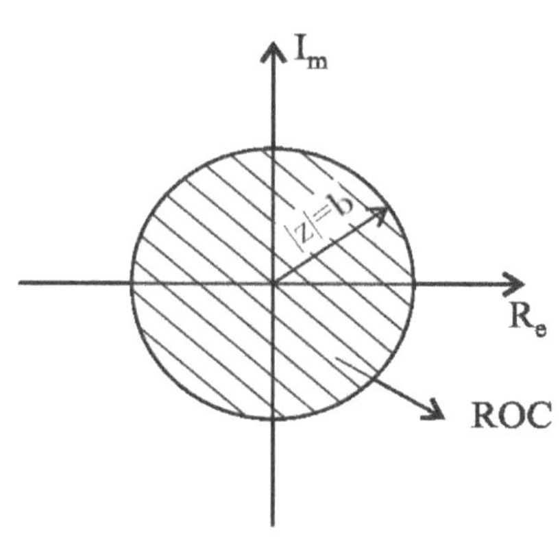

Fig. 3.4

Replace n with $-n$

$$= -\sum_{n=1}^{\infty}\left(bz^{-1}\right)^{-n} = -\sum_{n=1}^{\infty}\left(b^{-1}z\right)^{n} = -\left[\sum_{n=0}^{\infty}\left(b^{-1}z\right)^{n} - 1\right]$$

$$= -\left[\frac{1}{1-b^{-1}z} - 1\right] = \frac{z}{z-b}; \quad ROC:\left|b^{-1}z\right| < 1 \text{ or } \left|z\right| < \left|b\right|$$

Which is shown in Fig. 3.4.

Example 3.4: Determine the Z-transform and ROC of the signal
$x(n) = a^n u(n) + b^n u(-n-1)$

Solution:

The Z-transform is given by

$$X(z) = \sum_{n=0}^{\infty} a^n z^{-n} + \sum_{n=-\infty}^{-1} b^n z^{-n}$$

$$= \left[1 + \frac{a}{z} + \left(\frac{a}{z}\right)^z + \ldots\ldots\right] + \sum_{n=1}^{\infty}\left(b^{-1}z\right)^n$$

$$= \left[1 + \left(\frac{a}{z}\right) + \left(\frac{a}{z}\right)^2 + \ldots\right] + \left[\sum_{n=0}^{\infty}\left(b^{-1}z\right)^n - 1\right]$$

$$= \left[1 + \left(\frac{a}{z}\right) + \left(\frac{a}{z}\right)^2 + \ldots\right] + \left[1 + \frac{z}{b} + \left(\frac{z}{b}\right)^2 + \ldots\ldots - 1\right]$$

$$= \frac{1}{1-\left(\dfrac{a}{z}\right)} + \frac{1}{1-\dfrac{z}{b}} - 1$$

If

$$\left|\frac{a}{z}\right| < 1 \ \& \ \left|\frac{z}{b}\right| < 1,$$

ROC:

$$\left|z\right| > \left|a\right| \ \& \ \left|z\right| < \left|b\right|,$$

$$X(z) = \frac{z}{z-a} + \frac{b}{b-z} - 1 = \frac{z}{z-a} + \frac{z}{b-z}$$

ROC is shown in Fig. 3.5.

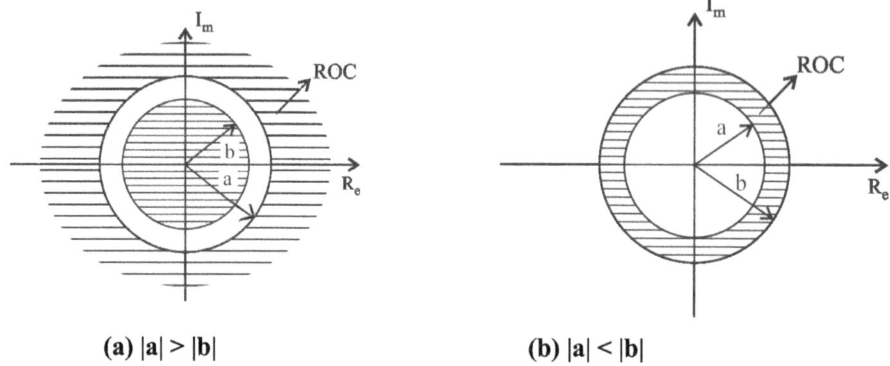

(a) |a| > |b| **(b) |a| < |b|**

Fig. 3.5

From Fig.3.5(a), we can say that $X(z)$ does not exist for $|a| > |b|$.

Example 3.5: Find the Z-transform of the sequence

$$x(n) = \frac{1}{n!} u(n)$$

Solution: The Z-transform is given by

$$X(z) = \sum_{n=0}^{\infty} \frac{1}{n!} z^{-n}$$

$$= 1 + \frac{\left(\dfrac{1}{z}\right)}{1!} + \frac{\left(\dfrac{1}{z}\right)^2}{2!} + \frac{\left(\dfrac{1}{z}\right)^3}{3!} + \ldots = e^{1/z}$$

$$\text{ROC is } \left|\frac{1}{z}\right| < \infty \quad \text{or} \quad |z| > 0.$$

Example 3.6: Find the Z-transform of the sequence

$$x(n) = \frac{1}{(n+1)!} u(n)$$

Solution: The Z-transform is given by

$$X(z) = \sum_{n=0}^{\infty} \frac{1}{(n+1)!} z^{-n}$$

$$= \frac{1}{1!} + \frac{z^{-1}}{2!} + \frac{z^{-2}}{3!} + \frac{z^{-3}}{4!} + \ldots\ldots$$

$$= z \left[\frac{z^{-1}}{1!} + \frac{z^{-2}}{2!} + \frac{z^{-3}}{3!} + \ldots\ldots \right]$$

$$= z \left[1 + \frac{\left(\dfrac{1}{z}\right)}{1!} + \frac{\left(\dfrac{1}{z}\right)^2}{2!} + \frac{\left(\dfrac{1}{z}\right)^3}{3!} + \ldots\ldots - 1 \right]$$

$$= z \left[e^{1/z} - 1 \right]$$

ROC is $\qquad |z| > 0.$

Example 3.7: Determine the Z-transform of the $x(n) = (\cos\omega_0 n)\, u(n)$

Solution:

The Z-transform is given by

$$X(z) = \sum_{n=-\infty}^{\infty} x(n)\, z^{-n} = \sum_{n=0}^{\infty} \cos\omega_0 n \; z^{-n}$$

$$= \sum_{n=0}^{\infty} \frac{e^{j\omega_0 n} + e^{-j\omega_0 n}}{2} \; z^{-n}$$

$$= \frac{1}{2} \left[\sum_{n=0}^{\infty} \left(e^{j\omega_0} z^{-1} \right)^n + \sum_{n=0}^{\infty} \left(e^{-j\omega_0} z^{-1} \right)^n \right]$$

ROC is $\left| e^{j\omega_0} z^{-1} \right| < 1$ and $\left| e^{-j\omega_0} z^{-1} \right| < 1 \quad$ i.e., $|z| > 1$

$$= \frac{1}{2} \left[\frac{1}{1 - e^{j\omega_0} z^{-1}} + \frac{1}{1 - e^{-j\omega_0} z^{-1}} \right] = \frac{1}{2} \left[\frac{2 - 2z^{-1} \cos\omega_0}{1 - 2z^{-1} \cos\omega_0 + z^{-2}} \right]$$

$$= \frac{1 - z^{-1} \cos\omega_0}{1 - 2z^{-1} \cos\omega_0 + z^{-2}}$$

The Z-transforms of common sequences are available, in closed form and are given in the form of table 3.1.

Table 3.1

S.No	Sequence	z-transform	ROC
1.	$\delta(n)$	1	Entire z-plane
2.	$u(n)$	$\dfrac{1}{1-z^{-1}}$	$\lvert z \rvert > 1$
3.	$-u(-n-1)$	$\dfrac{1}{1-z^{-1}}$	$\lvert z \rvert < 1$
4.	$\delta(n-k)$	z^{-k}	Entire z-plane except 0 (if $k > 0$) Entire z-plane except ∞ (if $k < 0$)
5.	$a^n u(n)$	$\dfrac{1}{1-az^{-1}}$	$\lvert z \rvert > \lvert a \rvert$
6.	$-a^n u(-n-1)$	$\dfrac{1}{1-az^{-1}}$	$\lvert z \rvert < \lvert a \rvert$
7.	$n\,a^n u(n)$	$\dfrac{az^{-1}}{\left(1-az^{-1}\right)^2}$	$\lvert z \rvert > \lvert a \rvert$
8.	$-n\,a^n u(-n-1)$	$\dfrac{az^{-1}}{\left(1-az^{-1}\right)^2}$	$\lvert z \rvert < \lvert a \rvert$
9.	$(\cos\omega_0 n)\,u(n)$	$\dfrac{1-\left(\cos\omega_0\right)z^{-1}}{1-2\left(\cos\omega_0\right)z^{-1}+z^{-2}}$	$\lvert z \rvert > 1$
10.	$(\sin\omega_0 n)\,u(n)$	$\dfrac{\left(\sin\omega_0\right)z^{-1}}{1-2\left(\cos\omega_0\right)z^{-1}+z^{-2}}$	$\lvert z \rvert > 1$
11.	$a^n \cos\omega_0 n$ $u(n)$	$\dfrac{1-\left(a\cos\omega_0\right)z^{-1}}{1-2\left(a\cos\omega_0\right)z^{-1}+a^2 z^{-2}}$	$\lvert z \rvert > \lvert a \rvert$
12.	$a^n \sin\omega_0 n\,u(n)$	$\dfrac{\left(a\sin\omega_0\right)z^{-1}}{1-2\left(a\cos\omega_0\right)z^{-1}+a^2 z^{-2}}$	$\lvert z \rvert > \lvert a \rvert$
13.	$\cosh(\omega_0 n)$	$\dfrac{z^2 - z\left(\cosh\omega_0\right)}{z^2 - 2z\left(\cosh\omega_0\right)+1}$	$\lvert z \rvert > \cosh\omega_0$
14.	$\sinh(\omega_0 n)$	$\dfrac{z\left(\sinh\omega_0\right)}{z^2 - 2z\left(\cosh\omega_0\right)+1}$	$\lvert z \rvert > \sinh\omega_0$

3.2 Properties of the Z-transform

3.2.1 Two Sided Z-transform Properties

(i) **Linearity:**

$$\text{If} \qquad x_1(n) \xleftrightarrow{\ z\ } X_1(z)$$

and $\qquad x_2(n)\xleftrightarrow{Z}X_2(z)$, then

$$a\, x_1(n) + b\, x_2(n)\xleftrightarrow{Z}a\, X_1(z) + b\, X_2(z) \qquad\qquad .....(3.2.1)$$

Proof: we know the z-transform of x(n)

i.e., $\qquad z\,\{x(n)\} = X(z) = \displaystyle\sum_{n=-\infty}^{\infty} x(n)\, z^{-n}$

$$z\{a\, x_1(n) + b\, x_2(n)\} = \sum_{n=-\infty}^{\infty}\left(a\, x_1(n) + b\, x_2(n)\right)z^{-n}$$

$$= a\sum_{n=-\infty}^{\infty} x_1(n)\, z^{-n} + b\sum_{n=-\infty}^{\infty} x_2(n)\, z^{-n}$$

$$= a\, X_1(z) + b\, X_2(z)$$

$$\therefore\quad a\, x_1(n) + b\, x_2(n)\xleftrightarrow{Z}a\, X_1(z) + b\, X_2(z)$$

(ii) Time Shifting:

If $\qquad x(n)\xleftrightarrow{Z}X(z)$, then

$$x(n-k)\xleftrightarrow{Z}z^{-k}\, X(z) \qquad\qquad .....(3.2.2)$$

Proof: $\qquad z\{x(n-k)\} = \displaystyle\sum_{n=-\infty}^{\infty} x(n-k)\, z^{-n}$

put $\quad n-k = l\ \rightarrow\ n = l + k$

$$n = -\infty \rightarrow \qquad l = -\infty$$

$$n = \infty \rightarrow \qquad l = \infty$$

$$= \sum_{l=-\infty}^{\infty} x(l)\, z^{-l}.\, z^{-k}$$

$$= z^{-k}\, X(z)$$

$\therefore \qquad x(n-k)\xleftrightarrow{Z}X(z)$

Similarly $\qquad x(n+k)\xleftrightarrow{Z}z^{k}\, X(z) \qquad\qquad .....(3.2.3)$

Example 3.8 Determine the Z-transform of

$$x(n) = \begin{cases} 1, & 0 \le n \le N-1 \\ 0, & \text{otherwise} \end{cases}$$

(or) $\qquad x(n) = u(n) - u(n-N)$

Solution:

$$X(z) = \sum_{n=0}^{N-1} 1.z^{-n} = 1 + z^{-1} + \dots + z^{-(N-1)}$$

$$= \begin{cases} N, & \text{if } z = 1 \\ \dfrac{1-z^{-N}}{1-z^{-1}}, & \text{if } z \ne 1 \end{cases} \qquad \dots\text{(3.2.4)}$$

It's ROC is entire z-transform except $z = 0$.

(iii) Scaling in the Z-domain:

If $\qquad x(n) \xleftrightarrow{\ Z\ } X(z), \quad \text{ROC: } l_1 < |z| < l_2$

then $\qquad a^n x(n) \xleftrightarrow{\ Z\ } X(z/a), \text{ROC: } |a| l_1 < |z| < |a| l_2 \qquad \dots\text{(3.2.5)}$

Proof: $\qquad z\{a^n x(n)\} = \sum_{n=-\infty}^{\infty} a^n x(n)\, z^{-n}$

$$= \sum_{n=-\infty}^{\infty} x(n)(a^{-1}z)^{-n}$$

$$= X(z/a)$$

Since the ROC of $X(z)$ is $l_1 < |z| < l_2$,

the ROC of $X(z/a)$ is $l_1 < \dfrac{|z|}{|a|} < l_2$

i.e., $\qquad |a| l_1 < |z| < |a| l_2$

Example 3.9: Determine the Z-transform of the signals.

(a) $x(n) = a^n (\cos \omega_0 n)\, u(n)$ (b) $x(n) = a^n (\sin \omega_0 n)\, u(n)$

Solution: From table 3.1

$$(\cos \omega_0 n)\, u(n) \xleftrightarrow{\ Z\ } \frac{1 - (\cos \omega_0) z^{-1}}{1 - 2(\cos \omega_0) z^{-1} + z^{-2}} \left.\vphantom{\begin{cases}1\\2\\3\\4\end{cases}}\right\}$$

$$(\sin \omega_0 n)\, u(n) \xleftrightarrow{\ Z\ } \frac{(\sin \omega_0) z^{-1}}{1 - 2(\cos \omega_0) z^{-1} + z^{-2}} \qquad |z| > 1$$

(a) From equation (3.2.5)

$$a^n (\cos \omega_0 n)\, u(n) \longleftrightarrow \frac{1 - (\cos \omega_0)a\, z^{-1}}{1 - 2(\cos \omega_0)a\, z^{-1} + a^2\, z^{-2}}$$

(b) Similarly

$$a^n (\sin \omega_0 n)\, u(n) \overset{Z}{\longleftrightarrow} \frac{(\sin \omega_0)a\, z^{-1}}{1 - 2(\cos \omega_0)a\, z^{-1} + a^2\, z^{-2}}$$

(iv) Scaling in the Time Domain:

If $\quad x(n)\overset{Z}{\longleftrightarrow}X(z)$, ROC: $l_1 < |z| < l_2$

then $\quad x(kn)\overset{Z}{\longleftrightarrow}X(z^{1/k})$, ROC: $l_1^{|k|} < |z| < l_2^{|k|}$ $\qquad\qquad$.....(3.2.6)

Proof: $\qquad z\{x(kn)\} = \displaystyle\sum_{n=-\infty}^{\infty} x(kn)\, z^{-n}$

Put $\qquad kn = l \to n = l/k \quad$ (l is different from l_1 and l_2)

$$n = -\infty \to \qquad l = -\infty$$

$$n = \infty \to \qquad l = \infty$$

$\therefore \qquad z\{x(kn)\} = \displaystyle\sum_{l=-\infty}^{\infty} x(l)\, z^{-l/k}$

$$= \sum_{l=-\infty}^{\infty} x(l)\, \left(z^{1/k}\right)^{-1}$$

$$= X\left(z^{1/k}\right)$$

Since the ROC of $X(z)$ is $l_1 < |z| < l_2$,

the ROC of $X(z^{1/k})$ is $l_1 < |z^{1/k}| < l_2$

$$l_1^{|k|} < |z| < l_2^{|k|}$$

Example 3.10: Obtain the Z-transform of $x(2n)$.

[SVU 2000]

Solution: $\qquad z\{x(2n)\} = \displaystyle\sum_{n=-\infty}^{\infty} x(2n)\, z^{-n}$

Put $\qquad 2n = l \to n = l/2$

$$n = -\infty \rightarrow \qquad l = -\infty$$

$$n = \infty \rightarrow \qquad l = \infty$$

$$\therefore \qquad z\{x(2n)\} = \sum_{l=-\infty}^{\infty} x(l)\, z^{-l/2}$$

$$= \sum_{l=-\infty}^{\infty} x(l)\, \left(z^{1/2}\right)^{-l}$$

$$= X\left(z^{1/2}\right)$$

$$\therefore \qquad \text{z-transform of } x(2n) \text{ is } X(z^{1/2})$$

(v) Differentiation in Z-domain:

If $x(n) \xleftrightarrow{\;Z\;} X(z)$ is a z-transform pair, then

$$nx(n) \longleftrightarrow -z\frac{d}{dz}\{X(z)\} \qquad\qquad .....(3.2.7)$$

Proof:

We know
$$X(z) = \sum_{n=-\infty}^{\infty} x(n)\, z^{-n}$$

Differentiate both sides w.r.t. z.

$$\frac{d\{X(z)\}}{dz} = \frac{d}{dz}\left\{ \sum_{n=-\infty}^{\infty} x(n)\, z^{-n} \right\}$$

$$= \sum_{n=-\infty}^{\infty} x(n)\frac{d}{dz}\{z^{-n}\}$$

$$= \sum_{n=-\infty}^{\infty} x(n)\,(-n).\, z^{-n-1}$$

$$= -z^{-1}\sum_{n=-\infty}^{\infty} n\, x(n)\, z^{-n}$$

$$-z\frac{d}{dz}\{X(z)\} = \sum_{n=-\infty}^{\infty} n\, x(n)\, z^{-n}$$

$\therefore$ From the above equation

$$n\, x(n) \xleftrightarrow{\;Z\;} -z\frac{d}{dz}\{X(z)\}$$

This can be generalized as

$$n^p\; x(n) \xleftrightarrow{\;z\;} -z\frac{d}{dz}\left\{z\left\{n^{p-1}x(n)\right\}\right\} \qquad\qquad \dots\dots (3.2.8)$$

where p is a positive integer.

Example 3.11: Obtain z-transform of $\;x_1(n) = n^2\; u(n)$

[SVU 2001]

Solution: General formula is $\;z\left\{n^p x(n)\right\} = -Z\dfrac{d}{dz}\left\{z\left\{n^{p-1}\; x(n)\right\}\right\}$

Here $\;p = 2$ and $x(n) = u(n)$

$$z\left\{n^2 u(n)\right\} = -z\frac{d}{dz}\left\{z\left\{n\; u(n)\right\}\right\} \qquad\qquad \dots\dots(3.2.9)$$

$$z\left\{n\; u(n)\right\} = -z\frac{d}{dz}\left\{z\left\{u(n)\right\}\right\} \qquad\qquad \dots\dots(3.2.10)$$

We know $\;z\left\{u(n)\right\} = \dfrac{z}{z-1}$

Substitute in (3.2.10)

$$\therefore \qquad z\left\{n\; u(n)\right\} = -z\frac{d}{dz}\left\{\frac{z}{z-1}\right\}$$

$$= -z\left[\frac{(z-1)1 - z.1}{(z-1)^2}\right]$$

$$= -z\;\frac{-1}{(z-1)^2}$$

$$z\left\{n\; u(n)\right\} = \frac{z}{(z-1)^2}$$

Substitute in (3.2.9)

$$\therefore \qquad z\left\{n^2\; u(n)\right\} = -z\frac{d}{dz}\left\{\frac{z}{(z-1)^2}\right\}$$

$$= -z\left\{\frac{(z-1)^2.1 - z.2(z-1)}{(z-1)^4}\right\}$$

$$= -z \, \frac{(1-z)^2}{(z-1)^4}$$

$$z\{n^2 u(n)\} = \frac{z(z^2-1)}{(z-1)^4} = \frac{z(z+1)}{(z-1)^3}$$

(vi) Time Reversal:

If $x(n) \xleftarrow{\ Z\ } X(z)$ is Z-transform pair with ROC: $l_1 < |z| < l_2$,

then $x(-n) \xleftarrow{\ Z\ } X(z^{-1})$, with ROC: $\dfrac{1}{l_2} < |z| < \dfrac{1}{l_1}$ .....(3.2.11)

is a pair

Proof: Consider $z\{x(-n)\} = \displaystyle\sum_{n=-\infty}^{\infty} x(-n) \, z^{-n}$

Put $m = -n$

$$n = -\infty \rightarrow m = \infty$$

$$n = \infty \rightarrow m = -\infty$$

$$Z\{x(-n)\} = \sum_{m=-\infty}^{\infty} x(m) \, z^{m}$$

$$= \sum_{m=-\infty}^{\infty} x(m) \, (z^{-1})^{-m}$$

$\therefore$ $z\{x(-n)\} = X\left(z^{-1}\right)$

The ROC of $x(z^{-1})$ is $l_1 < |z^{-1}| < l_2$ i.e., $\dfrac{1}{l_2} < |z| < \dfrac{1}{l_1}$

(vii) Convolution of Two Sequences:

If $x_1(n) \xleftarrow{\ Z\ } X_1(z)$ and $x_2(n) \xleftarrow{\ Z\ } X_2(z)$ are z-transform pairs, then

$$x(n) = x_1(n) * x_2(n) \xleftarrow{\ Z\ } X(z) = X_1(z)X_2(z)$$ .....(3.2.12)

Proof: We know

$$x_1(n) * x_2(n) = \sum_{m=-\infty}^{\infty} x_1(m) \, x_2(n-m)$$

Consider

$$z\{x_1(n) * x_2(n)\} = \sum_{n=-\infty}^{\infty} \left(x_1(n) * x_2(n)\right) z^{-n}$$

$$= \sum_{n=-\infty}^{\infty} \sum_{m=-\infty}^{\infty} x_1(m) \, x_2(n-m) \, z^{-n}$$

Interchange summations

$$= \sum_{m=-\infty}^{\infty} x_1(m) \sum_{n=-\infty}^{\infty} x_2(n-m) \, z^{-n}$$

Put $\quad n - m = l \rightarrow n = l + m$

$$n = -\infty \rightarrow l = -\infty$$

$$n = \infty \rightarrow l = \infty$$

$$Z\{x_1(n) * x_2(n)\} = \sum_{m=-\infty}^{\infty} x_1(m) \sum_{l=-\infty}^{\infty} x_2(l) \, z^{-l} \cdot z^{-m}$$

$$= \sum_{m=-\infty}^{\infty} x_1(m) \, z^{-m} \sum_{l=-\infty}^{\infty} x_2(l) \, z^{-l}$$

$$Z\{x_1(n) * x_2(n)\} = X_1(z) X_2(z)$$

Example 3.12 Compute the convolution of $x_1(n) = \{1, 1, 1\}$ and $x_2(n) = \{1, 1, 1\}$

Solution: $\quad X_1(z) = 1 + z^{-1} + z^{-2}$ and $\quad X_2(z) = 1 + z^{-1} + z^{-2}$

$$X(z) = X_1(z) \, X_2(z) = \left(1 + z^{-1} + z^{-2}\right)\left(1 + z^{-1} + z^{-2}\right)$$

$$= 1 + z^{-1} + z^{-2} + z^{-1} + z^{-2} + z^{-3} + z^{-2} + z^{-3} + z^{-4}$$

$$= 1 + 2z^{-1} + 3z^{-2} + 2z^{-3} + z^{-4}$$

$$\therefore \quad Z^{-1}\{X(z)\} = x(n) = \{1, 2, 3, 2, 1\}$$

3.2.2 One Sided Z-transform Properties

The two-sided Z-transform cannot be used to evaluate the output of non relaxed systems (systems with non zero initial values) where as the one sided Z-transform can be used to evaluate the output of non relaxed systems.

Almost all properties we have studied for the two-sided z–transform carry over to the one-sided Z-transform with the exception of the shifting property.

(i) Shifting Property

(a) Time delay: If $x(n) \xleftrightarrow{\;Z^+\;} X^+(z)$ is a pair then $\quad x(n-k)\xleftrightarrow{\;z^+\;}z^{-k}\left[X^+(z)+\sum_{n=1}^{k}x(-n)z^n\right],\ k>0$ .....(3.2.13)

In case x(n) is causal, then $x(n-k)\xleftrightarrow{\;z^+\;}z^{-k}X^+(z)$ .....(3.2.14)

Proof: Consider one sided Z-transform

$$Z^+\{x(n-k)\}=\sum_{n=0}^{\infty}x(n-k)\,z^{-n}$$

Put $\qquad n-k=l\rightarrow n=l+k$

$\qquad\qquad n=0\rightarrow l=-k$

$\qquad\qquad n=\infty\rightarrow l=\infty$

$\therefore\qquad Z^+\{x(n{-}k)\}=\sum_{l=-k}^{\infty}x(l)z^{-l}\times z^{-k}$

$$=z^{-k}\sum_{l=-k}^{\infty}x(l)z^{-l}$$

$$=z^{-k}\left[\sum_{l=-k}^{-1}x(l)z^{-l}+\sum_{l=0}^{\infty}x(l)z^{-l}\right]$$

$$=z^{-k}\left[\sum_{l=-k}^{-1}x(l)z^{-l}+X^+(z)\right]$$

Put $\qquad\qquad\qquad l=-n$

$$=z^{-k}\left[\sum_{n=1}^{k}x(-n)z^n+X^+(z)\right]$$

$\therefore\qquad Z^+\{n(n-k)\}=z^{-k}\left[X^+(z)+\sum_{n=1}^{k}x(-n)z^n\right]$

(b) Time advance: If $x(n)\xleftrightarrow{\;z^+\;}X^+(z)$ is a pair,

then $\quad x(n+k)\xleftrightarrow{\;z^+\;}z^{k}\left[X^+(z)-\sum_{n=0}^{k-1}x(n)z^{-n}\right],\ k>0$ .....(3.2.15)

Proof: Consider

$$Z^+\{x(n+k)\}=\sum_{n=0}^{\infty}x(n+k)z^{-n}$$

Put $\quad n+k=l \rightarrow n=l-k$

$\qquad n=0 \rightarrow l=k$

$\qquad n=\infty \rightarrow l=\infty$

$$Z^+\{x(n+k)\} = \sum_{l=k}^{\infty} x(l)z^{-l}.z^k$$

$$= z^k \sum_{l=k}^{\infty} x(l)z^{-l}$$

$$= z^k \left[\sum_{l=0}^{\infty} x(l)z^{-l} - \sum_{l=0}^{k-1} x(l)z^{-l}\right]$$

$$= z^k \left[X^+(z) - \sum_{l=0}^{k-1} x(l)z^{-l}\right]$$

Put $\qquad\qquad\qquad\qquad l=n$

$$Z\{x(n+k)\} = z^k \left[X^+(z) - \sum_{n=0}^{k-1} x(n)z^{-n}\right]$$

The theorems that are very much useful in the analysis of signals and systems are initial value and final value theorems, which are described as follows

(ii) Initial Value Theorem: If $x(n) \overset{Z^+}{\longleftrightarrow} X^+(z)$ is a pair,

then $\qquad\qquad x(0) = \underset{Z \to \infty}{\lim} X^+(z)$ $\qquad\qquad\qquad$ .....(3.2.16)

Proof: We know $\qquad X^+(z) = \sum_{n=0}^{\infty} x(n)z^{-n}$

$$= x(0) + x(1)z^{-1} + x(2)z^{-2} + \cdots$$

Take limits on both sides

$$\underset{Z \to \infty}{\lim} X^+(z) = \underset{Z \to \infty}{\lim}\left[x(0) + \frac{x(1)}{z} + \frac{x(2)}{z^2} + \cdots\right]$$

$$= x(0)$$

$\therefore \qquad\qquad x(0) = \underset{Z \to \infty}{\lim} X^+(z)$

i.e., the initial value of the signal x(n) can be determined with the above equation.

(iii) Final Value Theorem: If $x(n) \overset{Z^+}{\longleftrightarrow} X^+(z)$ is a pair,

then $\qquad\qquad \underset{n \to \infty}{\lim} x(n) = \underset{Z \to 1}{\lim}(z-1)X^+(z)$ $\qquad\qquad$ .....(3.2.17)

The limit exists if the ROC of $(Z-1)X^+(Z)$ includes the unit circle

Proof: Consider

$$Z^+\{x(n+1)-x(n)\} = \sum_{n=0}^{\infty}\left[x(n+1)-x(n)\right]Z^{-n}$$

$$Z^1 X^+(Z) - Z\,x(0) - X^+(Z) = \sum_{n=0}^{\infty}\left[x(n+1)-x(n)\right]Z^{-n}$$

$$X^+(Z)(Z-1) - Z\,x(0) = \sum_{n=0}^{\infty}\left[x(n+1)-x(n)\right]Z^{-n}$$

Take limits on both sides

$$\lim_{Z\to1}(z-1)X^+(z) - \lim_{Z\to1}z\,x(0) = \lim_{z\to1}\lim_{m\to\infty}\sum_{n=0}^{m}\left[x(n+1)-x(n)\right]z^{-n}$$

Interchange limits on right side

$$= \lim_{m\to\infty}\sum_{n=0}^{m}\left[x(n+1)-x(n)\right]$$

$$\therefore \qquad \lim_{Z\to1}(Z-1)X^+(z) = x(0) + \sum_{n=0}^{\infty}\left[x(n+1)-x(n)\right]$$

$$= x(\infty) = \lim_{n\to\infty}x(n)$$

$$\therefore \qquad \lim_{n\to\infty}x(n) = \lim_{Z\to1}(z-1)X^+(Z)$$

Example 3.13: If X (z) is the Z-transform of x (n), show that

(a) $Z\{x^*(n)\} = X^*(z^*)$, (conjugation property)

(b) $Z\{Re[x(n)]\} = \dfrac{1}{2}\left[X(z) + X^*(z^*)\right]$,

(c) $Z\{I_m[x(n)]\} = \dfrac{1}{2}\left[X(z) - X^*(z^*)\right]$

Solution: (a) We know $X(z) = \sum_{n=-\infty}^{\infty}x(n)\,z^{-n}$

Consider $X(z^*) = \sum_{n=-\infty}^{\infty}x(n)\,(z^*)^{-n}$

Take conjugation on both sides

$$X^*\left(z^*\right) = \sum_{n=-\infty}^{\infty} x^*(n)\left(\left(z^*\right)^{-n}\right)^*$$

$$= \sum_{n=-\infty}^{\infty} x^*(n)\, z^{-n}$$

$$\therefore \qquad Z\{x^*(n)\} = X^*\left(z^*\right)$$

(b) x(n) can be written as

$$x(n) = \mathrm{Re}\left[x(n)\right] + j I_m\left[x(n)\right]$$

$$\therefore \qquad X(z) = \sum_{n=-\infty}^{\infty} \left(\mathrm{Re}\left[x(n)\right] + j\, I_m\left[x(n)\right]\right) z^{-n}$$

$$X^*\left(z^*\right) = \sum_{n=-\infty}^{\infty} \left(\mathrm{Re}\left[x(n)\right] - j\, I_m\left[x(n)\right]\right) z^{-n}$$

Consider

$$\frac{1}{2}\left[X(z) + X^*\left(z^*\right)\right] = \sum_{n=-\infty}^{\infty} \mathrm{Re}\left[x(n)\right] z^{-n}$$

$$= z\{\mathrm{Re}\left[x(n)\right]\}$$

(c) Consider

$$\frac{1}{2}\left[X(z) - X^*\left(z^*\right)\right] = j\sum_{n=-\infty}^{\infty} I_m\left[x(n)\right] z^{-n}$$

$$= Z\{I_m\left[x(n)\right]\}$$

Example 3.14: If X(z) is the Z-transform of x(n), show that

If

$$x_k(n) = \begin{cases} x\left(\dfrac{n}{k}\right), & \text{if } n/k \text{ is integer} \\[2mm] 0, & \text{otherwise} \end{cases}$$

then

$$X_k(z) = X\left(z^k\right)$$

Solution: Consider

$$Z\{x_k(n)\} = z\left\{x\left(\frac{n}{k}\right)\right\} = \sum_{n=-\infty}^{\infty} x\left(\frac{n}{k}\right) z^{-n}$$

Put

$$\frac{n}{k} = l \rightarrow n = lk$$

$$n = -\infty \rightarrow l = -\infty$$

$$n = \infty \rightarrow l = \infty$$

$$Z\{x_k(n)\} = \sum_{l=-\infty}^{\infty} x(l) \, z^{-lk}$$

$$X_k(z) = \sum_{l=-\infty}^{\infty} x(l) \left(z^k\right)^{-l}$$

$$X_k(z) = X\left(z^k\right)$$

3.3 The Inverse Z-transform

The inverse Z-transform (IZT) allows us to recover the discrete-time sequence x(n), given its Z-transform. The inverse Z-transform may be defined as

$$x(n) = z^{-1}\left[X(z)\right] \qquad \qquad \dots\dots(3.3.1)$$

where $X(z)$ is the z-transform of x(n) and z^{-1} is the symbol for the inverse z–transform.

Assuming a causal sequence, the z-transform, $X(z)$ can be expanded into a power series as

$$X(z) = \sum_{n=0}^{\infty} x(n)z^{-n}$$

$$= x(0) + x(1)z^{-1} + x(2)z^{-2} + x(3)z^{-3} + \cdots\cdots$$

It is observed that the values of x(n) are the coefficients of Z^{-n} (n = 0, 1, ...) and so can be obtained directly by inspection. In practice, $X(z)$ is often expressed as a ratio of two polynomials in z^{-1} or equivalently in z.

$$X(z) = \frac{b_0 + b_1 z^{-1} + b_2 z^{-2} + \cdots + b_N z^{-N}}{a_0 + a_1 z^{-1} + a_2 z^{-2} + \cdots + a_M z^{-M}} \qquad \qquad \dots\dots(3.3.2)$$

In this form, the inverse z-transform, x(n), may be obtained using one of several methods including the following three

1. Power series expansion method (or) long division (or) synthetic division method;

2. Partial fraction expansion method;

3. Residue method (or) Cauchy's Residue method (or) Complex inversion integral method.

3.3.1 Power Series Method

Given the Z-transform, X(Z) of a sequence, it can be expanded into an infinite series in z^{-1} (if ROC specified is exterior to circle) or z (if ROC specified is interior to circle) by long division.

We will illustrate the method by an examples

Example 3.15: Determine the inverse Z-transform of

$$X(z) = \frac{1}{1 - 1.5z^{-1} + 0.5z^{-2}}$$

when

(a) $ROC : |z| > 1$ (b) $ROC : |z| < 0.5$

Solution: (a) Since the ROC is exterior of a circle, time domain sequence x(n) will be a causal. Thus we should divide the numerator of X(z) by its denominator, such that we get negative powers of z.

In this method, the numerator and denominator of X(z) are first expressed in either descending powers of z or ascending powers of z^{-1} and the quotient is then obtained by long division.

$$
\begin{array}{r}
1 + 1.5z^{-1} + 1.75z^{-2} + 1.875z^{-3} + 1.9375z^{-4} + \cdots \\
\hline
1 - 1.5z^{-1} + 0.5z^{-2} \,\big)\; 1 \\
1 - 1.5z^{-1} + 0.5z^{-2} \\
\hline
1.5z^{-1} - 0.5z^{-2} \\
1.5z^{-1} - 2.25z^{-2} + 0.75z^{-3} \\
\hline
1.75z^{-2} - 0.75z^{-3} \\
1.75z^{-2} - 2.625z^{-3} + 0.875z^{-4} \\
\hline
1.875z^{-3} - 0.875z^{-4} \\
1.875z^{-3} - 2.8125z^{-4} + 0.9375z^{-5} \\
\hline
1.9375z^{-4} - 0.9375z^{-5}
\end{array}
$$

$$\therefore \quad x(n) = \left\{\underset{\uparrow}{1},\ 1.5,\ 1.75,\ 1.875,\ 1.9375,\ \ldots\right\}$$

This long division approach can be reformulated so that the values of x(n) are obtained recursively

$$x(0) = b_0 / a_0$$

$$x(1) = \left[b_1 - x(0)a_1\right] / a_0$$

$$x(2) = \left[b_2 - x(1)a_1 - x(0)a_2\right] / a_0$$

$$\vdots \quad \vdots \quad \vdots \quad \vdots \quad \vdots \quad \vdots$$

$$x(n) = \left[b_n - \sum_{i=1}^{n} x(n-i)\, a_i \right] \Big/ a_o, \quad n = 1, 2, \ldots, \qquad \ldots\ldots(3.3.3)$$

where $x(0) = b_0 / a_0$ $\qquad\qquad\qquad\qquad\qquad\qquad\qquad\qquad\qquad$ $\ldots\ldots(3.3.4)$

$$X(z) = \frac{1}{1 - 1.5\, z^{-1} + 0.5\, z^{-2}} = \frac{b_0 + b_1 z^{-1} + b_2 z^{-2} + \ldots + b_N z^{-N}}{a_0 + a_1 z^{-1} + a_2 z^{-2} + \ldots + a_M z^{-M}}$$

comparing $\quad b_0 = 1,\ b_1 = b_2 = \ldots b_N = 0$

$$a_0 = 1,\ a_1 = -1.5,\ a_2 = 0.5$$

$$x(0) = \frac{b_0}{a_0} = \frac{1}{1} = 1$$

$$x(1) = \left[b_1 - \sum_{i=1}^{1} x(1-i) a_i \right] \Big/ a_0$$

$$= \left[b_1 - x(0) a_1 \right] / a_0$$

$$x(1) = \left[0 - 1 \times -1.5 \right] / 1 = 1.5$$

$$x(2) = \left[b_2 - \sum_{i=1}^{2} x(2-i) a_i \right] \Big/ a_0$$

$$= \left[0 - x(1) a_1 - x(0) a_2 \right] \Big/ a_0$$

$$x(2) = \left[-1.5 \times -1.5 - 1 \times 0.5 \right] \Big/ 1 = 1.75$$

$$x(3) = \left[b_3 - \sum_{i=1}^{3} x(3-i) a_i \right] \Big/ a_0$$

$$= \left[0 - x(2) a_1 - x(1) a_2 - x(0) a_3 \right] \Big/ a_0$$

$$= \left[-1.75 \times -1.5 - 1.5 \times 0.5 - 1 \times 0 \right] \Big/ 1$$

$$x(3) = 1.875$$

$$x(4) = \left[0 - x(3) a_1 - x(2) a_2 - x(1) a_3 - x(0) a_4 \right] \Big/ a_0$$

$$= \left[-1.875 \times -1.5 - 1.75 \times 0.5 - 1.5 \times 0 - 1 \times 0 \right] \Big/ 1$$

$$x(4) = 1.9375$$

$$x(n) = \left\{ \underset{\uparrow}{1},\ 1.5,\ 1.75,\ 1.875,\ 1.9375,\ \ldots \right\}$$

(b) In this case, the ROC is the interior of a circle, time domain sequence x(n) will be anticausal. Thus we should divide the numerator of X(z) by its denominator, such that we get positive powers of z.

In this method, the numerator and denominator of X(z) are first expressed in either ascending powers z or descending powers of z^{-1} and the quotient is then obtained by long division.

$$\frac{1}{2}z^{-2} - \frac{3}{2}z^{-1} + 1 \overline{\big)\; \begin{array}{l} 2z^2 + 6z^3 + 14z^4 + 30z^5 + 62z^6 + \ldots\ldots \\[4pt] 1 \\[2pt] \underline{1 - 3z + 2z^2} \\[2pt] \qquad 3z - 2z^2 \\[2pt] \qquad \underline{3z - 9z^2 + 6z^3} \\[2pt] \qquad\qquad 7z^2 - 6z^3 \\[2pt] \qquad\qquad \underline{7z^2 - 21z^3 + 14z^4} \\[2pt] \qquad\qquad\qquad 15z^3 - 14z^4 \\[2pt] \qquad\qquad\qquad \underline{15z^3 - 45z^4 + 30z^5} \\[2pt] \qquad\qquad\qquad\qquad 31z^4 - 30z^5 \end{array}}$$

$$\therefore \quad x(n) = \left\{ \ldots,\; 62,\, 30,\, 14,\, 6,\, 2,\, 0,\, \underset{\uparrow}{0} \right\}$$

3.3.2   Partial Fraction Expansion Method

In this method, the z-transform is expanded into a sum of simple partial fractions. The inverse z-transform of each partial fraction is then obtained from tables, such as Table 3.1, and then summed to obtain the overall inverse z-transform.

The z-transform is given as a ratio of polynomials in z or z^{-1}, as given below

$$X(z) = \frac{b_0 + b_1 z^{-1} + b_2 z^{-2} + \ldots + b_N z^{-N}}{a_0 + a_1 z^{-1} + a_2 z^{-2} + \ldots + a_M z^{-M}}$$

If N = M and X(z) has only first order poles, then it can be expanded as

$$X(z) = B_0 + \frac{C_1}{1 - P_1 z^{-1}} + \frac{C_2}{1 - P_2 z^{-1}} + \ldots + \frac{C_M}{1 - P_M z^{-1}}$$

$$= B_0 + \frac{C_1 z}{z - P_1} + \frac{C_2 z}{z - P_2} + \ldots + \frac{C_M z}{z - P_M}$$

$$= B_0 + \sum_{K=1}^{M} \frac{C_K z}{z - P_K} \qquad\qquad \ldots..(3.3.5)$$

where P_K are the poles of $X(z)$ (assumed distinct), C_K are the partial fraction coefficients and

$$B_0 = \frac{b_N}{a_N}$$

C_K can be obtained by

$$C_K = \frac{X(z)}{z}(z - P_K)\big|_{z=P_K} \qquad \dots\dots(3.3.6)$$

If $X(z)$ contains one or more multiple-order poles (i.e., poles that are coincident) then extra terms are required in equation (3.3.4). For example if $X(z)$ contains an mth – order pole at $z = P_l$, along with first order poles at $Z = P_K$, then $X(z)$ can be expanded as

$$X(z) = B_0 + \sum_{k=1}^{M} \frac{C_k z}{z - P_K} + \sum_{i=1}^{m} \frac{D_i z}{(z - P_l)^i} \qquad \dots\dots (3.3.7)$$

The coefficients, D_i may be obtained by

$$D_i = \frac{1}{(m-i)!} \frac{d^{m-i}}{dz^{m-i}} \left[(z - p_l)^m \frac{X(z)}{z} \right]_{Z=P_l} \qquad \dots\dots (3.3.8)$$

If $N < M$, then B_0 will be zero.

If $N > M$, then $X(z)$ must be reduced first, to make $N \leq M$, by long division with the numerator and denominator polynomials written in descending powers of z^{-1}. The remainder can then be expressed either as in equation (3.3.4) (if there are no multiple order poles) or as in equation (3.3.7) (if there are multiple order poles).

While writing inverse z-transform, if ROC is $|z| > P_{max}$, where $P_{max} = \max \{p_1, p_2, \dots, p_k\}$, then the time domain sequence is causal.

i.e., $\qquad\qquad \dfrac{z}{z - p_k} \xleftarrow{\ z\ } (p_k)^n u(n), \ \text{ROC} : |z| > P_{max}, \ k=1, 2, \dots \ \dots\dots (3.3.9)$

if ROC: $|z| < P_{min}$, where $P_{min} = \min \{p_1, p_2, \dots, p_k\}$, then the time domain sequence is anticausal.

i.e., $\qquad\qquad \dfrac{z}{z - p_k} \xleftarrow{\ z\ } -(p_k)^n u(-n-1), \ \text{ROC}: |z| < P_{min} \quad \dots\dots(3.3.10)$

Example 3.16: Determine the inverse z-transform of

$$X(z) = \frac{1}{1 - 1.5\, z^{-1} + 0.5\, z^{-2}}$$

when   (a) ROC: $|z| > 1$ (b) ROC: $|z| < 0.5$

Solution:

$$X(z) = \frac{z^2}{z^2 - 1.5\,z + 0.5} = \frac{z^2}{(z-1)\,(z-0.5)}$$

Divide both sides with z

$$\frac{X(z)}{z} = \frac{z}{(z-1)\,(z-0.5)}$$

$$= \frac{C_1}{z-1} + \frac{C_2}{z-0.5}$$

$$C_1 = (z-1)\frac{z}{(z-1)(z-0.5)}\bigg|_{z=1} = \frac{1}{1-0.5} = 2$$

$$C_2 = (z-0.5)\frac{z}{(z-1)\,(z-0.5)}\bigg|_{z=0.5} = \frac{0.5}{0.5-1} = -1$$

$$\therefore \quad \frac{X(z)}{z} = \frac{2}{z-1} - \frac{1}{z-0.5}$$

$$X(z) = \frac{2\,z}{z-1} - \frac{z}{z-0.5}$$

(a) ROC: $|z| > 1$

$$\therefore \quad x(n) = 2\,u(n) - (0.5)^n\,u(n)$$

$$x(n) = \left[2 - (0.5)^n\right]u(n)$$

(b) ROC: $|z| < 0.5$

$$x(n) = -2\,u(-n-1) - \left[-(0.5)^n\,u(-n-1)\right]$$

$$= \left[(0.5)^n - 2\right]u(-n-1)$$

Example 3.17: Obtain the inverse z-transform of the following function.

$$X(z) = \frac{(z^3 - 1)}{z(z-1)\,(z-0.5)} \quad \text{for } |z| > \tfrac{1}{2} \qquad \text{[SVU 2000]}$$

Solution:

$$X(z) = \frac{(z-1)\,(z^2 + z + 1)}{z(z-1)\,(z-0.5)} = \frac{z^2 + z + 1}{z(z-0.5)}$$

Divide both sides with z

$$\frac{X(z)}{z} = \frac{z^2 + z + 1}{z^2(z - 0.5)}$$

$$\frac{X(z)}{z} = \frac{D_1}{z} + \frac{D_2}{z^2} + \frac{C_1}{z - 0.5}$$

$$m = 2,$$

$i = 1;$
$$D_1 = \frac{1}{(2-1)!} \frac{d^{2-1}}{dz^{2-1}} \left[z^2 \frac{(z^2 + z + 1)}{z^2(z - 0.5)} \right]\Bigg|_{z=0}$$

$$= \frac{d}{dz} \left[\frac{z^2 + z + 1}{z - 0.5} \right]\Bigg|_{z=0}$$

$$= \frac{(z - 0.5)(2z + 1) - (z^2 + z + 1).1}{(z - 0.5)^2}\Bigg|_{z=0}$$

$$D_1 = \frac{-0.5 - 1}{(-0.5)^2} = \frac{-1.5}{0.5 \times 0.5} = -6$$

$i = 2,$
$$D_2 = \frac{1}{(2-2)!} \frac{d^{2-Z}}{dz^{2-Z}} \left[z^2 \frac{(z^2 + z + 1)}{z^2(z - 0.5)} \right]\Bigg|_{z=0}$$

$$= \frac{z^2 + z + 1}{z - 0.5}\Bigg|_{z=0}$$

$$D_2 = -2$$

$$C_1 = (z - 0.5).\frac{(z^2 + z + 1)}{z^2(z - 0.5)}\Bigg|_{z=0.5}$$

$$C_1 = \frac{(0.5)^2 + 0.5 + 1}{(0.5)^2} = 1 + \frac{1}{0.5} + \frac{1}{(0.5)^2} = 1 + 2 + 4 = 7$$

$$\therefore \quad \frac{X(z)}{z} = \frac{-6}{z} - \frac{2}{z^2} + \frac{7}{z - 0.5}$$

$$X(z) = -6 - 2z^{-1} + \frac{7z}{z - 0.5}$$

$$x(n) = -6\,\delta(n) - 2\delta(n)z^{-1} + 7(0.5)^n\, u(n)$$

$$x(n) = -6\,\delta(n) - 2\delta(n - 1) + 7(0.5)^n\, u(n)$$

Example 3.18: Find the inverse z-transform of the following function

$$X(z) = \frac{z-2}{(z-0.8)\,(z+1.2)} \quad \text{for } |z| > 1.2$$

[JNTU 2000]

Solution:
$$X(z) = \frac{z-2}{(z-0.8)\,(z+1.2)}$$

Method I:

$$X(z) = \frac{C_1}{z-0.8} + \frac{C_2}{z+1.2}$$

$$C_1 = \left.\frac{z-2}{z+1.2}\right|_{z=0.8} = -0.6$$

$$C_2 = \left.\frac{z-2}{z-0.8}\right|_{z=-1.2} = 1.6$$

$$\therefore \quad X(z) = \frac{-0.6}{z-0.8} + \frac{1.6}{z+1.2}$$

$$X(z) = \frac{-0.6\,z\,z^{-1}}{z-0.8} + \frac{1.6\,z\,z^{-1}}{z+1.2}$$

$$\therefore \quad x(n) = -0.6(0.8)^n\,u(n)z^{-1} + 1.6(-1.2)^n\,u(n)z^{-1}$$

$$= -0.6(0.8)^{n-1}u(n-1) + 1.6(-1.2)^{n-1}u(n-1)$$

$$= -\frac{0.6}{0.8}(0.8)^n\,u(n-1) - \frac{1.6}{1.2}(-1.2)^n\,u(n-1)$$

$$x(n) = -\frac{3}{4}(0.8)^n\,u(n-1) - \frac{4}{3}(-1.2)^n\,u(n-1)$$

Method II:

Given
$$X(z) = \frac{z-2}{(z-0.8)\,(z+1.2)}$$

Divide both sides with z

$$\frac{X(z)}{z} = \frac{z-2}{z(z-0.8)\,(z+1.2)}$$

$$\frac{X(z)}{z} = \frac{C_1}{z} + \frac{C_2}{z-0.8} + \frac{C_3}{z+1.2}$$

$$C_1 = z.\frac{(z-2)}{z(z-0.8)\,(z+1.2)}\bigg|_{z=0}$$

$$= \frac{-2}{-0.8 \times 1.2} = \frac{25}{12}$$

$$C_2 = (z-0.8)\frac{(z-2)}{z(z-0.8)\,(z+1.2)}\bigg|_{z=0.8}$$

$$= \frac{0.8-2}{0.8(0.8+1.2)} = -\frac{1.2}{1.6} = -\frac{12}{16} = -\frac{3}{4}$$

$$C_3 = (z+1.2)\frac{(z-2)}{z(z-0.8)\,(z+1.2)}\bigg|_{z=-1.2}$$

$$= \frac{-1.2-2}{-1.2(-1.2-0.8)} = -\frac{3.2}{2.4} = -\frac{4}{3}$$

$$\frac{X(z)}{z} = \frac{25}{12}\cdot\frac{1}{z} - \frac{\left(\dfrac{3}{4}\right)}{(z-0.8)} - \frac{\left(\dfrac{4}{3}\right)}{(z+1.2)}$$

$$X(z) = \frac{25}{12} - \left(\frac{3}{4}\right)\frac{z}{(z-0.8)} - \left(\frac{4}{3}\right)\frac{z}{(z+1.2)}$$

$$x(n) = \frac{25}{12}\,\delta(n) - \frac{3}{4}(0.8)^n\,u(n) - \frac{4}{3}(-1.2)^n\,u(n)$$

For $n = 0$

$$x(0) = \frac{25}{12} - \frac{3}{4} - \frac{4}{3} = 0$$

$$x(n) = -\frac{3}{4}(0.8)^n\,u(n-1) - \frac{4}{3}(-1.2)^n\,u(n-1)$$

which is same as **Method I** result.

Example 3.19 Determine the inverse z-transform of

$$X(z) = \frac{1}{(z+2)^2}; \quad |z| < \frac{1}{2} \hspace{3cm} \text{[JNTU 2000]}$$

Solution:

Method I:

$$X(z) = \frac{1}{(z+2)^2}$$

divide both sides with z

$$\frac{X(z)}{z} = \frac{1}{z(z+2)^2}$$

$$\frac{X(z)}{z} = \frac{C_1}{z} + \frac{D_1}{z+2} + \frac{D_2}{(z+2)^2}$$

$$C_1 = z\frac{1}{z(z+2)^2}\bigg|_{z=0} = \frac{1}{4}$$

$$D_1 = \frac{d}{dz}\left[(z+2)^2 \frac{1}{z(z+2)^2}\right]\bigg|_{z=-2}$$

$$= \frac{d}{dz}\left(z^{-1}\right)\bigg|_{z=-2}$$

$$= -1\, z^{-2}\big|_{z=-2}$$

$$= -\frac{1}{4}$$

$$D_2 = (z+2)^2 \frac{1}{z(z+2)^2}\bigg|_{z=-2} = -\frac{1}{2}$$

$$\frac{X(z)}{z} = \frac{\left(\dfrac{1}{4}\right)}{z} + \frac{\left(-\dfrac{1}{4}\right)}{z+2} + \frac{\left(-\dfrac{1}{2}\right)}{(z+2)^2}$$

$$X(z) = \frac{1}{4} + \left(-\frac{1}{4}\right)\frac{z}{z+2} + \left(-\frac{1}{2}\right)\frac{z}{(z+2)^2}$$

Since ROC:
$$|z| < \frac{1}{2}$$

$$x(n) = \frac{1}{4}\delta(n) - \frac{1}{4}\left(-(-2)^n u(-n-1)\right) - \frac{1}{2}\left(\frac{n}{2}(-2)^n u(-n-1)\right)$$

$$= \frac{1}{4}\delta(n) + \frac{1}{4}(-2)^n u(-n-1) - \frac{n}{4}(-2)^n u(-n-1)$$

$$= \frac{1}{4}\delta(n) + \frac{1}{4}(-2)^n u(-n-1)(1-n)$$

$$x(n) = \frac{1}{4}\delta(n) + \frac{(1-n)}{4}(-2)^n u(-n-1)$$

Method II:

ROC: $|z| < \dfrac{1}{2}$

$\therefore$ we know

$$-(-2)^n\, u(-n-1) \longleftrightarrow \frac{z}{z+2}$$

Multiply with n in time domain, it results differentiation and multiplication with z in z-domain.

$$-n(-2)^n\, u(-n-1) \longleftrightarrow -z\frac{d}{dz}\left[\frac{z}{z+2}\right]$$

$$-n(-2)^n\, u(-n-1) \longleftrightarrow \frac{-2\,z}{(z+2)^2}$$

$$\frac{n}{2}(-2)^n\, u(-n-1)z^{-1} \longleftrightarrow \frac{1}{(z+2)^2}$$

$$\frac{(n-1)}{2}(-2)^{n-1}\, u(-n) \longleftrightarrow \frac{1}{(z+2)^2}$$

$$\frac{(1-n)}{4}(-2)^n\, u(-n) \longleftrightarrow \frac{1}{(z+2)^2}$$

i.e., $x(n) = \dfrac{(1-n)}{4}(-2)^n\, u(-n)$

for $n = 0$

Method I result $x(0) = \dfrac{1}{4}\delta(0) + \dfrac{(1-0)}{4}(-2)^0\, u(-0-1) = \dfrac{1}{4}$

Method II result $x(0) = \dfrac{(1-0)}{4}(-2)^0\, u(0) = \dfrac{1}{4}$

for $n = -1$

Method I result $x(-1) = \dfrac{1}{4}\delta(-1) + \dfrac{2}{4}(-2)^{-1}\, u(0) = -\dfrac{1}{4}$

Method II result $x(-1) = \dfrac{1-(-1)}{4}(-2)^{-1}\, u(0) = -\dfrac{1}{4}$

for $n = 1$

Method I result $x(1) = \dfrac{1}{4}\delta(1) + 0 = 0$

Method II result $x(1) = 0$

$\therefore$ both results are same

Example 3.20: Find the inverse z-transform for the following function

$$H(z) = \frac{2}{(z+2)^2} \text{ given ROC: } |z| > 2$$

Solution:

Method I:

$$H(z) = \frac{2}{(z+2)^2}$$

divide both sides with z

$$\frac{H(z)}{z} = \frac{2}{z(z+2)^2}$$

$$\frac{H(z)}{z} = \frac{C_1}{z} + \frac{D_1}{(z+2)} + \frac{D_2}{(z+2)^2}$$

$$C_1 = z.\frac{2}{z(z+2)^2}\Bigg|_{z=0} = \frac{1}{2}$$

$$D_1 = \frac{d}{dz}\left[(z+2)^2 \frac{2}{z(z+2)^2}\right]\Bigg|_{z=-2}$$

$$= -2\,z^{-2}\Big|_{z=-2} = -\frac{2}{4} = -\frac{1}{2}$$

$$D_2 = (z+2)^2 \frac{2}{z(z+2)^2}\Bigg|_{z=-2} = -1$$

$$\frac{H(z)}{z} = \frac{\left(\frac{1}{2}\right)}{z} + \frac{\left(-\frac{1}{2}\right)}{z+2} + \frac{(-1)}{(z+2)^2}$$

$$H(z) = \frac{1}{2} - \frac{1}{2}\frac{z}{z+2} - \frac{z}{(z+2)^2}$$

$\because \qquad$ ROC: $|z| > 2$

$$h(n) = \frac{1}{2}\delta(n) - \frac{1}{2}(-2)^n u(n) + \frac{n}{2}(-2)^n u(n)$$

$$h(n) = \frac{1}{2}\delta(n) + \frac{(n-1)}{2}(-2)^n u(n)$$

Method II:

$\because$ ROC: $|z| > 2$

We know

$$(-2)^n u(n) \longleftrightarrow \frac{z}{z+2}$$

$$n(-2)^n u(n) \longleftrightarrow -z\frac{d}{dz}\left(\frac{z}{z+2}\right)$$

$$n(-2)^n u(n) \longleftrightarrow \frac{-2z}{(z+2)^2}$$

$$-n(-2)^n u(n)z^{-1} \longleftrightarrow \frac{2}{(z+2)^2}$$

$$-(n-1)(-2)^{n-1} u(n-1) \longleftrightarrow \frac{2}{(z+2)^2}$$

$$\frac{(n-1)}{2}(-2)^n u(n-1) \longleftrightarrow \frac{2}{(z+2)^2}$$

$\therefore$ $x(n) = \frac{(n-1)}{2}(-2)^n u(n-1)$

both results are same

3.3.3   Residue Method

The z-transform of a sequence x(n) is given by $X(z) = \displaystyle\sum_{n=-\infty}^{\infty} x(n)z^{-n}$

Multiplying both sides by z^{m-1} and integrating w.r.t. z about a closed contour C in the region of convergence of X(z) we have

$$\oint_c X(z)\, z^{m-1}\, dz = \oint_c \sum_{n=-\infty}^{\infty} x(n)\, z^{-n+m-1}\, dz$$

$$= \sum_{n=-\infty}^{\infty} x(n)\oint_c z^{m-n-1}\, dz \qquad\qquad \dots(3.3.11)$$

Assume that C encloses the origin of the z-plane, then by Cauchy's residue theorem,

$$\oint_c z^{m-n-1}\, dz = 2\pi j\, \delta_{mn} \qquad\qquad .....(3.3.12)$$

where $\quad \delta_{mn} = \begin{cases} 1, & \text{if } m = n \\ 0, & \text{if } m \neq n \end{cases} \qquad\qquad .....(3.3.13)$

hence $\quad \oint_c X(z)\, z^{m-1} dz = 2\pi j \sum_{n=-\infty}^{\infty} x(n)\, \delta_{mn}$

when $\quad m = n,\ \delta_{mn} = 1$

$\therefore \qquad \oint_c X(z)\, z^{m-1} dz = 2\pi j\, x(m)$

or $\qquad x(m) = \dfrac{1}{2\pi j} \oint_c X(z)\, z^{m-1}\, dz$

$\therefore$ The inverse z-transform is given by the Contour integral as

$$x(n) = \frac{1}{2\pi j} \oint_c X(z)\, z^{n-1}\, dz \qquad\qquad ..... (3.3.14)$$

where C is the path of integration enclosing all the poles of $X(z)$

For rational polynomials, the Contour integral is evaluated using a fundamental result in complex variable theory known as Cauchy's residue theorem.

$$x(n) = \frac{1}{2\pi j} \oint_c z^{n-1}\, X(z)\, dz$$

= sum of the residues of $z^{n-1}\, X(z)$ at all the poles inside C.

The residue of $z^{n-1}\, X(z)$ at the pole p_k is given by

$$\text{Res}\left[F(z),\, P_k\right] = \frac{1}{(m-1)!} \frac{d^{m-1}}{dz^{m-1}} \left[(z - p_k)^m\, F(z) \right] \Bigg|_{z=p_k} \qquad .....(3.3.15)$$

where $F(z) = Z^{n-1}\, X(z)$, m is the order of the pole at p_k and $\text{Res}\left[F(z),\, P_k\right]$ is the residue of $F(z)$ at $z = p_k$.

For a simple (distinct) pole, equation (3.3.15) reduces to

$$\text{Res}\left[F(z),\, P_k\right] = (z - p_k) F(z)\Big|_{z=p_k} \qquad\qquad .....(3.3.16)$$

Example 3.21: Determine the inverse z-transform of

$$X(z) = \frac{1}{1 - 1.5\, z^{-1} + 0.5\, z^{-2}}$$

when　　(a)　ROC: $|z| > 1$

　　　　　(b)　ROC: $|z| < 0.5$

using Residue method

Solution:

Given

$$X(z) = \frac{1}{1 - 1.5\, z^{-1} + 0.5\, z^{-2}} = \frac{z^2}{(z-1)(z-0.5)}$$

Let

$$F(z) = z^{n-1}\, X(z), \text{ then}$$

$$F(z) = \frac{z^{n-1}\, z^2}{(z-1)(z-0.5)} = \frac{z^{n+1}}{(z-1)(z-0.5)}$$

(a)　Given ROC: $|z| > 1$, i.e., time domain sequence to be causal

$$x(n) = \mathrm{Res}\left[F(z),\ 1\right] + \mathrm{Res}\left[F(z),\ 0.5\right]$$

$$\mathrm{Res}\left[F(z),\ 1\right] = (z-1)F(z)\Big|_{z=1}$$

$$= (z-1)\frac{z^{n+1}}{(z-1)(z-0.5)}\Bigg|_{z=1}$$

$$= \frac{1^{n+1}}{0.5} = 2$$

$$\mathrm{Res}\left[F(z),\ 0.5\right] = (z-0.5)\frac{z^{n+1}}{(z-1)(z-0.5)}\Bigg|_{z=0.5}$$

$$= \frac{(0.5)^{n+1}}{-0.5} = \frac{(0.5)^n \cdot 0.5}{-0.5} = -(0.5)^n$$

∴

$$x(n) = 2 - (0.5)^n \quad n \geq 0$$

$$= \left[2 - (0.5)^n\right]u(n)$$

(b)　Given ROC: $|z| < 0.5$, i.e., time domain sequence to be anticausal.

∴

$$x(n) = -\mathrm{Res}\left[F(z),\ 1\right] - \mathrm{Res}\left[F(z),\ 0.5\right]$$

$$= \left[-2 - \left(-(0.5)^n\right)\right]u(-n-1)$$

$$= \left[(0.5)^n - 2\right]u(-n-1)$$

Example 3.22: Obtain the inverse z-transform of the following function

$$X(z) = \frac{\left(z^3 - 1\right)}{z(z-1)(z-0.5)} \quad \text{for } |z| > \frac{1}{2}, \text{ using Residue method.}$$

Solution:

$$X(z) = \frac{(z-1)\left(z^2 + z + 1\right)}{z(z-1)(z-0.5)} = \frac{z^2 + z + 1}{z(z-0.5)}$$

Since ROC is $|z| > \dfrac{1}{2}$, time domain sequence is causal.

Let

$$F(z) = z^{n-1} X(z)$$

$$F(z) = \frac{z^{n-1}\left(z^2 + z + 1\right)}{z(z-0.5)}$$

For n = 0,

$$F(z) = \frac{z^2 + z + 1}{z^2(z-0.5)}$$

there are two poles at z = 0 (origin)

$$x(0) = \text{Res}\big[F(z),\, 0\big] + \text{Res}\big[F(z),\, 0.5\big]$$

$$\text{Res}\big[F(z),\, 0\big] = \frac{1}{(2-1)!}\frac{d^{2-1}}{dz^{2-1}}\left[z^2 \cdot \frac{\left(z^2 + z + 1\right)}{z^2(z-0.5)}\right]\Bigg|_{z=0}$$

$$= \frac{d}{dz}\left[\frac{z^2 + z + 1}{z - 0.5}\right]\Bigg|_{z=0}$$

$$= \frac{(z-0.5)(z-z+1) - \left(z^2 + z + 1\right)}{(z-0.5)^2}\Bigg|_{z=0}$$

$$= \frac{-0.5 - 1}{(-0.5)^2} = -2 - 4 = -6$$

$$\text{Res}\big[F(z),\, 0.5\big] = (z-0.5) \cdot \frac{\left(z^2 + z + 1\right)}{z^2(z-0.5)}\Bigg|_{z=0.5}$$

$$= \frac{(0.5)^2 + (0.5) + 1}{(0.5)^2} = 1 + 2 + 4 = 7$$

$$\therefore \qquad\qquad x(0) = -6 + 7 = 1$$

For n = 1, $\qquad F(z) = \dfrac{z^2 + z + 1}{z(z - 0.5)}$

there exist one pole at origin.

$$x(1) = \operatorname{Res}\left[F(z),\ 0\right] + \operatorname{Res}\left[F(z),\ 0.5\right]$$

$$\operatorname{Res}\left[F(z),\ 0\right] = z.\dfrac{\left(z^2 + z + 1\right)}{z(z - 0.5)}\bigg|_{z=0}$$

$$= \dfrac{1}{-0.5} = -2$$

$$\operatorname{Res}\left[F(z),\ 0.5\right] = \dfrac{(z - 0.5)\left(z^2 + z + 1\right)}{z(z - 0.5)}\bigg|_{z=0.5}$$

$$= \dfrac{(0.5)^2 + 0.5 + 1}{0.5} = 0.5 + 1 + 2 = 3.5$$

$$\therefore \qquad x(1) = -2 + 3.5 = 1.5$$

For n = 2 onwards, there won't be pole at origin

$$\therefore \quad \text{for } n > 1, \quad F(z) = \dfrac{z^{n-1}\left(z^2 + z + 1\right)}{z(z - 0.5)} = \dfrac{z^n\left(z^2 + z + 1\right)}{z^2(z - 0.5)}$$

$$x(n) = \operatorname{Res}\left[F(z),\ 0\right] + \operatorname{Res}\left[F(z),\ 0.5\right]$$

$$\operatorname{Res}\left[F(z),\ 0.5\right] = (z - 0.5)\dfrac{z^{n-1}\left(z^2 + z + 1\right)}{z(z - 0.5)}\bigg|_{z=0.5}$$

$$= (0.5)^{n-1}\dfrac{(0.5)^2 + 0.5 + 1}{0.5}$$

$$= \dfrac{(0.5)^n}{0.5}\dfrac{(0.5)^2 + 0.5 + 1}{0.5}$$

$$= (0.5)^n\left[1 + 2 + 4\right] = 7(0.5)^n$$

$$\text{Res}[F(z),\ 0] = \frac{d}{dz}\left[z^2\ \frac{z^n\left(z^2+z+1\right)}{z^2\left(z-0.5\right)}\right]\Bigg|_{z=0}$$

$$= \frac{\left(z-0.5\right)\left[\left(n+2\right)z^{n+1}+\left(n+1\right)z^n+n\ z^{n-1}\right]-z^n\left(z^2+z+1\right).1}{\left(z-0.5\right)^2}\Bigg|_{z=0}$$

$$= 0$$

$$\therefore \qquad x(n) = 7(0.5)^n\ \text{ for } n > 1$$

Compare the result of example 3.17, with the above result for different values of n, both are same.

Example 3.23: Find inverse z-transform of

$$X(z) = \frac{z}{(z-1)(z-2)}\ \text{ for ROC: } 1 < |z| < 2$$

Using (a) Residue method (b) Partial fraction method

Solution:

(a) Here the Contour of integration lies in the annular region of ROC, hence the inverse z-transform is

$$x(n) = \text{Res}[F(z),\ 1] + \left[-\text{Res}[F(z),\ 2]\right]$$

First term is for $n \ge 0$, because ROC: i.e., $|z| > 1$

Second term is for $n < 0$, because ROC: i.e., $|z| < 2$

where $\qquad F(z) = z^{n-1}\ X(z)$

$$= \frac{z^{n-1}.z}{(z-1)(z-2)} = \frac{z^n}{(z-1)(z-2)}$$

Here there is no pole at origin for any value of n $n \ge 0$,

$$\text{Res}[F(z),\ 1] = (z-1)\frac{z^n}{(z-1)(z-2)}\Bigg|_{z=1}$$

$$= \frac{1^n}{-1} = -1u(n)$$

$$n < 0, \qquad \text{Res}\big[F(z),\ 2\big] = (z-2)\frac{z^n}{(z-1)(z-2)}\bigg|_{z=2}$$

$$= \frac{2^n}{1} = 2^n\, u(-n-1)$$

$$\therefore \qquad x(n) = -u(n) - 2^n\, u(-n-1)$$

(b) $\quad X(z) = \dfrac{z}{(z-1)(z-2)}$

Divide with z on both sides

$$\frac{X(z)}{z} = \frac{1}{(z-1)(z-2)}$$

$$\frac{X(z)}{z} = \frac{c_1}{z-1} + \frac{c_2}{z-2}$$

$$c_1 = (z-1)\frac{1}{(z-1)(z-2)}\bigg|_{z=1} = -1$$

$$c_2 = (z-2)\frac{1}{(z-1)(z-2)}\bigg|_{z=2} = 1$$

$$\therefore \qquad \frac{X(z)}{z} = \frac{-1}{z-1} + \frac{1}{z-2}$$

$$X(z) = \frac{-z}{z-1} + \frac{z}{z-2}$$

For first term $|z| > 1$; for second term $|z| < 2$

$$x(n) = -u(n) - (2)^n\, u(-n-1)$$

3.4 System Function

The response of a system y(n), whose impulse response h(n) for the input x(n) is given by:

$$y(n) = x(n) * h(n) \qquad\qquad .....(3.4.1)$$

by applying convolution property of z-transform

$$Y(z) = X(z)\, H(z) \qquad\qquad .....(3.4.2)$$

where $Y(z)$ is the z-transform of the output sequence $y(n)$, $X(z)$ is the z-transform of the input sequence $x(n)$, and $H(z)$ is the z-transform of the unit sample response $h(n)$.

If we know $h(n)$ and $x(n)$, we can determine their corresponding z-transforms $H(z)$ and $X(z)$, multiply them to obtain $Y(z)$, and then we can determine $y(n)$ by evaluating the inverse z-transform of $Y(z)$. Alternatively, if we know $x(n)$ and output $y(n)$ of the system, we can determine the unit sample response by first solving for $H(z)$ from the relation

$$H(z) = \frac{Y(z)}{X(z)} \qquad\qquad .....(3.4.3)$$

and then evaluating the inverse Z-transform of $H(z)$

Since
$$H(z) = \sum_{n=-\infty}^{\infty} h(n)z^{-n} \qquad\qquad .....(3.4.4)$$

it is clear that $H(z)$ represents the z-domain characterization of a system, whereas $h(n)$ is the corresponding time-domain characterization of the system. The transform $H(z)$ is called the system function or Transfer function.

3.5 Solution of Difference Equations of Digital Filters

The Nth order system or digital filters are described by a general form of Linear constant coefficient difference equation as

$$\sum_{k=0}^{N} a_k y(n-k) = \sum_{k=0}^{M} b_k x(n-k) \qquad\qquad .....(3.5.1)$$

where $y(n)$ is the output, $x(n)$ is the input and a_k, b_k are linear constant coefficients.
By taking $a_0 = 1$

$$y(n) = -\sum_{k=1}^{N} a_k y(n-k) + \sum_{k=0}^{M} b_k x(n-k) \qquad\qquad .....(3.5.2)$$

3.5.1 Response of System with Zero Initial Conditions

Let us consider the system function $H(z)$ of system described by the eqn. 3.5.2. We represent $H(z)$ as a ratio of two polynomials $B(z)/A(z)$, where $B(z)$ is the numerator polynomial that contains the zeros of $H(z)$ and $A(z)$ is the denominator polynomial that determines the poles of $H(z)$. Furthermore, let us assume that the input signal $x(n)$ has a rational z-transform $X(z)$ of the form.

$$X(z) = \frac{N(z)}{Q(z)} \qquad\qquad .....(3.5.3)$$

system is relaxed i.e., initial conditions for the difference equation are zero, $y(-1) = y(-2) = \ldots = y(-N) = 0$.

The z-transform of the output of the system has the form

$$Y(z) = H(z)\,X(z) = \frac{B(z)N(z)}{A(z)Q(z)} \qquad\qquad .....(3.5.4)$$

Now suppose that the system contains simple poles P_1, P_2, ... P_N and the z-transform of the input signal contains poles q_1, q_2, ... q_L, where $P_k \neq q_m$ for all $k = 1, 2, ...$ N and m $= 1, 2, ..., L$. Then partial fraction expansion of $Y(z)$ yields as

$$Y(z) = \sum_{k=1}^{N} \frac{A_k}{1 - P_k z^{-1}} + \sum_{k=1}^{L} \frac{Q_k}{1 - q_k z^{-1}} \qquad\qquad .....(3.5.5)$$

The inverse transform of $Y(z)$ is

$$y(n) = \sum_{k=1}^{N} A_k \left(P_k\right)^n u(n) + \sum_{k=1}^{L} Q_k \left(q_k\right)^n u(n) \qquad\qquad .....(3.5.6)$$

Output sequence $y(n)$ can be subdivided into two parts. The first part is a function of the poles (P_k) of the system and is called the natural response of the system. The second part of the response is a function of the poles (q_k) of the input signal and is called the forced response of the system. As initial conditions are zero the response $y(n)$ is called zero-state response.

$$\therefore \qquad y_{zs}(n) = y_n(n) + y_f(n) \qquad\qquad .....(3.5.7)$$

where $\quad y_{zs}(n)$ = zero-state response

$\qquad\quad y_n(n)$ = natural response

$\qquad\quad y_f(n)$ = forced response

3.5.2 Response of System with Nonzero Initial Conditions

If a causal signal $x(n)$ is applied to the system, the effects of all previous input signals to the system are reflected in the initial conditions $y(-1)$, $y(-2)$, ... $y(-N)$. As we are interested in determining the output $y(n)$ for $n \geq 0$, we may use the one-sided Z-transform, which allows us to deal with the initial conditions.

Thus the one-sided z-transform of eqn. (3.5.2) is

$$Y^+(z) = -\sum_{k=1}^{N} a_k z^{-k} \left[Y^+(z) + \sum_{n=1}^{k} y(-n)z^n \right] + \sum_{k=0}^{M} b_k Z^{-k} X^+(z) \qquad .....(3.5.8)$$

as $x(n)$ is causal $X^+(z) = X(z)$

$$Y^+(z) = \frac{\displaystyle\sum_{k=0}^{M} b_k z^{-k}}{1 + \displaystyle\sum_{k=1}^{N} a_k z^{-k}} X(z) - \frac{\displaystyle\sum_{k=1}^{N} a_k z^{-k} \sum_{n=1}^{k} y(-n)z^n}{1 + \displaystyle\sum_{k=1}^{N} a_k z^{-k}}$$

$$= H(z)\,X(z) + \frac{N_0(z)}{A(z)} \qquad\qquad .....(3.5.9)$$

where
$$N_0(z) = -\sum_{k=1}^{N} a_k z^{-k} \sum_{n=1}^{k} y(-n) z^{n} \qquad \ldots\ldots(3.5.10)$$

From eq., (3.5.9), the response of the system with non zero initial conditions can be subdivided into two parts. The first part is the zero-state response of the system which is in z-domain as
$$Y_{zs}(z) = H(z) \, X(z) \qquad \ldots\ldots(3.5.11)$$

The second part of eq. (3.5.9) is the response due to initial values. This is called zero-input response of the system, which is defined as
$$Y_{zi}^{+}(z) = \frac{N_0(z)}{A(z)} \qquad \ldots\ldots(3.5.12)$$

Hence the total response is the sum of $Y_{zs}(z)$ and $Y_{zi}^{+}(z)$

In time domain it is
$$y(n) = y_{zs}(n) + y_{zi}(n) \qquad \ldots\ldots(3.5.13)$$

Example 3.24: Compute the response of the system $y(n) = 0.7\, y(n-1) - 0.12\, y(n-2) + x(n-1) + x(n-2)$ to the input $x(n) = nu(n)$ using z-transform. $\hfill$ [JNTU 2000]

Solution:

Given difference equation
$$y(n) = 0.7\, y(n-1) - 0.12\, y(n-2) + x(n-1) + x(n-2)$$

As the system is zero initial values (relaxed) system, apply shifting properties of two-sided z-transform.
$$Y(z) = 0.7z^{-1} Y(z) - 0.12\, z^{-2} Y(z) + X(z)\, z^{-1} + X(z)\, z^{-2}$$
$$Y(z)\, [1 - 0.7\, z^{-1} + 0.12\, z^{-2}] = X(z)\, [z^{-1} + z^{-2}]$$
$$\frac{Y(z)}{X(z)} = \frac{z^{-1} + z^{-2}}{1 - 0.7z^{-1} + 0.12z^{-2}}$$

which is the system function
$$H(z) = \frac{z+1}{z^2 - 0.7z + 0.12}$$
$$H(z) = \frac{z+1}{(z-0.4)(z-0.3)}$$

Given input sequence $x(n) = nu(n)$
$$\therefore \qquad z\{x(n)\} = \frac{z}{(z-1)^2} = X(z)$$

Response of the system $y(n) = x(n) * h(n)$

where $\qquad h(n) = z^{-1}\{H(z)\}$

According to convolution property of z-transform

$\therefore \qquad\qquad Y(z) = X(z)\, H(z)$

$$Y(z) = \frac{z}{(z-1)^2}\frac{(z+1)}{(z-0.4)(z-0.3)}$$

divide both sides with z

$$\frac{Y(z)}{z} = \frac{z+1}{(z-0.4)(z-0.3)(z-1)^2}$$

$$\frac{Y(z)}{z} = \frac{C_1}{z-0.4} + \frac{C_2}{z-0.3} + \frac{D_1}{z-1} + \frac{D_2}{(z-1)^2}$$

$$C_1 = (z-0.4)\frac{(z+1)}{(z-0.4)(z-0.3)(z-1)^2}\bigg|_{z=0.4}$$

$$= \frac{(0.4+1)}{(0.4-0.3)(0.4-1)^2} = 38.9$$

$$C_2 = (z-0.3)\frac{(z+1)}{(z-0.4)(z-0.3)(z-1)^2}\bigg|_{z=0.3}$$

$$C_2 = \frac{(0.3+1)}{(0.3-0.4)(0.3-1)^2} = -26.5$$

$$D_1 = \frac{1}{(2-1)!}\frac{d^{2-1}}{dz^{2-1}}\left[(z-1)^2\frac{(z+1)}{(z-0.3)(z-0.4)(z-1)^2}\right]\bigg|_{z=1}$$

$$= \frac{d}{dz}\left[\frac{z+1}{(z-0.3)(z-0.4)}\right]\bigg|_{z=1}$$

$$= \frac{(z-0.3)(z-0.4).1-(2z-0.7).1}{\left[(z-0.3)(z-0.4)\right]^2}\bigg|_{z=1}$$

$$= \frac{(1-0.3)(1-0.4)-(2-0.7)}{\left[(1-0.3)(1-0.4)^2\right]}$$

$$= -4.9$$

$$D_2 = \frac{1}{(2-2)!}\frac{d^{2-2}}{dz^{2-2}}\left[(z-1)^2\frac{(z+1)}{(z-0.3)(z-0.4)(z-1)^2}\right]_{z=1}$$

$$= \frac{z+1}{(z-0.3)(z-0.4)}\Bigg|_{z=1}$$

$$= \frac{1+1}{(1-0.3)(1-0.4)}$$

$$= 4.7$$

$$\frac{Y(z)}{z} = \frac{38.9}{z-0.4} - \frac{26.5}{z-0.3} - \frac{4.9}{z-1} + \frac{4.7}{(z-1)^2}$$

$$Y(z) = \frac{38.9z}{z-0.4} - 26.5\frac{z}{z-0.3} - 4.9\frac{z}{z-1} + 4.7\frac{z}{(z-1)^2}$$

Inverse z-transform is

$$y(n) = 38.9(0.4)^n u(n) - 26.5(0.3)^n u(n) - 4.9\, u(n) + 4.7\, n\, u(n)$$

which is the response of the system for the input nu(n).

Example 3.25: Solve the following difference equation using z-transform method:

$$S(n+2) + 3\, S(n+1) + 2\, S(n) = 0$$

initial conditions are $S(0) = 0$ and $S(1) = 1$ [JNTU 2002]

Solution: Given the difference equation

$$S(n+2) + 3S(n+1) + 2S(n) = 0$$

As the system is non relaxed (system with initial values), apply shifting properties of one-sided z-transform.

$$z^2[S^+(z) - S(0)\, z^{-0} - S(1)\, z^{-1}] + 3z^1\,[S^+(z) - S(0)\, z^{-0}] + 2\, S^+(z) = 0$$

$$z^2\, S^+(z) - S(0)\, z^2 - S(1)\, z^1 + 3z\, S^+(z) - 3\, S(0)\, z + 2\, S^+(z) = 0$$

given $S(0) = 0$ and $S(1) = 1$

$$\therefore \qquad z^2\, S^+(z) - z + 3\, z\, S^+(z) + 2S^+(z) = 0$$

$$S^+(z)\,[z^2 + 3z + 2] = z$$

$$S^+(z) = \frac{z}{z^2 + 3z + 2} = \frac{z}{(z+1)(z+2)}$$

divide with z on both sides

$$\frac{S^+(z)}{z} = \frac{1}{(z+1)(z+2)}$$

$$\frac{S^+(z)}{z} = \frac{c_1}{z+1} + \frac{c_2}{z+2}$$

$$C_1 = (z+1)\frac{1}{(z+1)(z+2)}\bigg|_{z=-1} = 1$$

$$C_2 = (z+2)\frac{1}{(z+1)(z+2)}\bigg|_{z=-2} = -1$$

$$\frac{S^+(z)}{z} = \frac{1}{z+1} - \frac{1}{z+2}$$

$$S^+(z) = \frac{z}{z+1} - \frac{z}{z+2}$$

Inverse z-transform

$$S(n) = (-1)^n\, u(n) - (-2)^n\, u(n)$$

3.6 Poles and Zeros of the System

Zeros: The zeros of a z-transform $X(z)$ are the values of z for which $x(z) = 0$

Poles: The poles of a z-transform are the values of z for which $X(z) = \infty$

If $X(z)$ is a rational function, then

$$X(z) = \frac{b_0 + b_1 z^{-1} + \ldots + b_N z^{-N}}{a_0 + a_1 z^{-1} + \ldots + a_M z^{-M}} = \frac{\displaystyle\sum_{k=0}^{N} b_k z^{-k}}{\displaystyle\sum_{k=0}^{M} a_k z^{-k}} \qquad \ldots\text{(3.6.1)}$$

If $a_0 \neq 0$ and $b_0 \neq 0$, we can avoid the negative powers of z by factoring out the terms $b_0\, z^{-N}$ and $a_0\, z^{-M}$ as follows.

$$X(z) = \frac{b_0 z^{-N}}{a_0 z^{-M}} \cdot \frac{z^N + \left(\dfrac{b_1}{b_0}\right)z^{N-1} + \ldots + \left(\dfrac{b_N}{b_0}\right)}{z^M + \left(\dfrac{a_1}{a_0}\right)z^{M-1} + \ldots + \left(\dfrac{a_M}{a_0}\right)}$$

Since numerator and denominator are polynomials in z, they can be expressed in factored form as:

$$X(z) = \frac{b_0}{a_0} z^{-N+M} \frac{(z-z_1)(z-z_2)...(z-z_N)}{(z-P_1)(z-P_2)...(z-P_M)}$$

$$X(z) = G\, z^{M-N}\, \frac{\prod\limits_{k=1}^{N}(z-z_k)}{\prod\limits_{k=1}^{M}(z-P_k)} \qquad\qquad .....(3.6.2)$$

where $\qquad\qquad G = \dfrac{b_0}{a_0}$

Thus $X(z)$ has N finite zeros at $z = z_1$, z_2, ... z_N (the roots of the numerator polynomial), M finite poles at $Z = P_1$, P_2,...P_M (the roots of the denominator polynomial), and $|M - N|$ zeros (if $M > N$) or poles (if $M < N$) occur at the origin $Z = 0$. Poles or zeros may also occur at $z = \infty$. A zero exists at $z = \infty$ if $X(\infty) = 0$ and a pole exists at $z = \infty$ if $X(\infty) = \infty$. If we count the poles and zeros at zero and infinity, we find that $X(z)$ has the same number of poles as zeros.

We can represent $X(z)$ graphically by a pole-zero plot (or pattern) in the complex plane, which shows that location of poles by cross (X) and the location of zeros by circles (0). The multiplicity of multiple-order poles or zeros is indicated by a number close to the corresponding cross or circle. The ROC does not contain poles.

Example 3.26: A causal linear shift invariant system is described by the difference equation

$$y(n) = y(n-1) + y(n-2) + x(n-1)$$

(i) Find the system function $H(z)$. Plot the poles and zeros indicating ROC

(ii) Find the unit – sample response of the system.

[JNTU 2002]

Solution:

(i) Given difference equation

$$y(n) = y(n-1) + y(n-2) + x(n-1)$$

apply shifting property of two-sided Z-transform

$$Y(z) = Y(z)z^{-1} + Y(z)z^{-2} + X(z)z^{-1}$$

$$Y(z)\left[1 - z^{-1} - z^{-2}\right] = X(z)z^{-1}$$

$$H(z) = \frac{Y(z)}{X(z)} = \frac{z^{-1}}{1 - z^{-1} - z^{-2}}$$

$$H(z) = \frac{z}{z^2 - z - 1}$$

$$z = \frac{1 \pm \sqrt{1^2 + 4}}{2} = \frac{1 \pm \sqrt{5}}{2}$$

$$z = 1.6,\ -0.6$$

$$\therefore \qquad H(z) = \frac{z}{(z - 1.6)(z + 0.6)}$$

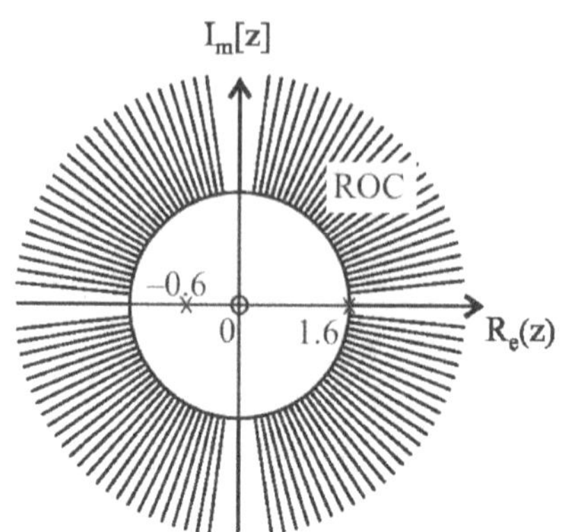

Poles are z = 1.6 and z = – 0.6

Zero is z = 0

Since the system is causal, ROC is $|z| > 1.6$

Pole – zero plot is

(ii) $\quad H(z) = \dfrac{z}{(z - 1.6)(z + 0.6)}$

divide with Z on both sides

$$\frac{H(z)}{z} = \frac{1}{(z - 1.6)(z + 0.6)}$$

$$\frac{H(z)}{z} = \frac{C_1}{z - 1.6} + \frac{C_2}{z + 0.6}$$

$$C_1 = (z - 1.6)\frac{1}{(z - 1.6)(z + 0.6)}\Bigg|_{z=1.6} = 0.45$$

$$C_2 = (z + 0.6)\frac{1}{(z + 0.6)(z + 0.6)}\Bigg|_{Z=-0.6} = -\,0.45$$

$$\frac{H(z)}{z} = \frac{0.45}{z - 1.6} - \frac{0.45}{z + 0.6}$$

$$H(z) = 0.45\frac{z}{z - 1.6} - 0.45\frac{z}{z + 0.6}$$

Unit – sample response is inverse z-transform of H(z)

$$h(n) = 0.45(1.6)^n u(n) - 0.45(-0.6)^n u(n)$$

Example 3.27: Consider the system

$$H(z) = \frac{1 - 2z^{-1} + 2z^{-2} - z^{-3}}{(1 - z^{-1})(1 - 0.5z^{-1})(1 - 0.2z^{-1})}$$

$$ROC : 0.5 < |z| < 1$$

(i) Sketch the pole zero pattern

(ii) Determine the impulse response of the system

[JNTU 2000]

Solution: Given $H(z) = \dfrac{1 - 2z^{-1} + 2z^{-2} - z^{-3}}{(1 - z^{-1})(1 - 0.5z^{-1})(1 - 0.2z^{-1})}$

$$= \frac{z^3 - 2z^2 + 2z - 1}{(z-1)(z-0.5)(z-0.2)}$$

$$\text{(i)} \quad = \frac{(z-1)(z^2 - z + 1)}{(z-1)(z-0.5)(z-0.2)}$$

$$= \frac{(z-1)(z-0.5 - j\,0.87)(z-0.5 + j\,0.87)}{(z-1)(z-0.5)(z-0.2)}$$

Poles are   z = 1, 0.5, 0.2

Zeros are   z = 1, 0.5 + j 0.87, 0.5 – j 0.87,

Pole – Zero pattern is

$$\text{(i)} \qquad H(z) = \frac{(z-1)\,(z^2 - z + 1)}{(z-1)(z-0.5)\,(z-0.2)}$$

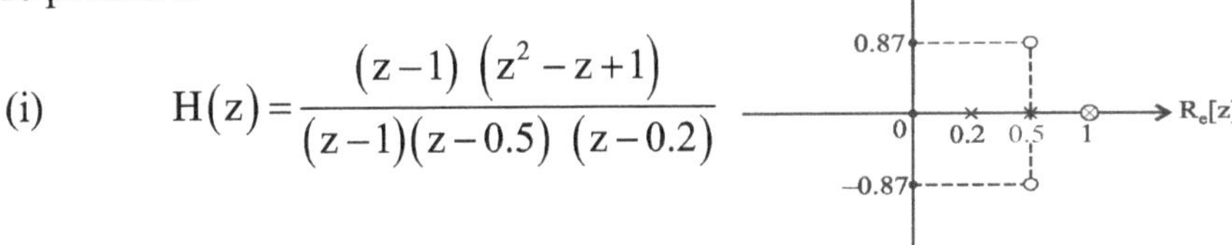

divide both sides with Z

$$\frac{H(z)}{z} = \frac{z^2 - z + 1}{z(z-0.5)\,(z-0.2)}$$

$$\frac{H(z)}{z} = \frac{C_1}{z} + \frac{C_2}{(z-0.5)} + \frac{C_3}{(z-0.2)}$$

$$C_1 = z \frac{\left(z^2 - z + 1\right)}{z(z-0.5)\,(z-0.2)}\Bigg|_{Z=0} = 10$$

$$C_2 = (z-0.5) \frac{\left(z^2 - z + 1\right)}{z(z-0.5)\,(z-0.2)}\Bigg|_{Z=0.5} = 5$$

$$C_3 = (z-0.2) \frac{\left(z^2 - z + 1\right)}{z(z-0.5)\,(z-0.2)}\Bigg|_{Z=0.2} = -14$$

$$\frac{H(z)}{z} = \frac{10}{z} + \frac{5}{(z-0.5)} - \frac{14}{(z-0.2)}$$

$$H(z) = 10 + 5\frac{z}{(z-0.5)} - 14\frac{z}{(z-0.2)}$$

Impulse response h(n) is the inverse Z-transform of H(Z)

$$h(n) = 10\delta(n) + 5(0.5)^n\, u(n) - 14(0.2)^n\, u(n)$$

3.7　　Stability Criterion

Recall from our discussion in the first unit, a necessary and sufficient condition for a linear time-invariant system to be BIBO stable is

$$\sum_{n=-\infty}^{\infty} |h(n)| < \infty \qquad\qquad\qquad .....(3.7.1)$$

In turn, this condition implies that H(z) most contain the unit circle within its ROC.

since $\quad H(z) = \sum\limits_{n=-\infty}^{\infty} h(n)\, z^{-n}$

Take modulus on both sides

$$|H(z)| \le \sum_{n=-\infty}^{\infty} \left|h(n)z^{-n}\right| = \sum_{n=-\infty}^{\infty} |h(n)|\,\left|z^{-n}\right|$$

when it is evaluated on the unit circle i.e., $|z| = 1$,

$$|H(z)| \le \sum_{n=-\infty}^{\infty} |h(n)| \qquad\qquad\qquad .....(3.7.2)$$

Hence, if the system is BIBO stable, the unit circle is contained in the ROC of H(z). This also can be stated like "A linear time-invariant system is BIBO stable if and only if the ROC of the system function includes the unit circle".

Example 3.28: Show that all poles of an LTI discrete system need to lie inside the unit circle for the system to be stable.

(or)

prove that the necessary and sufficient condition for stability of a system is $|Z| < 1$

[JNTU 2002]

Solution: Consider an LTI discrete system with transfer function

$$H(z) = \sum_{k=1}^{M} \frac{a_k}{1 - b_k z^{-1}} = \sum_{k=1}^{M} \frac{a_k z}{z - b_k}$$

Here poles are

$$|z| = b_k, \qquad k = 1, 2, \dots M$$

Unit sample response h(n) is

$$h(n) = \sum_{k=1}^{M} h_k(n)$$

where $\quad h_k(n) = a_k (b_k)^n u(n)$

For the system to be stable, each component of the sequence $h_k(n)$ must satisfy the condition

$$\sum_{n=0}^{\infty} |h_k(n)| < \infty$$

$$\Rightarrow \sum_{n=0}^{\infty} |h_k(n)|$$

$$= \sum_{n=0}^{\infty} |a_k (b_k)^n|$$

$$= a_k \sum_{n=0}^{\infty} |(b_k)^n|$$

For the above system to be finite, the magnitude of each term must be less than unity.

i.e., each $|b_k| < 1$, where b_k is a pole

i.e., $|z|$ must be less than 1

So, all poles of the system must lie inside the unit circle, for the system to be stable.

3.7.1 Schür-Cohn Stability Test

When the denominator polynomial of a transfer function of the system is large and which can not be factorised, it is not possible to find the poles of the system. Consequently we cannot decide whether the system is stable or not. In such cases stability can be decided by using Schür – Cohn stability test.

Let us consider transfer function of a system, whose stability to be decided,

$$H(z) = \frac{1}{1 - \frac{7}{4}z^{-1} - \frac{1}{2}z^{-2}}$$

Consider only the denominator polynomial, here order of the denominator polynomial is 2

So denote the polynomial as

$$D_2(z) = 1 - \frac{7}{4}z^{-1} - \frac{1}{2}z^{-2},$$

Let $k_2 = -\frac{1}{2}$ and $|k_2| = \frac{1}{2}$

If k_2 is greater than or equal to 1, system is unstable

If k_2 is less than 1, then find k_1 by forming reverse polynomial $R_2(z)$ from which $D_1(z)$ can be found by using $D_1(z) = \dfrac{D_2(z) - k_2 R_2(z)}{1 - k_2^2}$

Here $|k_2|$ is less than one

So form reverse polynomial $R_2(z) = -\frac{1}{2} - \frac{7}{4}z^{-1} + 1\,z^{-2}$

$$\therefore \quad D_1(z) = \frac{D_2(z) - k_2 R_2(z)}{1 - k_2^2}$$

$$= \frac{1 - \frac{7}{4}z^{-1} - \frac{1}{2}z^{-2} + \frac{1}{2}\left(-\frac{1}{2} - \frac{7}{4}z^{-1} + 1z^{-2}\right)}{1 - \frac{1}{4}}$$

$$D_1(z) = 1 - \frac{7}{2}z^{-1}$$

$$k_1 = -\frac{7}{2}; \quad |k_1| = \frac{7}{2}$$

Here $|k_1|$ is greater than 1

So, the system is unstable.

If $D_N(z)$ is given, form $R_N(z)$ and use recursive equation

$$D_{N-1}(z) = \frac{D_N(z) - k_N R_N(z)}{1 - k_N^2}; \quad \text{to get } k_{N-1}, \ldots k_1 \qquad \ldots\ldots(3.7.3)$$

If any one of k_N, k_{N-1}, $\ldots k_1$ is greater than or equal to one, stop calculating remaining k values and decide the system as unstable.

Example 3.29: Find the stability of the following transfer function

[JNTU 2000]

$$H(z) = \frac{z^2 + z + 1}{z^4 + 2z^3 + 3z^2 + 4z + 6}$$

Solution: Given

$$H(z) = \frac{z^2 + z + 1}{z^4 + 2z^3 + 3z^2 + 4z + 6}$$

Write in the form of negitive powers of Z

$$H(z) = \frac{z^2 + z + 1}{z^4 \left(1 + 2z^{-1} + 3z^{-2} + 4z^{-3} + 6z^{-4}\right)}$$

$$= \frac{z^{-2} + z^{-3} + z^{-4}}{1 + 2z^{-1} + 3z^{-2} + 4z^{-3} + 6z^{-4}}$$

Here $D_4(z) = 1 + 2z^{-1} + 3z^{-2} + 4z^{-3} + 6z^{-4}$

$$k_4 = 6$$

which is greater than 1

Hence the system is unstable.

3.8 Frequency Response of Stable Systems

In the design of discrete filters, it is often necessary to examine the spectrum (frequency response) of the filter to ensure that the desired specifications are satisfied. The frequency response of a system can be readily obtained from its Z-transform.

If we set $z = e^{j\omega T}$, i.e., evaluate the Z-transform around the unit circle, we get the Fourier transform of the system, where T is the sampling period.

$$H(z) = \sum_{n=-\infty}^{\infty} h(n)z^{-n} \bigg|_{z=e^{j\omega T}}$$

If we assume $T = 1$ sec, Frequency response becomes

$$H(e^{j\omega}) = H(\omega) = \sum_{n=-\infty}^{\infty} h(n)e^{-jn\omega} \qquad \qquad .....(3.8.1)$$

$H(\omega)$ is the frequency response of the system, $H(\omega)$ is complex. Its modulus gives the magnitude response and its phase is the phase response of the system.

Note:

  (i) At the pole the magnitude response exhibits a peak whereas, at the zero, the magnitude response falls to zero.

  (ii) The magnitude response, $|H(\omega)|$, is symmetrical about half the sampling frequency, and the phase response is anti symmetrical about the same frequency.

Example 3.30: The output $y(n)$ of an LTI discrete system to the input $x(n)$ is given by

$$y(n) = x(n) - 2x(n-1) + y(n-1)$$

Compute and sketch the magnitude and phase of the frequency response of the system as a function of frequency for $w \leq \pi$.

Solution: Given difference equation

$$y(n) = x(n) - 2x(n-1) + y(n-1)$$

Z-domain of the above equation is

$$Y(z) = X(z) - 2X(z)z^{-1} + Y(z)z^{-1}$$

$$Y(z)(1-z^{-1}) = X(z)(1-2z^{-1})$$

$$H(z) = \frac{Y(z)}{X(z)} = \frac{1-2z^{-1}}{1-z^{-1}}$$

Frequency response can be obtained by putting $z = e^{j\omega}$

$$\therefore \ H(e^{j\omega}) = H(\omega) = \frac{1-2e^{-j\omega}}{1-e^{-j\omega}}$$

Magnitude response:

$$|H(\omega)| = \sqrt{H(\omega).H^*(\omega)}$$

$$= \left[\frac{1-2e^{-j\omega}}{1-e^{-j\omega}} \cdot \frac{1-2e^{j\omega}}{1-e^{j\omega}} \right]^{1/2}$$

$$|H(\omega)| = \left[\frac{5 - 4\cos\omega}{2 - 2\cos\omega}\right]^{1/2}$$

Phase response: $\angle H(\omega)$

$$H(\omega) = \frac{1 - 2e^{-j\omega}}{1 - e^{-j\omega}} \cdot \frac{1 - e^{j\omega}}{1 - e^{j\omega}}$$

$$= \frac{1 - 2e^{-j\omega} - e^{j\omega} + 2}{2 - \left(e^{j\omega} + e^{-j\omega}\right)}$$

$$= \frac{3 - 2\left(\cos\omega - j\sin\omega\right) - \left(\cos\omega + j\sin\omega\right)}{2 - 2\cos\omega}$$

$$= \frac{3 - 3\cos\omega + j\sin\omega}{2 - 2\cos\omega}$$

$$\theta(\omega) = \angle H(\omega) = \tan^{-1}\left(\frac{H_I(\omega)}{H_R(\omega)}\right)$$

$$= \tan^{-1}\left(\frac{\sin\omega}{3 - 3\cos\omega}\right)$$

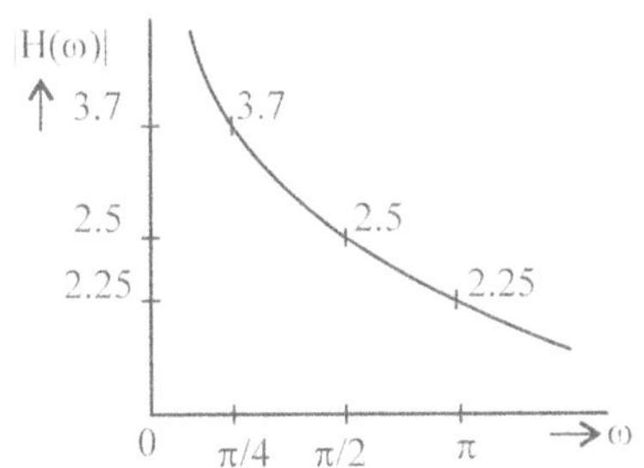

Magnitude response

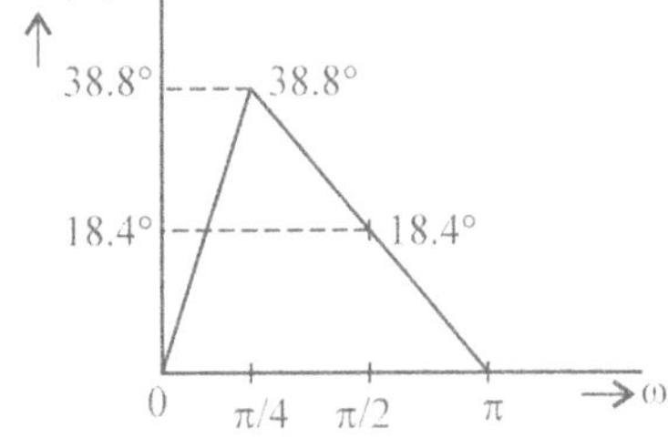

Phase response

Example 3.31: A discrete-time system S gives an output sequence which is the same as the input sequence except that it is delayed by K samples. The sampling period is T sec. sketch the magnitude and phase responses of the system.

Solution: Let x(n) is the input and y(n) is the output of a system then

$$y(n) = S\left[x(n)\right]$$

$$y(n) = x(n - kT)$$

$$Y(z) = X(z) z^{-kT}$$

$$H(z) = \frac{Y(z)}{X(z)} = z^{-kT}$$

Frequency response: Put $z = e^{jw}$

$$H(\omega) = e^{-j\omega kT}$$

Magnitude response: $\left|H(\omega)\right| = 1$

Phase response: $\theta(\omega) = \angle H(\omega) = -\omega kT$

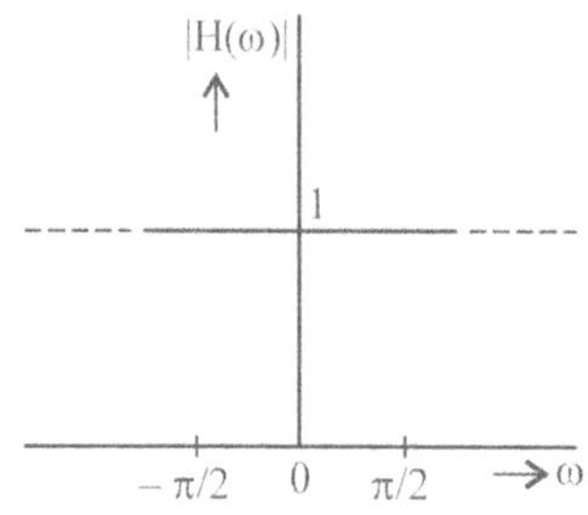

Magnitude plot

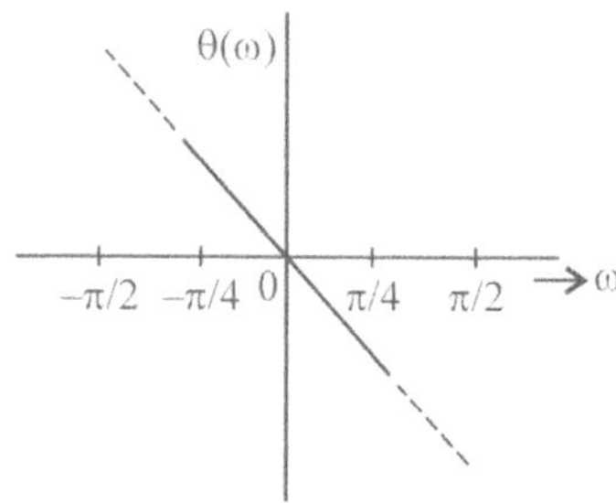

Phase plot

3.9   Realization of Digital Filters

A digital filter transfer function can be realized by using the following methods.

  (i)   Direct form I

 (ii)   Direct form II (or) Canonic form

(iii)   Cascade form

(iv)   Parallel form

3.9.1   Direct Form I Realization

Let us consider N^{th} order Linear constant coefficient difference equation which represents digital filter is

$$y(n) = -\sum_{k=1}^{N} a_k y(n-k) + \sum_{k=0}^{M} b_k x(n-k) \qquad .....(3.9.1)$$

$$\begin{aligned} y(n) = \ & -a_1 \, y\,(n-1) - a_2 \, y\,(n-2) - ....- a_{N-1} \, y(n-N+1) - a_N \, y\,(n-N) + \\ & b_0 \, x(n) + b_1 x\,(n-1) + ... + b_M \, x(n-M) \qquad .....(3.9.2) \end{aligned}$$

Let $b_0 \, x\,(n) + b_1 \, x\,(n-1) + ...+ b_M \, x\,(n-M) = v(n) \qquad .....(3.9.3)$

Then $\quad y(n) = -a_1 y(n-1) - a_2 y(n-2) - \ldots - a_N\, y(n-N) + v(n)$ $\qquad$ …..(3.9.4)

The equation (3.9.3) can be realized as shown in Fig. 3.6.

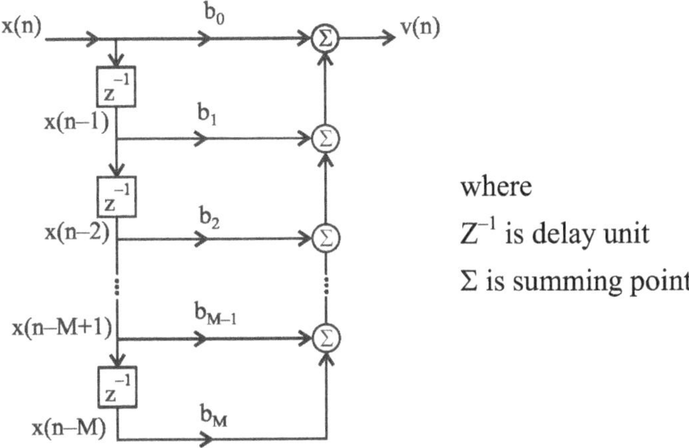

where

Z^{-1} is delay unit

Σ is summing point

Fig. 3.6 Realization structure of Eqn. (3.9.3)

Similarly the Eqn. (3.9.4) can be realized as shown in Fig. 3.7.

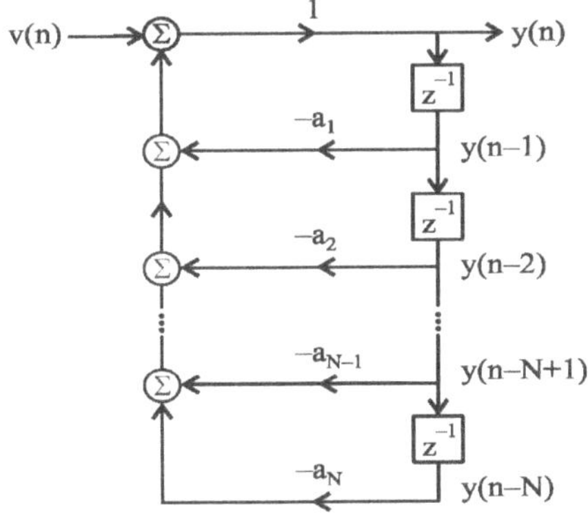

Fig. 3.7 Realization structure of eqn. (3.9.4)

The complete realization of Eqn. (3.9.2) can be obtained by combining Fig. 3.6 and 3.7.

The structure shown in Fig. 3.8 is called Direct form I. This requires $M + N + 1$ multiplications, $M + N$ additions and $M + N + 1$ memory locations.

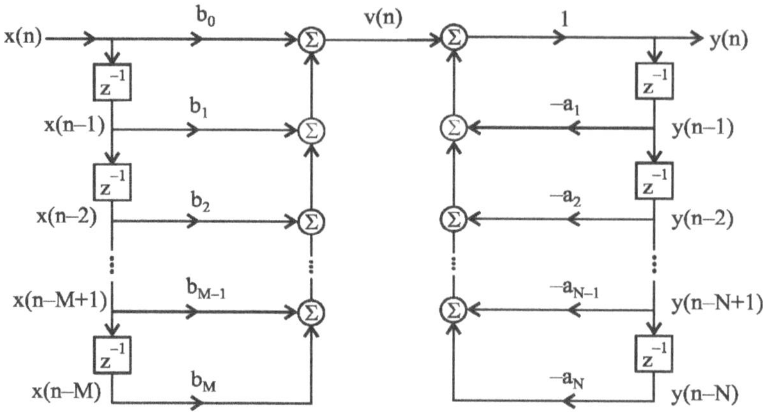

Fig. 3.8 Direct form I realization of eqn. 3.9.3

Example: 3.32: Realize the system described by the following difference equation $y(n) - y(n-1) + 0.5y(n-2) = x(n) + x(n-1)$ using Direct form I method.

Solution: Given difference equation is

$$y(n) - y(n-1) + 0.5y(n-2) = x(n) + x(n-1)$$

$$y(n) = y(n-1) - 0.5\,y(n-2) + x(n) + x(n-1)$$

Let $\quad x(n) + x(n-1) = v(n),$ then

$$y(n) = y(n-1) - 0.5\,y(n-2) + v(n)$$

$$x(n) + x(n-1) = v(n) \text{ can be realized as}$$

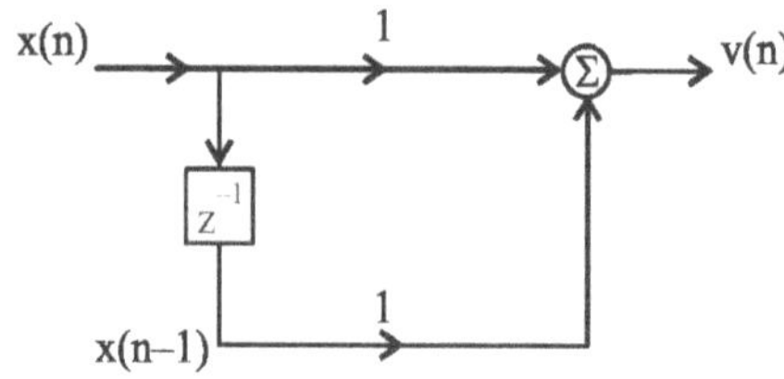

$$y(n) = y(n-1) - 0.5y(n-2) + v(n) \text{ can be realized as}$$

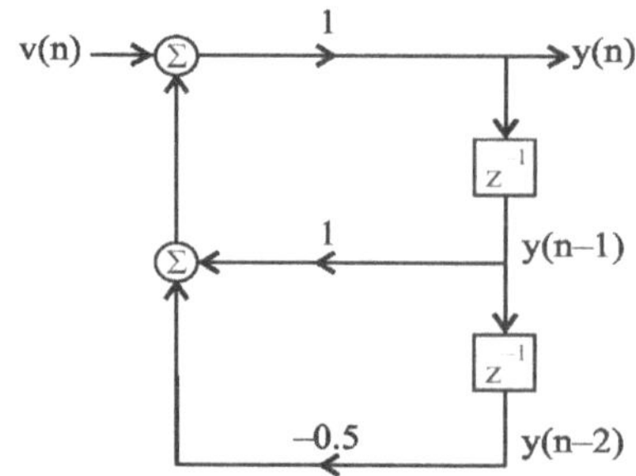

Direct form I can be obtained by combining above two figures.

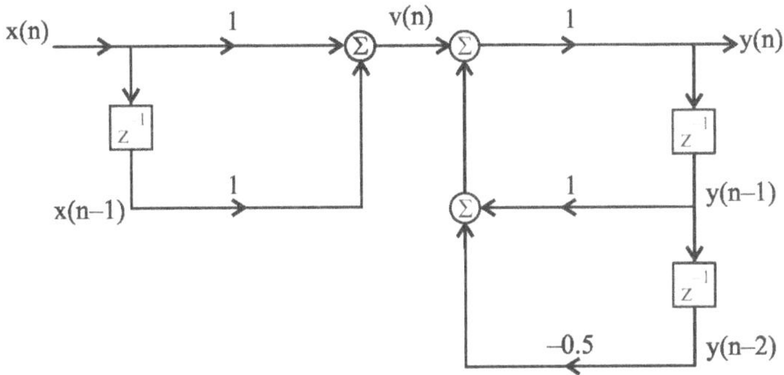

3.9.2 Direct Form II (or) Canonic Form Realization

Eqn (3.9.1) can be expressed as

$$H(z) = \frac{Y(z)}{X(z)} = \frac{\displaystyle\sum_{k=0}^{M} b_k z^{-k}}{1 + \displaystyle\sum_{k=1}^{N} a_k z^{-k}} \qquad \text{.....(3.9.5)}$$

Let

$$\frac{Y(z)}{X(z)} = \frac{Y(z)}{V(z)} \cdot \frac{V(z)}{X(z)}$$

Where

$$\frac{V(z)}{X(z)} = \frac{1}{1 + \displaystyle\sum_{k=1}^{N} a_k z^{-k}}$$

which can be written as

$$V(z) = X(z) - a_1 z^{-1} V(z) - a_z z^{-2} V(z) - ... - a_N z^{-N} V(z) \qquad \text{.....(3.9.6)}$$

and $\dfrac{Y(z)}{V(z)} = \displaystyle\sum_{k=0}^{M} b_k z^{-k}$ which can be written as

$$Y(z) = b_0 V(z) + b_1 z^{-1} V(z) + b_2 z^{-2} V(z) + ... + b_M z^{-M} V(z) \qquad \text{.....(3.9.7)}$$

Convert equations (3.9.6) and (3.9.9) into time domain

$$v(n) = x(n) - a_1 v(n-1) - a_2 v(n-2) - ... - a_N v(n-N) \qquad \text{.....(3.9.8)}$$

and

$$y(n) = b_0 v(n) + b_1 v(n-1) + b_2 v(n-2) + ... + b_M v(n-M) \qquad \text{.....(3.9.9)}$$

If we observe equation (3.9.8) and (3.9.9), both use the same delay terms v(n–1), v(n–2) ... etc., to express v(n) and y(n).

The realization of equations (3.9.8) and (3.9.9) are shown in Figs. 3.8 and 3.9.

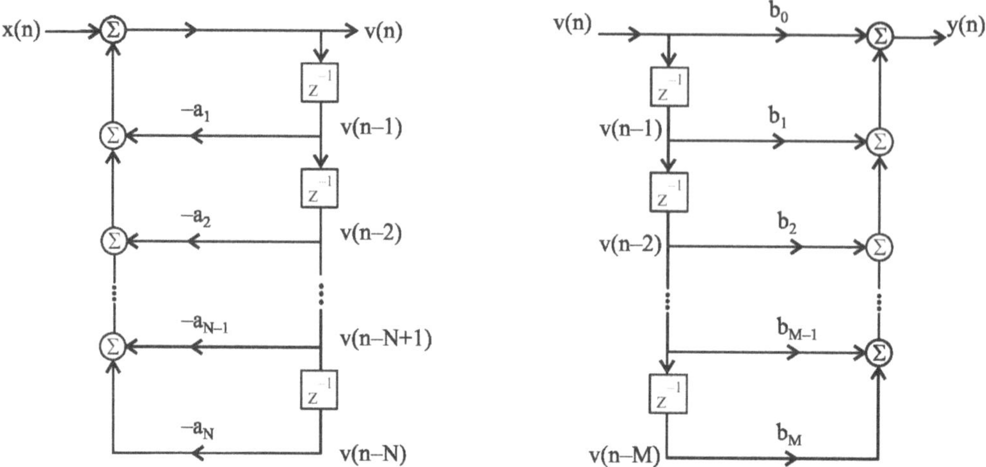

Fig. 3.9 Realization of eqn.(3.9.8).   **Fig. 3.10** Realization of eqn.(3.9.9).

The realization of difference equation 3.9.8 can be obtained by combining Figs. 3.9 and 3.10.

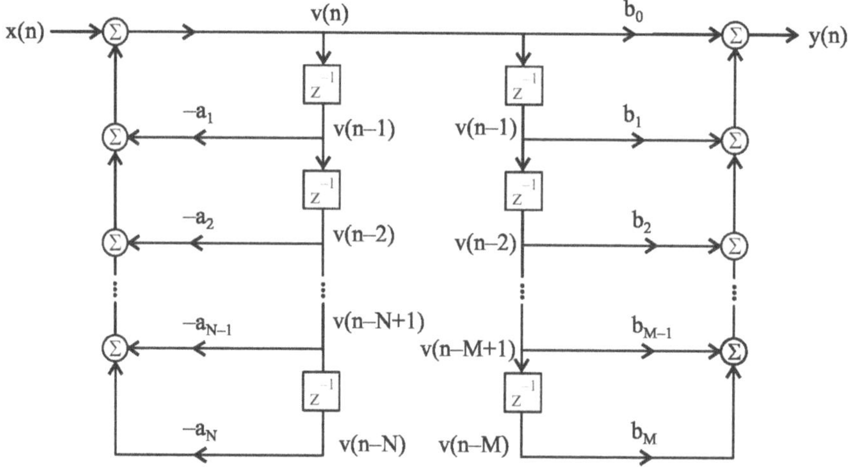

Fig. 3.11 Realization of equation 3.9.1.

From Fig 3.11, we observe that the two delay elements contain the same input v(n) and the output v(n − 1). So we can merge these delays into one delay and can be redrawn as shown in Fig 3.11.

Let M = N,

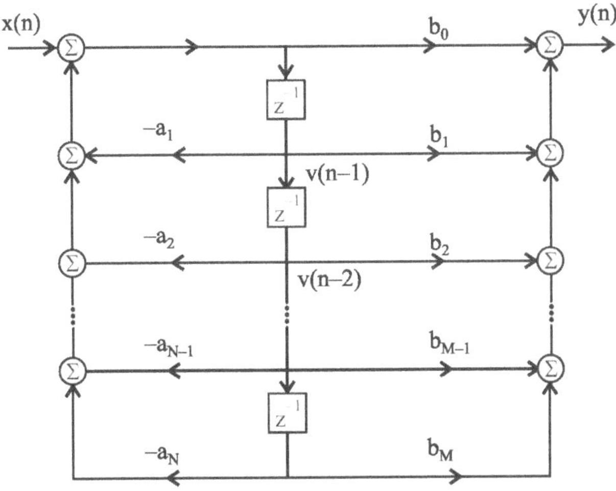

Fig. 3.12 Direct form II realization.

The realization structure shown in Fig 3.12 is called a Direct form II realization. This structure requires M + N + 1 multiplications, M + N additions and the maximum of (M, N) memory locations. Since this realization minimizes the number of delay units (memory locations), it is called Canonic.

Example 3.33: Realize the following digital filter in Canonical form

$$H(z) = \frac{16z^2(z+1)}{(4z+3)(4z^2 - 2z + 1)}$$

Solution: Given

$$H(z) = \frac{16z^2(z+1)}{(4z+3)(4z^2 - 2z + 1)}$$

$$= \frac{16\, z^3\left(1 + z^{-1}\right)}{z^3\left(4 + 3z^{-1}\right)\left(4 - 2z^{-1} + z^{-2}\right)}$$

$$H(z) = \frac{16\left(1 + z^{-1}\right)}{3z^{-3} - 2z^{-2} + 4z^{-1} + 16}$$

$$= \frac{Y(z)}{X(z)} = \frac{1 + z^{-1}}{1 + \dfrac{4}{16}z^{-1} - \dfrac{2}{16}z^{-2} + \dfrac{3}{16}z^{-3}}$$

Here numerator order is 1 and denominator order is 3. So this realization requires 3 delay elements. If we observe Direct form II realization (Fig 3.11), denominator coefficients with opposite signs are written in the left side and numerator coefficients with the same signs are written in the right side of the structure.

So direct form II realization of the given H(z) is

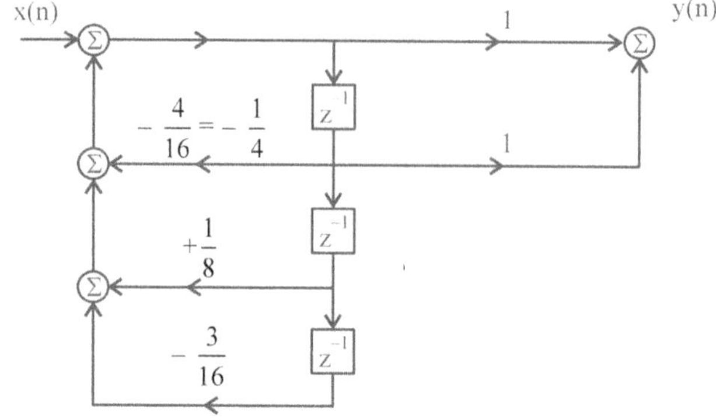

where   x(n) is the time domain representation of X(z) and

y(n) is the time domain representation of Y(z)

Example 3.34: Realize the difference equation using Canonic form

$$y(n)=\frac{1}{2}y(n-1)+\frac{1}{4}y(n-2)+x(n)+x(n-1)$$

Solution: Given

$$y(n)=\frac{1}{2}y(n-1)+\frac{1}{4}y(n-2)+x(n)+x(n-1)$$

Z–domain is

$$Y(z)=\frac{1}{2}Y(z)z^{-1}+\frac{1}{4}Y(z)z^{-2}+X(z)+X(z)z^{-1}$$

$$Y(z)\left[1-\frac{1}{2}z^{-1}-\frac{1}{4}z^{-2}\right]=X(z)\left[1+z^{-1}\right]$$

$$H(z)=\frac{Y(z)}{X(z)}=\frac{1+z^{-1}}{1-\frac{1}{2}z^{-1}-\frac{1}{4}z^{-2}}$$

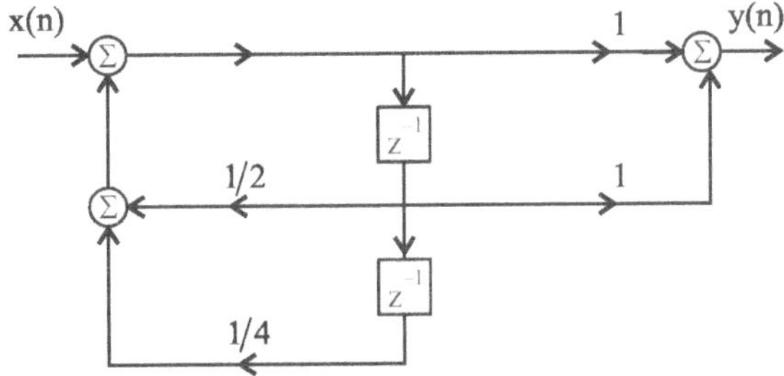

3.9.3 Cascade Form Realization

Let us consider the transfer function of a digital filter as

$$H(z) = H_1(z)H_2(z) \ldots H_k(z) \qquad\qquad \ldots\ldots (3.9.10)$$

This can be represented as

$$x(n) = x_1(n) \xrightarrow{\quad} \boxed{H_1(z)} \xrightarrow{y_1(n)\ \ x_2(n)} \boxed{H_2(z)} \xrightarrow{y_2(n)\quad x_k(n)} \cdots \xrightarrow{\quad} \boxed{H_k(z)} \xrightarrow{y_k(n) = y(n)}$$

Fig. 3.12 Block diagram representation of equation (3.9.10).

Then realize each $H_k(z)$ in Canonic form and cascade all structures. $H_k(z)$ may be either second order sections or first order sections.

For example, consider a system whose transfer function

$$H(z) = \frac{\left(b_{k0} + b_{k1}z^{-1} + b_{k2}z^{-2}\right)\left(b_{m0} + b_{m1}z^{-1} + b_{m2}z^{-2}\right)}{\left(1 + a_{k1}z^{-1} + a_{k2}z^{-2}\right)\left(1 + a_{m1}z^{-1} + a_{m2}z^{-2}\right)} \qquad\qquad \ldots\ldots (3.9.11)$$

$$= H_1(z)H_2(z)$$

where $\qquad H_1(z) = \dfrac{b_{k0} + b_{k1}z^{-1} + b_{k2}z^{-2}}{1 + a_{k1}z^{-1} + a_{k2}z^{-2}}$ and

$$H_2(z) = \frac{b_{m0} + b_{m1}z^{-1} + b_{m2}z^{-2}}{1 + a_{m1}z^{-1} + a_{m2}z^{-2}}$$

Realizing $H_1(z)$ and $H_2(z)$ in Canonic form, and cascading, we get cascade form of the system.

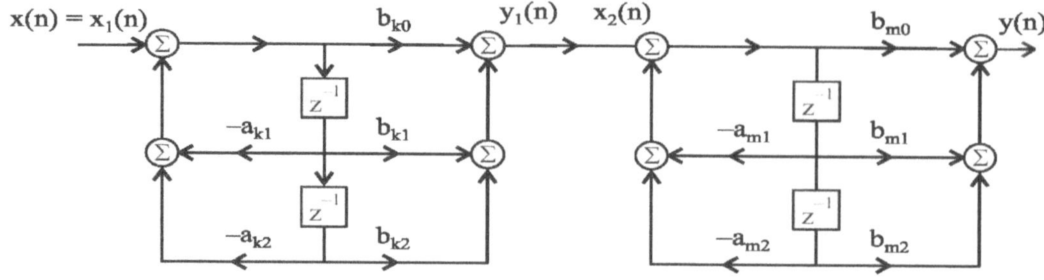

Fig. 3.13 Cascade realization of equation (3.9.11).

Example 3.35: Realize the following transfer function of a digital filter using cascade form realization.

$$H(z) = \frac{\left(1 - \dfrac{1}{2}z^{-1}\right)\left(1 - \dfrac{1}{2}z^{-1} + \dfrac{1}{4}z^{-2}\right)}{\left(1 + \dfrac{1}{4}z^{-1}\right)\left(1 + z^{-1} + \dfrac{1}{2}z^{-2}\right)\left(1 - \dfrac{1}{4}z^{-1} + \dfrac{1}{4}z^{-2}\right)}$$

Solution:

Let $\qquad H_1(z) = \dfrac{1 - \dfrac{1}{2}z^{-1}}{1 + \dfrac{1}{4}z^{-1}}$; $\qquad H_2(z) = \dfrac{1 - \dfrac{1}{2}z^{-1} + \dfrac{1}{4}z^{-2}}{1 + z^{-1} + \dfrac{1}{2}z^{-2}}$

and $\qquad H_3(z) = \dfrac{1}{1 - \dfrac{1}{4}z^{-1} + \dfrac{1}{4}z^{-2}}$

$\therefore \qquad H(z) = H_1(z) \, . \, H_2(z) \, . \, H_3(z) = \dfrac{Y(z)}{X(z)}$

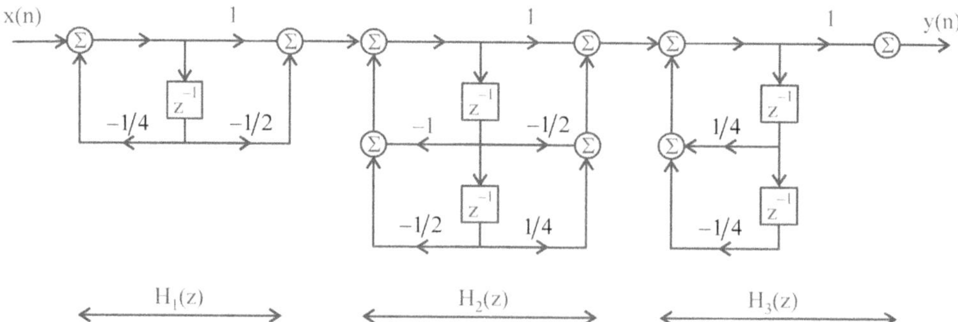

where $\quad$ x(n) is the time domain version of X(z) and

$\qquad\qquad$ y(n) is the time domain version of Y(z)

3.9.4 Parallel Form Realization

A parallel form realization can be obtained by expressing the transfer function H(z) (equation 3.9.5) of a system as a sum of first and second order sections.

So
$$H(z) = B_0 + \sum_{k=1}^{N} H_k(z) \qquad\qquad \dots\dots (3.9.12)$$

where
$$H_k(z) = \frac{b_{k0} + b_{k1}\, z^{-1}}{1 + a_{k1}\, z^{-1} + a_{k2}\, z^{-2}} \qquad \text{(second order section)}$$

or
$$H_k(z) = \frac{b_{k0}}{1 + a_{k1}\, z^{-1}} \qquad \text{(first order section)}$$

The equation (3.9.12) can be realized in parallel form by realizing individual sections using Direct form II, as shown in Fig 3.14.

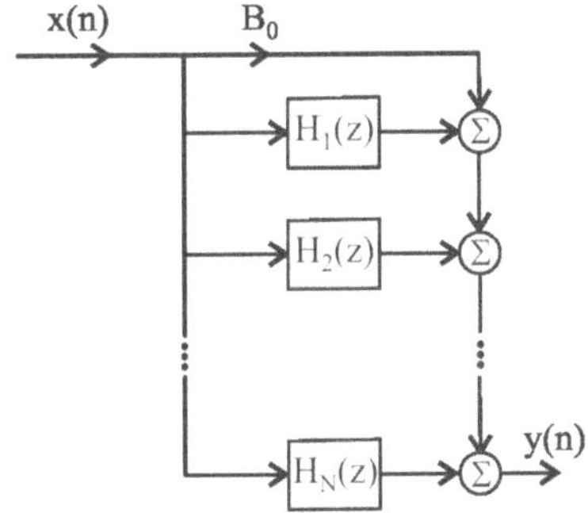

Fig 3.14 Parallel form realization.

Here   x(n) is the time domain representation of X(z) and

y(n) is the time domain representation of Y(z)

Example 3.36: Show the parallel realization of the following transfer function

$$H(z) = \frac{z^2 + 4z + 4}{(z - 0.8)(z^2 - 0.9\, z + 0.14)} \qquad\qquad \text{[JNTU 2000/s]}$$

Solution:
$$H(z) = \frac{(z + 2)(z + 2)}{(z - 0.8)(z - 0.7)(z - 0.2)}$$

Method I:

Divide both sides with Z

$$\frac{H(z)}{z} = \frac{(z + 2)(z + 2)}{z(z - 0.8)(z - 0.7)(z - 0.2)}$$

$$\frac{H(z)}{z} = \frac{C_1}{z} + \frac{C_2}{z-0.8} + \frac{C_3}{z-0.7} + \frac{C_4}{z-0.2}$$

$$C_1 = z\frac{(z+2)^2}{z(z-0.8)(z-0.7)(z-0.2)}\Bigg|_{z=0} = -35.7$$

$$C_2 = (z-0.8)\frac{(z+2)^2}{z(z-0.8)(z-0.7)(z-0.2)}\Bigg|_{z=0.8} = 163.3$$

$$C_3 = (z-0.7)\frac{(z+2)^2}{z(z-0.8)(z-0.7)(z-0.2)}\Bigg|_{z=0.7} = -208.2$$

$$C_4 = (z-0.2)\frac{(z+2)^2}{z(z-0.8)(z-0.7)(z-0.2)}\Bigg|_{z=0.2} = 80.6$$

$$\therefore \qquad \frac{H(z)}{z} = \frac{-35.7}{z} + \frac{163.3}{z-0.8} - \frac{208.2}{z-0.7} + \frac{80.6}{z-0.2}$$

$$\frac{Y(z)}{X(z)} = H(z) = -35.7 + \frac{163.3\,z}{z-0.8} - 208.2\frac{z}{z-0.7} + 80.6\frac{z}{z-0.2}$$

$$= -35.7 + 163.3\frac{1}{1-0.8z^{-1}} - 208.2\frac{1}{1-0.7z^{-1}} + 80.6\frac{1}{1-0.2z^{-1}}$$

Parallel form realization is

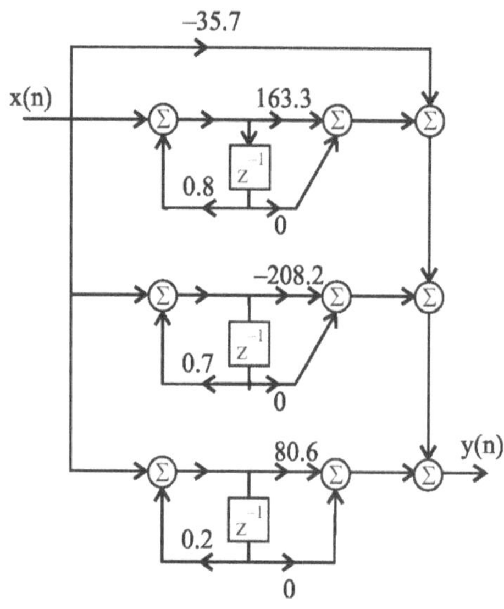

Method II   without dividing with z.

$$H(z) = \frac{C_1}{z-0.8} + \frac{C_2}{z-0.7} + \frac{C_3}{z-0.2}$$

$$C_1 = (z-0.8)\frac{(z+2)^2}{(z-0.8)(z-0.7)(z-0.2)}\Bigg|_{z=0.8} = 130.6$$

$$C_2 = (z-0.7)\frac{(z+2)^2}{(z-0.8)(z-0.7)(z-0.2)}\Bigg|_{z=0.7} = -145.8$$

$$C_3 = (z-0.2)\frac{(z+2)^2}{(z-0.8)(z-0.7)(z-0.2)}\Bigg|_{z=0.2} = 26.8$$

$$\therefore \quad H(z) = \frac{130.6}{z-0.8} - \frac{145.8}{z-0.7} + \frac{26.8}{z-0.2}$$

$$= 130.6\frac{z^{-1}}{1-0.8z^{-1}} - 145.8\frac{z^{-1}}{1-0.7z^{-1}} + 26.8\frac{z^{-1}}{1-0.2z^{-1}}$$

Parallel realization is

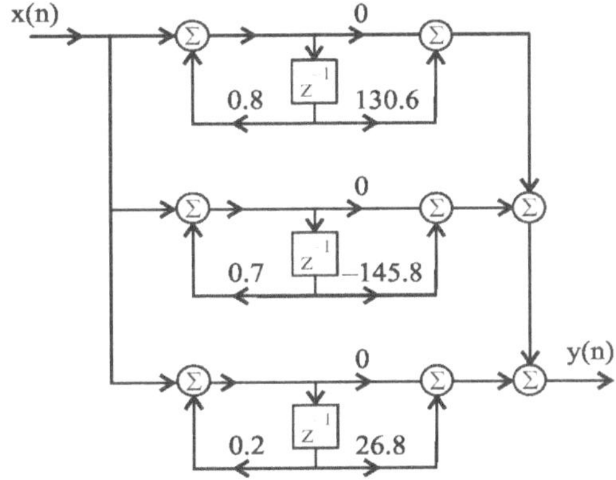

3.9.5   Comparison of Different Realizations of Discrete Filters

(i)   When the order of H(z) is low, Direct form I and Direct form II are suitable.

(ii)   When the order of H(z) is high, as Direct form I and Direct form II are more sensitive to finite word length effects i.e., large errors will result if a small number of bits are used to represent the coefficients, cascade and parallel form realizations are more suitable.

(iii)   The difficulty in cascade structure is pairing of poles and zeros.

Example 3.37: Realize the following transfer function using parallel, cascade, Direct form I and Canonical methods. Use only first order sections.

$$H(z) = \frac{(z-1)(z+2)}{(z^2+4z+4)(z^2+6z+9)}$$

Solution: $$H(z) = \frac{(z-1)(z+2)}{(z+2)^2(z+3)^2}$$

$$= \frac{z-1}{(z+2)(z+3)^2}$$

Parallel form:

$$H(z) = \frac{C_1}{z+2} + \frac{D_1}{z+3} + \frac{D_2}{(z+3)^2}$$

$$C_1 = (z+2)\frac{(z-1)}{(z+2)(z+3)^2}\bigg|_{z=-2} = -3$$

$$D_1 = \frac{d}{dz}\left[(z+3)^2\frac{(z-1)}{(z+2)(z+3)^2}\right]_{z=-3}$$

$$D_1 = \frac{(z+2)1-(z-1)1}{(z+2)^2}\bigg|_{z=-3} = 3$$

$$D_2 = (z+3)^2\frac{(z-1)}{(z+2)(z+3)^2}\bigg|_{z=-3} = 4$$

$$\therefore \quad H(z) = \frac{-3}{z+2} + \frac{3}{z+3} + \frac{4}{(z+3)^2}$$

$$H(z) = \frac{-3z^{-1}}{1+2z^{-1}} + \frac{3z^{-1}}{1+3z^{-1}} + \frac{4z^{-2}}{1+6z^{-1}+9z^{-2}}$$

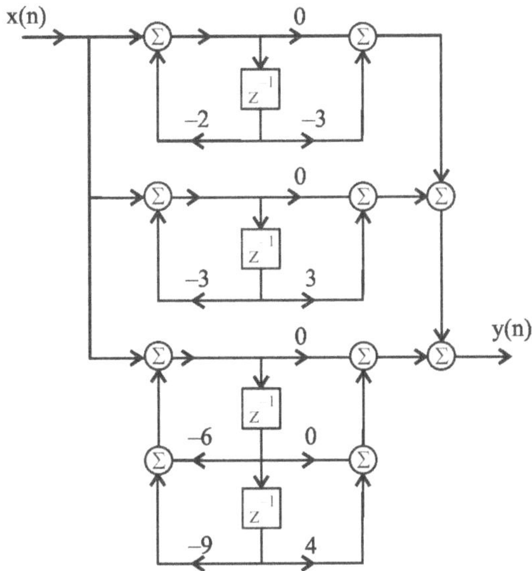

Cascade form:

$$H(z) = \frac{z-1}{(z+2)(z+3)^2}$$

$$= \frac{z-1}{z+2} \cdot \frac{1}{z+3} \cdot \frac{1}{z+3}$$

$$= H_1(z)\, H_2(z)\, H_3(z)$$

$$= \frac{1-z^{-1}}{1+2z^{-1}} \cdot \frac{z^{-1}}{1+3z^{-1}} \cdot \frac{z^{-1}}{1+3z^{-1}}$$

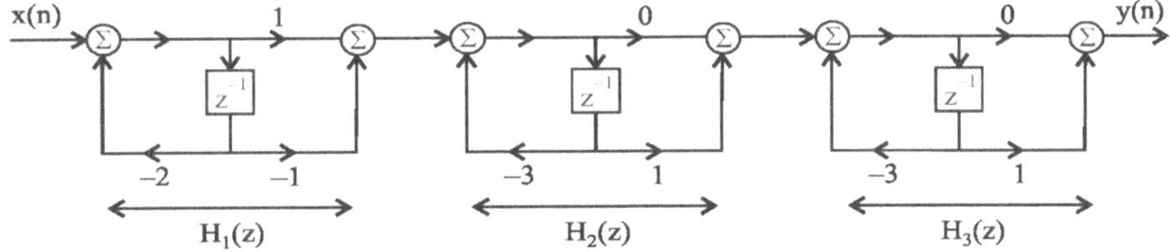

Direct form I:

$$H(z) = \frac{z-1}{(z+2)(z+3)^2}$$

$$= \frac{z-1}{(z+2)(z^2+6z+9)}$$

$$= \frac{z-1}{z^3 + 8z^2 + 21z + 18}$$

$$= \frac{z\left(1 - z^{-1}\right)}{z^3\left(1 + 8z^{-1} + 21z^{-2} + 18z^{-3}\right)}$$

$$H(z) = \frac{Y(z)}{X(z)} = \frac{z^{-2} - z^{-3}}{1 + 8z^{-1} + 21z^{-2} + 18z^{-3}}$$

Direct form II (or) Canonic form:

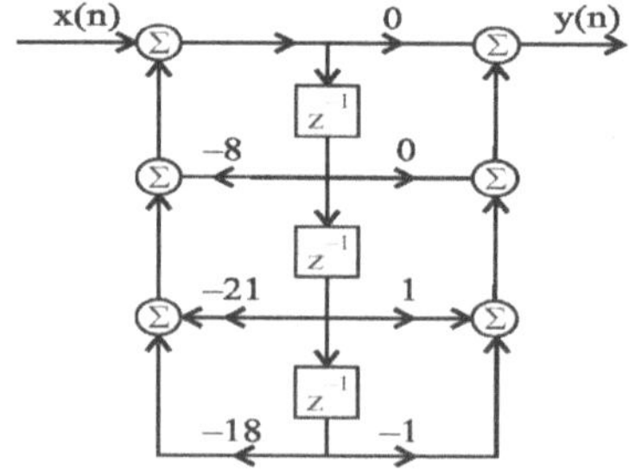

Review Questions

1. If a time domain sequence x(n) is delayed by m, its corresponding DFS pair, when x(n) and X(k) is DFS pair, is

   (a) $e^{j2\pi km/N} \cdot X(k)$ (b) $e^{-j2\pi km/N} \cdot X(k)$

   (c) $\dfrac{e^{-j2\pi km/N}}{X(k)}$ (d) $e^{-j2\pi km/N} - X(k)$ *Ans :* [b]

2. If a time domain sequence x(n) is advanced by m, its corresponding DFS pair, when x(n) and X(k) is DFS pair, is

   (a) $e^{j2\pi km/N} \cdot X(k)$ (b) $e^{-j2\pi km/N} \cdot X(k)$

   (c) $\dfrac{e^{-j2\pi km/N}}{X(k)}$ (d) $e^{-j2\pi km} - X(k)$ *Ans:* [a]

3. If X(k) is delayed by 1, its corresponding DFS pair, when x(n) and X(k) is DFS pair, is

    (a) $e^{j2\pi nl/N} \cdot x(n)$ (b) $e^{-j2\pi nl/N} \cdot x(n)$

    (c) $\dfrac{e^{-j2\pi nl/N}}{x(n)}$ (d) $e^{j2\pi nl/N} - x(n)$ *Ans:* [a]

4. If X(k) is advanced by 1, its corresponding DFS pair, when x(n) and X(k) is DFS pair, is

    (a) $e^{j2\pi nl/N} \cdot x(n)$ (b) $e^{-j2\pi nl/N} \cdot x(n)$

    (c) $\dfrac{e^{-j2\pi nl/N}}{x(n)}$ (d) $e^{j2\pi nl/N} - x(n)$ *Ans:* [b]

5. If x(n) and X(k) is a DFS pair, then X(–k) corresponding DFS pair is

    (a)   N X(n)   (b) X(–n)   (c)   x(n)   (d) X(–n)   *Ans:* [a]

6. If x(n) (complex) and X(k) is a DFS pair, then even part of x(n) DFS pair is

    (a)   Even part of X(k)   (b)   Odd part of X(k)

    (c)   Imaginary part of X(k)   (d)   Real part of X(k)

                                                  Ans: [d]

7. If x(n) (real) and X(k) is a DFS pair, then real part of X(k) is

    (a)   Odd symmetry   (b)   Even symmetry

    (c)   Zero   (d)   None   *Ans:* [b]

8. If x(n) (real) and X(k) is a DFS pair, then imaginary part of X(k) is

    (a)   Odd symmetry   (b)   Even symmetry

    (c)   Zero   (d)   None   *Ans:* [a]

9. The other name(s) of circular convolution is

    (a)   Linear convolution   (b)   Periodic convolution

    (c)   Both (a) and (b)   (d)   None   *Ans:* [b]

10. The DFS coefficients cannot be found for the signal $x(n) = \cos \sqrt{3}\pi\, n$ because

    (a)   Period is rational number

    (b)   Period is irrational

    (c)   Period is zero

    (d)   None   *Ans:* [b]

11. Period of the signal $x(n) = \cos(\pi n / 2)$ is

    (a) Zero   (b) 4   (c) 6   (d) None   *Ans:* [b]

12. If a signal is continuous and non periodic its spectrum will be
    (a)   Discrete and periodic
    (b)   Continuous and periodic
    (c)   Continuous and aperiodic
    (d)   Discrete and aperiodic   *Ans:* [c]

13. If a signal is continuous and non periodic its spectrum can be found by
    (a)   Fourier transform
    (b)   Fourier series
    (c)   Discrete fourier series
    (d)   Discrete fourier transform   *Ans:* [a]

14. If a signal is continuous and periodic its spectrum will be
    (a)   Discrete and periodic
    (b)   Continuous and periodic
    (c)   Continuous and aperiodic
    (d)   Discrete and aperiodic   *Ans:* [d]

15. If a signal is continuous and periodic its spectrum can be found by
    (a)   Fourier transform   (b)   Fourier series
    (c)   Discrete fourier series   (d)   Discrete fourier transform   *Ans:* [b]

16. If a signal is discrete and aperiodic its spectrum can be found by
    (a) Fourier transform   (b) DTFT   (c) Fourier series   (d) DFT   *Ans:* [b]

17. If a signal is discrete and aperiodic its spectrum will be
    (a)   Discrete and aperiodic
    (b)   Continuous and periodic
    (c)   Continuous and aperiodic
    (d)   Discrete and aperiodic   *Ans:* [b]

18. If a signal is discrete and periodic its spectrum will be
    (a)   Discrete and periodic
    (b)   Continuous and periodic
    (c)   Continuous and aperiodic
    (d)   Discrete and aperiodic   *Ans:* [a]

19. If x(n) (complex) and x(k) is a DFT pair, then $x^*(n)$ corresponding DFT pair is

   (a) X(k)   (b) $X^*(k)$ (c) $X^*(-k)$ (d) Zero   *Ans:* [c]

20. If x(n) and x(k) is a DFT pair, then x(N–n) corresponding DFT pair is

   (a) X(k)   (b) X(k–N)   (c) X(–k)   (d) X(N–k)   *Ans:* [d]

21. The following are the properties of DFT

   (a)   Linear convolution   (b)   Periodic convolution

   (c)   Circular convolution   (d)   Both b and c   *Ans:* [d]

22. The other name of twiddle factor is

   (a)   Phase factor   (b)   Imaginary factor

   (c)   Real factor   (d)   None   *Ans:* [a]

23. The symmetry properties of phase factor is

   (a) $W_N^{K+N} = W_N^K$ (b) $W_N^{K/N} = W_N^K$

   (c) $W_N^{K+\frac{N}{2}} = -W_N^K$ (d)   None   *Ans:* [c]

24. The periodicity property of phase factor is

   (a) $W_N^{K+N} = W_N^K$ (b) $W_N^{K/N} = W_N^K$

   (c) $W_N^{K+\frac{N}{2}} = -W_N^K$ (d)   None   *Ans:* [a]

25. W_N (phase factor) is equal to

   (a) $e^{j2\pi/N}$ (b) $e^{j2\pi N}$ (c) $e^{-j2\pi N}$ (d) $e^{-j2\pi/N}$ *Ans:* [d]

26. Linear convolution can be obtained by using

   (a) DFT   (b) Circular convolution   (c) Both a and b   (d) None

   Ans: [c]

27. Circular convolution can be obtained by using

   (a) DFT   (b) Linear convolution   (c) Both a and b   (d) None

   Ans: [c]

28. The response of a system whose impulse response h(n) = {1, 1, 1}, for the input x(n) = {1, 1, 1} is

   (a)   {1, 3, 2, 1, 0}   (b)   {1, 2, 3, 2, 1}

   (c)   {3, 2, 1, 2, 3}   (d)   {1, 2, 3, 1, 2}   *Ans:* [b]

29. The length of output sequence of linear convolution of two sequences with lengths M and N is

   (a) M + N   (b) max (M & N)   (c) M + N – 1   (d) min (M & N)   *Ans:* [c]

30. The length of output sequence of circular convolution of two sequences with lengths M and N is

    (a) $M + N$ (b) max (M & N)   (c) $M + N - 1$ (d) min (M & N)   *Ans:* [b]

31. The purpose of FFT is

    (a) To increase number of computations

    (b) To reduce number of computations

    (c) To increase speed

    (d) Both b and c   *Ans:* [d]

32. The N-pt DFT, N must be __________ to use Radix-2 FFT algorithms

    (a) An integer power of 2   (b) Odd number

    (c) Irrational number   (d) None   *Ans:* [a]

33. The number of computations required to compute N-pt DFT directly

    (a) N   (b) N–1   (c) N/2   (d) N^2 *Ans:* [d]

34. The number of multiplications required to compute N-pt DFT using Radix-2 DIT FFT algorithm

    (a) $\dfrac{N}{2}\log_2^N$ (b) $N\log_2^N$ (c) N   (d) N^2 *Ans :* [a]

35. The number of additions required to compute N-pt DFT using Radix-2 DIF FFT algorithm

    (a) $\dfrac{N}{2}\log_2^N$ (b) $N\log_2^N$ (c) N   (d) N^2 *Ans:* [b]

36. In DIT FFT algorithm

    (a) x(n) is decimated into even and odd numbered samaples

    (b) x(k) is decimated into even and odd numbered samples

    (c) Both (a) and (b)

    (d) None   *Ans:* [a]

37. In DIF FFT algorithm

    (a) x(n) is decimated into even and odd numbered samaples

    (b) x(k) is decimated into even and odd numbered samples

    (c) Both (a) and (b)

    (d) None   *Ans:* [b]

38. The DIT FFT algorithm input samples will be in

    (a) Scrambled order   (b) Normal order

    (c) Bitreversal order   (d) Both (a) and (c)   *Ans:* [d]

39. The DIF FFT algorithm input samples will be in

    (a) Scrambled order   (b) Normal order

    (c) Bitreversal order   (d) Both (a) and (c)   *Ans:* [b]

40. The DIT FFT algorithm output samples will be in

    (a) Scrambled order   (b) Normal order

    (c) Bitreversal order   (d) Both (a) and (c)   *Ans:* [b]

41. The DIF FFT algorithm output samples will be in

    (a) Scrambled order   (b) Normal order

    (c) Bitreversal order   (d) Both (a) and (c)   *Ans:* [d]

42. Number of stages required to comput 16-pt DFT using Radix-2 DIT or DIF FFT algorithm

    (a) 3   (b) 4   (c) 6   (d) 1   *Ans:* [b]

43. Since shape of the 2-pt DFT flow graph is butterfly, FFT algorithms are also called <u>butterfly algorithms</u>.

44. Number of butterflies required for each stage of N-pt DFT using FFT algorithms are

    (a) N   (b) N^2 (c) $\dfrac{N}{2}$ (d) N–1   *Ans:* [c]

45. Inverse DFT can be computed efficiently by using FFT algorithms

    (a) True (b) False (c) Cannot say   *Ans:* [a]

Chapter 4

IIR Digital Filters

4.0 Introduction

Basically a "filter is a system or an algorithm which allows certain frequencies and attenuates the remaining frequencies".

It is known that, the response of a system is obtained by convolving input sequence with impulse response of the system as shown in Fig. 4.1.

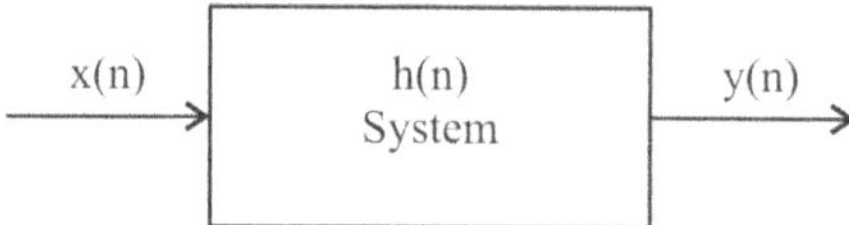

Fig. 4.1 Discrete System.

where

$$x(n) = \text{input or excitation to the system}$$

$$y(n) = \text{output or response of the system}$$

$$h(n) = \text{Impulse response of the system}$$

so

$$y(n) = h(n) * x(n) \qquad \qquad \text{.....(4.0.1)}$$

$$= \sum_{k=-\infty}^{\infty} h(k)x(n-k) \qquad \qquad \text{.....(4.0.2)}$$

where k is a dummy variable.

Consider a causal system, then

$$y(n) = \sum_{k=0}^{\infty} h(k)x(n-k) \qquad \qquad \text{.....(4.0.3)}$$

Here impulse response h(k) ranges from 0 to ∞. Hence this equation represents an Infinite Impulse Response (IIR) system or filter.

As Eqn. (4.0.3) requires infinite number of memory locations to store samples, which is not practically possible, we go for N^{th} order Linear Constant coefficient difference equation which represents IIR filter.

$$y(n) = -\sum_{k=1}^{N} a_k y(n-k) + \sum_{k=0}^{M} b_k x(n-k) \qquad \text{.....(4.0.4)}$$

The transfer function or system function of an IIR filter is

$$H(Z) = \frac{Y(Z)}{X(Z)} = \frac{\displaystyle\sum_{k=0}^{M} b_k Z^{-k}}{1 + \displaystyle\sum_{k=1}^{N} a_k Z^{-k}} \qquad \text{.....(4.0.5)}$$

Designing of an IIR filter from the given specifications is equivalent to finding filter coefficients a_k and b_k of the Eqn. (4.0.5) or finding the transfer function of the filter.

4.0.1 Filter Types

Magnitude response of filter contains different bands or regions namely passband, transitionband and stopband.

Passband: The range of frequencies of signal that are passed through the filter is called passband.

Stopband: The range of frequencies that are rejected is called stopband.

Transition band: The region between pass band and stop band is called transition band.

Different types of filters are

    1.   Lowpass Filter,   2.   Highpass filter,

    3.   Bandpass filter and   4.   Band reject filter.

1. **Lowpass Filter:** The magnitude response of an ideal low pass filter allows low frequencies in the passband $0 < \Omega < \Omega_c$, whereas the higher frequencies are attenuated in the stopband $\Omega > \Omega_c$. The frequency Ω_c is the 3dB cutoff frequency at which the magnitude $|H(j\Omega)| = 1/\sqrt{2}$

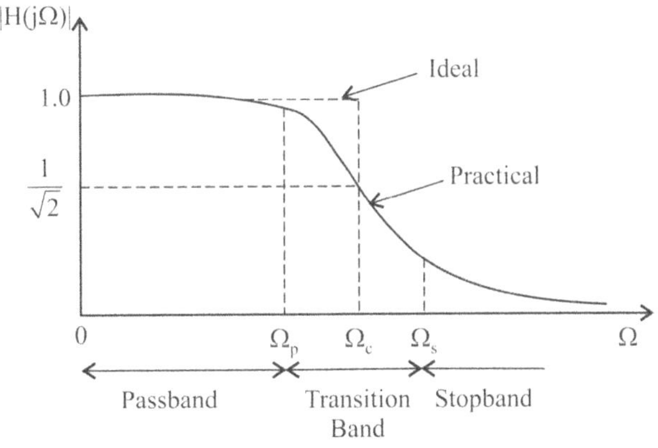

(a)

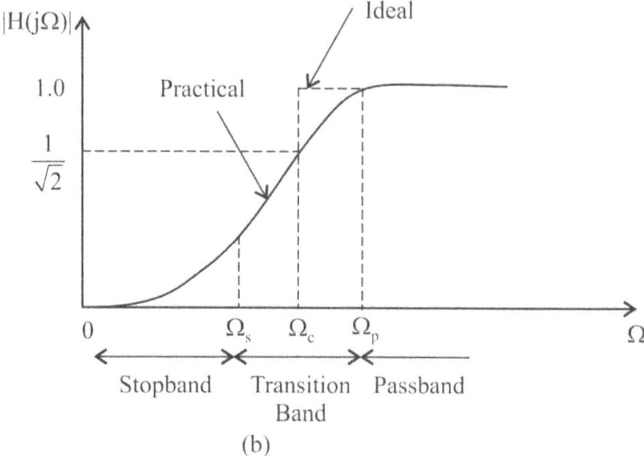

Fig. 4.2 Magnitude response of filters (a) Lowpass (b) Highpass.

But practically ideal filters are not realizable. The magnitude responses of ideal (dotted lines) and practical lowpass filters are shown in fig. 4.2(a).

2. **Highpass Filter:** It allows frequencies above $\Omega > \Omega_c$ and attenuates the frequencies between 0 and Ω_c. The magnitude responses of ideal (dotted lines) and practical high pass filters are shown in Fig. 4.2 (b).

3. **Bandpass Filter:** It allows only a band of frequencies from Ω_l to Ω_u and attenuates the remaining frequencies. The magnitude responses of ideal and practical are shown in Fig. 4.3(a) and (b).

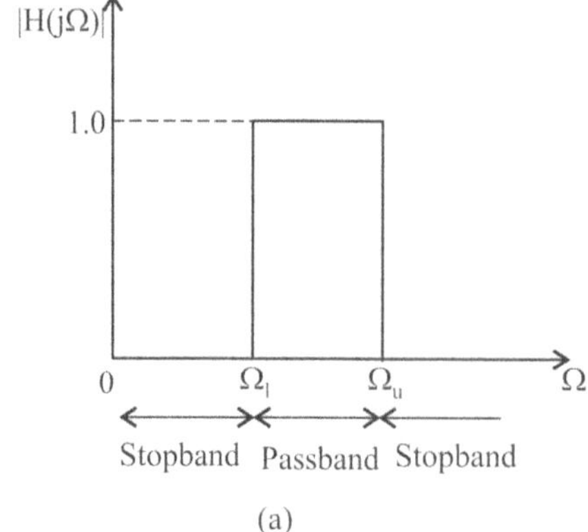

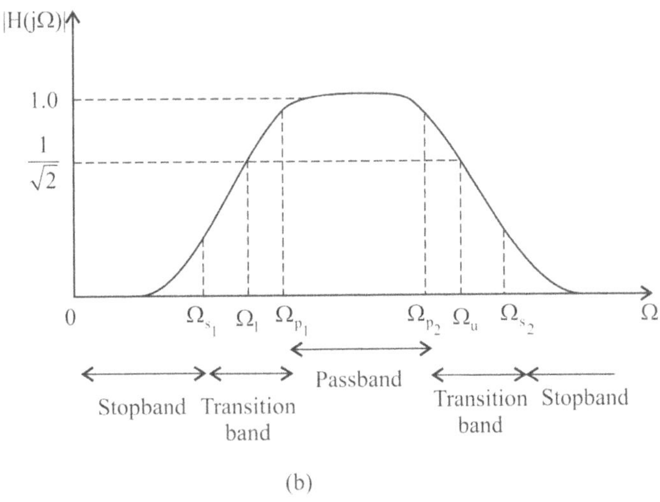

(b)

Fig. 4.3 Magnitude response of bandpass filter (a) Ideal (b) Practical.

4. Band Reject Filter: It rejects a band of frequencies from Ω_l to Ω_u and allows the remaining frequencies. The magnitude responses of ideal and practical are shown in Fig. 4.4(a) and (b).

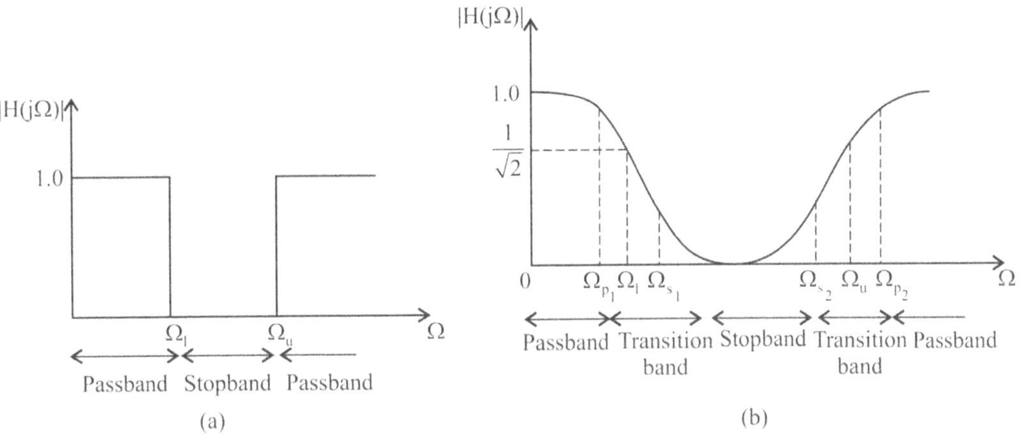

Fig. 4.4 Magnitude response of band reject filter (a) Ideal (b) Practical.

4.1 Analog Filter Approximations

The most common way of designing IIR digital filter (Lowpass) is first designing an analog prototype filter either by Butterworth Approximation or Chebyshev Approximation and then transforming the prototype to a digital filter. To design Highpass, Bandpass and Band reject IIR digital filters, first design an analog prototype

Lowpass filter by transforming desired filter specifications into lowpass filter, then transform to desired filter using spectral transformations. The Digital version of desired filter can be obtained by transforming analog prototype to a digital filter.

4.1.1 Filter Specifications

Magnitude specifications of an analog Lowpass are shown in Fig. 4.5(a) and (b).

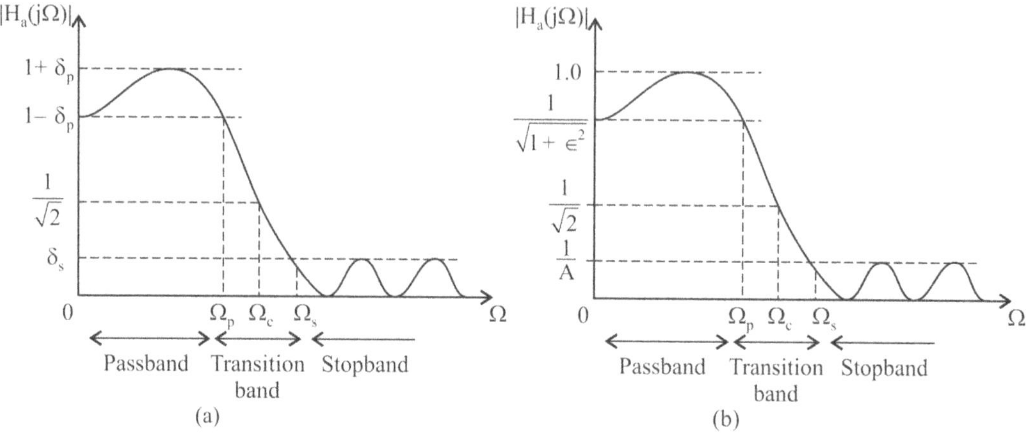

Fig. 4.5 (a) Typical magnitude specifications of an analog lowpass filter

(b) Normalized magnitude specifications of an analog lowpass filter.

The various parameters in the Fig. 4.5(a) and (b) are

Ω_p = Passband edge frequency in Hz

Ω_s = stopband edge frequency in Hz

Ω_c = 3db cutoff frequency in Hz.

$\in$ = parameter specifying allowable passband

The limits of the tolerances in the passband and stopband, δ_p and δ_s, are called ripples. Usually these ripples are specified in dB in terms of the peak passband ripple α_p and the minimum stopband attenuation α_s, defined by

$$\alpha_p = -20\log_{10}\left(1-\delta_p\right)\,\text{dB} \qquad\qquad .....(4.1.1)$$

$$= -20\log_{10}\frac{1}{\sqrt{1+\in^2}}\,\text{dB}$$

$$\alpha_s = -20\log_{10}\left(\delta_s\right)\text{dB}$$

$$= -20\log_{10}\left(\frac{1}{A}\right)\text{dB} \qquad\qquad .....(4.1.2)$$

The ratio of passband edge frequency Ω_P and the stopband edge frequency Ω_s is called Transition ratio or Selectivity parameter and is denoted by K.

$$K = \frac{\Omega_p}{\Omega_s} \qquad\qquad .....(4.1.3)$$

The discrimination parameter is denoted by K_1 and is defined as

$$K_1 = \frac{\epsilon}{\sqrt{A^2 - 1}} \qquad\qquad .....(4.1.4)$$

Usually $K < 1$, for lowpass filter and $K_1 \ll 1$

The magnitude specifications of digital lowpass filter are shown in Fig 4.6

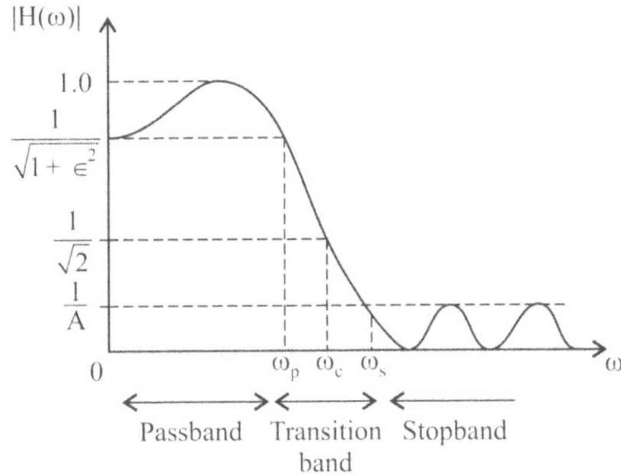

Fig. 4.6 Magnitude specification of digital lowpass filter.

ω_p = passband edge frequency in radians

ω_s = stopband edge frequency in radians

ϵ = parameter specifying allowable passband

A = parameter specifying allowable stopband

ω_c = 3dB – cutoff frequency in radians

The analog prototype of Lowpass filter may be designed either by Butterworth Approximation or by Chebyshev Approximation, which are discussed in the next sections.

4.2 Butterworth Approximation

The magnitude – squared response of an analog Lowpass Butterworth filter $H_a(s)$ of Nth order is given by

$$\left|H_a\left(j\Omega\right)\right|^2 = \frac{1}{1+\left(\Omega/\Omega_C\right)^{2N}} \qquad\qquad .....(4.2.1a)$$

$$= \frac{1}{1+\in^2\left(\Omega/\Omega_P\right)^{2N}} \qquad\qquad .....(4.2.1b)$$

where N is the order of the filter, Ω_C is its $-$ 3dB cutoff frequency, Ω_P is the passband edge frequency and $1/\left(1+\in^2\right)$ is the band-edge value of $\left|H\left(j\Omega\right)\right|^2$. The magnitude response of the Butterworth Lowpass filter is shown in Fig 4.7. The magnitude response has a maximally flat passband and stopband.

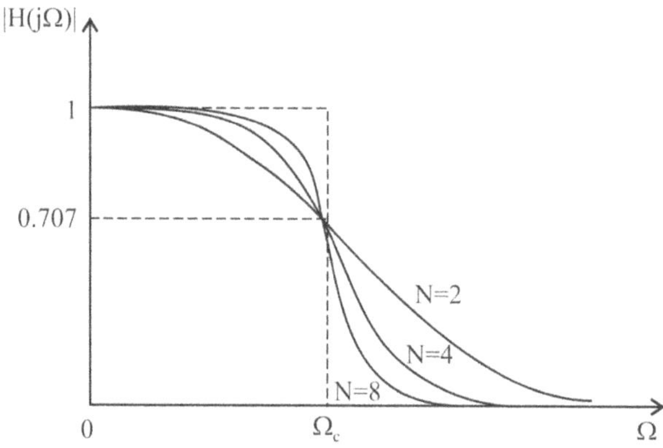

Fig. 4.7 Magnitude response of butterworth lowpass filter.

From Fig 4.7, it is observed that by increasing the filter order N, the response approximates the ideal response.

The two parameters completely characterizing Butterworth filter are Ω_C and N. These are determined by considering the low-pass filter with the following specifications.

$$1 \leq \left|H_a\left(j\Omega\right)\right| \leq \frac{1}{\sqrt{1+\in^2}} \text{ for } \left|\Omega\right| \leq \Omega_P \qquad\qquad .....(4.2.2)$$

and $\qquad \left|H_a\left(j\Omega\right)\right| \leq \frac{1}{A} ; \qquad \text{for } \Omega_S \leq \left|\Omega\right| \leq \infty \qquad\qquad .....(4.2.3)$

at $\Omega = \Omega_P$ from equation (4.2.1a) and (4.2.2)

$$\left|H_a\left(j\Omega_P\right)\right|^2 = \frac{1}{1+\left(\Omega_P/\Omega_C\right)^{2N}} = \frac{1}{1+\in^2} \qquad\qquad .....(4.2.4a)$$

at $\Omega = \Omega_S$ from equation (4.2.1a) and (4.2.3)

$$\left|H_a\left(j\Omega_S\right)\right|^2 = \frac{1}{1+\left(\Omega_S/\Omega_C\right)^{2N}} = \frac{1}{A^2} \qquad\qquad .....(4.2.4b)$$

by eliminating Ω_C from equations (4.2.4a) and (4.2.4b), we get equation for order of the filter N.

Equation (4.2.4a) can be written as

$$1+\left(\Omega_P/\Omega_C\right)^{2N} = 1 + \epsilon^2$$

$$\frac{\left(\Omega_P\right)^{2N}}{\left(\Omega_C\right)^{2N}} = \epsilon^2$$

$$\frac{\left(\Omega_P\right)^{N}}{\left(\Omega_C\right)^{N}} = \epsilon$$

$$\left(\Omega_C\right)^{N} = \frac{\left(\Omega_P\right)^{N}}{\epsilon}$$

$$\therefore\qquad \Omega_C = \frac{\Omega_P}{\epsilon^{1/N}} \qquad\qquad .....(4.2.5)$$

Equation (4.2.4b) can be written as

$$1+\left(\Omega_S/\Omega_C\right)^{2N} = A^2$$

$$\frac{\left(\Omega_S\right)^{2N}}{\left(\Omega_C\right)^{2N}} = A^2 - 1$$

$$\frac{\left(\Omega_S\right)^{N}}{\left(\Omega_C\right)^{N}} = \sqrt{A^2 - 1}$$

$$\left(\Omega_C\right)^{N} = \frac{\left(\Omega_S\right)^{N}}{\sqrt{A^2 - 1}}$$

$$\Omega_C = \frac{\Omega_S}{\left[\sqrt{A^2 - 1}\right]^{1/N}} \qquad\qquad .....(4.2.6)$$

from equations (4.2.5) and (4.2.6)

$$\frac{\Omega_P}{\in^{1/N}} = \frac{\Omega_S}{\left[\sqrt{A^2 - 1}\right]^{1/N}}$$

$$\frac{\Omega_P}{\Omega_S} = \left[\frac{\in}{\sqrt{A^2 - 1}}\right]^{1/N}$$

Taking log on both sides

$$\log_{10}\left(\Omega_P/\Omega_S\right) = \frac{1}{N}\log_{10}\left(\frac{\in}{\sqrt{A^2 - 1}}\right)$$

$$\therefore \qquad N = \frac{\log_{10}\left(\dfrac{\in}{\sqrt{A^2 - 1}}\right)}{\log_{10}\left(\Omega_P/\Omega_S\right)}$$

$$N = \frac{\log_{10} K_1}{\log_{10} K} \qquad\qquad .....(4.2.7)$$

N is rounded up to the next higher integer.

This value of N can be used next in equation (4.2.5) or equation (4.2.6) to solve for the 3-dB cutoff frequency Ω_C.

The expression for the normalized transfer function of the Butterworth lowpass filter is given by

$$H_a\left(s\right) = \frac{1}{\displaystyle\prod_{k=1}^{N}\left(S - P_K\right)} \qquad\qquad .....(4.2.8)$$

where $\qquad\qquad P_K = e^{j\left(\pi(N + 2K - 1)/2N\right)},\ K = 1, 2, ..., N \qquad\qquad .....(4.2.9)$

Unnormalized transfer function can be obtained by replacing S with $\dfrac{S}{\Omega_C}$ in equation (4.2.8)

4.2.1 Poles of Butterworth Filter

From equation (4.2.1a), for a normalized filter $\Omega_C = 1$, then

$$\left|H_a\left(j\Omega\right)\right|^2 = \frac{1}{1 + \Omega^{2N}}$$

The normalized poles in the S-domain can be obtained by substituting $\Omega = S/j$ and equating the denominator polynomial to zero, i.e.,

$$1 + \left(\frac{S}{j}\right)^{2N} = 0 \ \text{ or } \ 1 + \left(-S^2\right)^N = 0 \qquad\qquad .....(4.2.10)$$

The solution to the above equation gives the poles of the filter.

equation (4.2.10) can be written as

$$\left(-1\right)^N S^{2N} = -1$$

expressing -1 in the polar form, i.e., $-1 = e^{j\,(2K-1)\pi}$, $\qquad$ K = 1, 2, ... N

$\therefore \qquad\qquad\qquad \left(-1\right)^N S^{2N} = e^{j\,(2K-1)\pi}, \qquad\qquad$ K = 1, 2, ... N

$$-S^2 = e^{j\,(2K-1)\pi/N}$$

$$S^2 = -\,e^{j\,(2K-1)\pi/N}$$

$$S = \left(-1\right)^{1/2} e^{j\,(2K-1)\pi/2N}$$

$$S = j\,e^{j\,(2K-1)\pi/2N}, \qquad\qquad \text{K = 1, 2, ... N}$$

Let us indicate S with S_K

$\therefore \qquad\qquad\qquad S_K = e^{j\pi/2}\ e^{j(2K-1)\pi/2N}$

Poles are $\qquad\qquad\quad S_K = e^{j(2K+N-1)\pi/2N}, \qquad\qquad$ K = 1, 2, ... N $\qquad .....(4.2.11)$

For N = 4, Butterworth normalized (i.e., $\Omega_C = 1$) poles are

$$S_1 = e^{j(2+4-1)\pi/8} = -\,0.383 + j\,0.924$$

$$S_2 = e^{j(4+4-1)\pi/8} = -\,0.924 + j\,0.383$$

$$S_3 = e^{j(6+4-1)\pi/8} = -\,0.924 - j\,0.383$$

$$S_4 = e^{j(8+4-1)\pi/8} = -\,0.383 - j\,0.924$$

So normalized poles lie on the unit circle in the S-plane as shown in Fig 4.8.

Unnormalized poles lie on the circle in the S-plane, but radius depends on 3dB cutoff frequency. S_1^1, S_2^1, S_3^1 and S_4^1 are corresponding poles of S_1, S_2, S_3 and S_4 on the right side of S-plane.

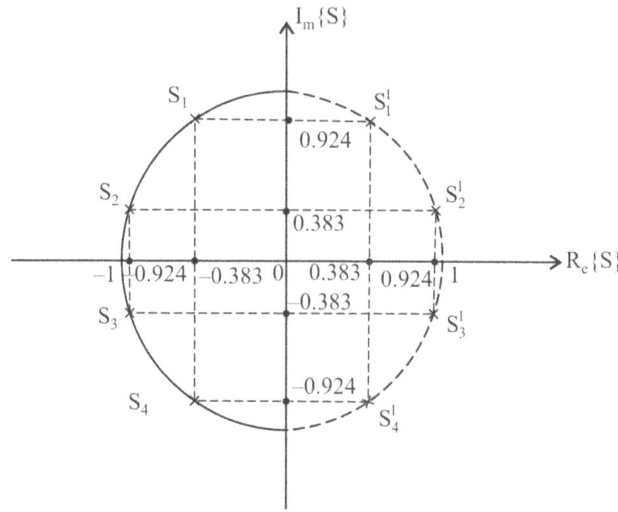

Fig. 4.8 Butterworth poles location in the S-plane.

Example 4.1 Determine the lowest order of a transfer function $H_a(s)$ having a maximally flat lowpass characteristic with a 1dB cutoff frequency at 1 kHz and a minimum attenuation of 40 dB at 5 kHz.

Solution:

Given

$$\alpha_P = 1 \text{dB}$$

$$\alpha_S = 40 \text{ dB}$$

$$\Omega_P = 1 \text{ kHz}$$

$$\Omega_S = 5 \text{ kHz}$$

We know

$$\alpha_P = -20\log_{10}\frac{1}{\sqrt{1+\epsilon^2}}$$

$$1 = -20\log_{10}\frac{1}{\sqrt{1+\epsilon^2}}$$

$$\frac{1}{\sqrt{1+\epsilon^2}} = 10^{-1/20}$$

$$\epsilon^2 = 0.25893 \;\rightarrow\; \epsilon = 0.50885$$

and

$$\alpha_S = -20\log_{10}\frac{1}{A}$$

$$40 = -20 \ \log_{10} \frac{1}{A}$$

$$10^{-40/20} = \frac{1}{A}$$

$$A = 100$$

$$A^2 = 10,000$$

$$K = \frac{\Omega_P}{\Omega_S} = \frac{1K}{5K} = 0.2$$

$$K_1 = \frac{\in}{\sqrt{A^2 - 1}} = \frac{0.50885}{\sqrt{10,000 - 1}} = 5.0887 \times 10^{-3}$$

$$\therefore \ \text{Order of the filter} \quad N = \frac{\log_{10} K_1}{\log_{10} K} = 3.28$$

By rounding up to next higher integer, $\quad N = 4$

4.3 Chebyshev Approximation

Here approximation error, which is the difference between the ideal response and the actual response, is minimized over a prescribed band of frequencies. There are two types of Chebyshev filters. In the Type I Chebyshev approximation, the magnitude characteristic is equiripple in the passband and monotonic in the stopband, whereas in Type II approximation, the magnitude response is monotonic in the passband and equiripple in the stopband.

4.3.1   Type I Chebyshev Approximation

The magnitude-squared response of an analog lowpass Type I Chebyshev filter $H_a(s)$ of Nth order is given by

$$\left| H_a \left(j\Omega \right) \right|^2 = \frac{1}{1 + \in^2 T_N^2 \left(\Omega / \Omega_P \right)} \qquad \qquad .....(4.3.1)$$

Where Ω_P is the 3-dB cutoff frequency, $\in$ is a parameter of the filter that is related to the ripple in the passband and $T_N(x)$ is the Nth – order Chebyshev polynomial defined as

$$T_N \left(x \right) = \begin{cases} \cos \left(N \cos^{-1} \left(x \right) \right), & |x| \leq 1 \\ \cosh \left(N \cosh^{-1} \left(x \right) \right), & |x| > 1 \end{cases} \qquad .....(4.3.2)$$

The Chebyshev polynomials can be generated by the recursive equation

$$T_{N+1}(x) = 2xT_N(x) - T_{N-1}(x), \qquad N = 1, 2, \dots \qquad \dots\dots(4.3.3)$$

where $T_0(x) = 1$ and $T_1(x) = x$. From equation (4.3.3) we obtain

$$T_2(x) = 2x^2 - 1, \; T_3(x) = 4x^3 - 3x, \text{ and so on.}$$

Some of the properties of these polynomials are

(i) $\left| T_N(x) \right| \le 1$ for all $|x| \le 1$

(ii) $T_N(1) = 1$ for all N

(iii) All the roots of the polynomial $T_N(x)$ occur in the interval $-1 \le x \le 1$.

The magnitude response of the Type I Chebyshev filter is shown in fig 4.9.

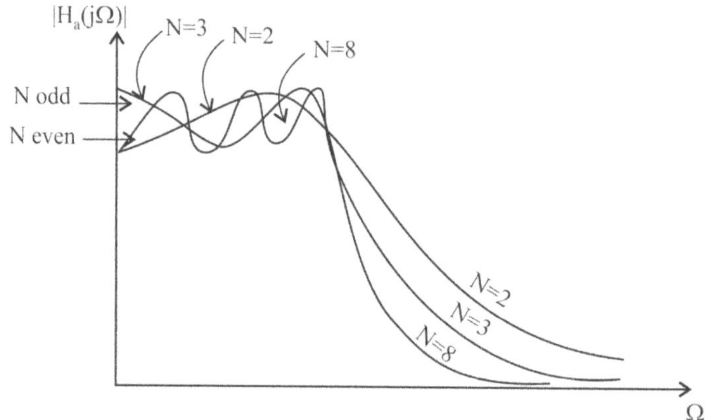

Fig. 4.9 Magnitude response of Type I Chebyshev lowpass filter.

The order N of the filter is determined as at $\Omega = \Omega_s$

$$\left| H_a(j\Omega_S) \right|^2 = \frac{1}{1 + \epsilon^2\, T_N^2(\Omega_S/\Omega_P)} = \frac{1}{A^2}$$

$$= \frac{1}{1 + \epsilon^2 \left\{ \cosh\left(N\cosh^{-1}(\Omega_S/\Omega_P) \right) \right\}^2} = \frac{1}{A^2}$$

$$1 + \epsilon^2 \left\{ \cosh\left(N\cosh^{-1}(\Omega_S/\Omega_P) \right) \right\}^2 = A^2$$

$$\left[\cosh\left(N\cosh^{-1}(\Omega_S/\Omega_P) \right) \right]^2 = \frac{A^2 - 1}{\epsilon^2}$$

$$\cosh\left(N\cosh^{-1}\left(\Omega_S/\Omega_P\right)\right) = \frac{\sqrt{A^2 - 1}}{\in}$$

apply $\cosh^{-1}$ on both sides

$$N\cosh^{-1}\left(\Omega_S/\Omega_P\right) = \cosh^{-1}\left(\frac{\sqrt{A^2-1}}{\in}\right)$$

$$\therefore \quad N = \frac{\cosh^{-1}\left(\frac{1}{k_1}\right)}{\cosh^{-1}\left(\frac{1}{k}\right)} \qquad \qquad .....(4.3.4)$$

The Normalized transfer function $H_a(s)$ is given by

$$H_a(s) = \frac{H_0}{\prod\limits_{k=1}^{N}\left(S - P_k\right)} \qquad \qquad .....(4.3.5)$$

where

$$P_K = \sigma_K + j\Omega_K, \quad k = 1, 2, \ldots N \qquad .....(4.3.6)$$

$$\sigma_K = A\sin\left[\frac{(2K-1)\pi}{2N}\right], \Omega_K = B\cos\left[\frac{(2K-1)\pi}{2N}\right] \quad .....(4.3.7)$$

$$A = \frac{Y^2 - 1}{2Y}, \quad B = \frac{Y^2 + 1}{2Y}, \quad Y = \left[\frac{1 + \sqrt{1 + \in^2}}{\in}\right]^{1/N} \quad .....(4.3.8)$$

or

$$P_K = \sinh(\alpha)\cos(\beta_K) + j\cosh(\alpha)\sin(\beta_K) \qquad .....(4.3.9)$$

where

$$\alpha = \frac{1}{N}\sinh^{-1}\left(\frac{1}{\in}\right) \qquad \qquad .....(4.3.10a)$$

$$\beta_K = \frac{(2K + N - 1)\pi}{2N}, \quad k = 1, 2, \ldots N \qquad .....(4.3.10b)$$

and

$$H_0 = \begin{cases} \dfrac{1}{\sqrt{1 + \in^2}}\prod\limits_{k=1}^{N}\left(-P_K\right) & \text{for even } N \\[4mm] \prod\limits_{k=1}^{N}\left(-P_K\right) & \text{for odd } N \end{cases} \qquad .....(4.3.11)$$

To get unnormalized transfer function, replace S with $\dfrac{S}{\Omega_P}$ in equation (4.3.5)

4.3.2   Poles of Chebyshev Filter

From equation (4.3.1)

$$\left|H_a\left(j\Omega\right)\right|^2 = \frac{1}{1 + \epsilon^2\, T_N^2\left(\Omega/\Omega_P\right)}$$

The poles can be obtained by substituting $\Omega = s\,/\,j$ and equating the denominator to zero. i.e.,

$$1 + \epsilon^2\, T_N^2\left(\frac{-jS}{\Omega_P}\right) = 0$$

$$\epsilon^2\, T_N^2\left(\frac{-jS}{\Omega_P}\right) = -1$$

$$\epsilon\, T_N\left(\frac{-jS}{\Omega_P}\right) = \sqrt{-1}$$

$$T_N\left(\frac{-jS}{\Omega_P}\right) = \pm\frac{j}{\epsilon} = \cos\left[N\cos^{-1}\left(\frac{-jS}{\Omega_P}\right)\right] \qquad .....(4.3.12)$$

We define

$$\cos^{-1}\left(-\frac{jS}{\Omega_P}\right) = \phi - j\theta \qquad .....(4.3.13)$$

$\therefore$ Equation (4.3.9) gives $\pm\dfrac{j}{\epsilon} = \cos\left(N\phi - jN\theta\right)$

$$\pm\frac{j}{\epsilon} = \cos\left(N\phi\right)\cos\left(jN\theta\right) + \sin\left(N\phi\right)\sin\left(jN\theta\right)$$

$\because$ $\qquad\qquad\qquad \cos\theta = \dfrac{e^{j\theta} + e^{-j\theta}}{2}$

$$\cos j\theta = \frac{e^{j(j\theta)} + e^{-j(j\theta)}}{2}$$

$$= \frac{e^{-\theta} + e^{\theta}}{2} = \cosh\theta$$

$$\pm \frac{j}{\in} = \cos\left(N\phi\right)\cosh\left(N\theta\right) + j\sin\left(N\phi\right)\sinh\left(N\theta\right) \qquad .....(4.3.14)$$

Equating the real and imaginary parts of both sides equation (4.3.14) results in

$$\cos\left(N\phi\right)\cosh\left(N\theta\right) = 0 \qquad .....(4.3.15a)$$

$$\sin\left(N\phi\right)\sinh\left(N\theta\right) = \pm\frac{1}{\in} \qquad .....(4.3.15b)$$

Since $\cosh\left(N\theta\right) > 0$ for θ real, to satisfy equation 4.35a, we have

$$\phi = \frac{\left(2k-1\right)\pi}{2N}, \quad k = 1, 2, \dots N \qquad .....(4.3.16)$$

Now solve for θ, where $\sin\left(N\phi\right) = \pm 1$ from equation 4.35b

$$\therefore \qquad \theta = \pm\frac{1}{N}\sinh^{-1}\left(\frac{1}{\in}\right) \qquad .....(4.3.17)$$

From equations 4.33, 4.36 and 4.37

$$S_K = j\Omega_p \cos\left(\phi - j\theta\right)$$

$$= j\Omega_P \left[\cos\phi\cosh\theta + j\sin\phi\sinh\theta\right]$$

$$= \Omega_P \left[-\sin\phi\sinh\theta + j\cos\phi\cosh\theta\right] \qquad .....(4.3.18)$$

We know the identity

$$\sinh^{-1} x = \ln_e\left(x + \sqrt{1+x^2}\right)$$

$$\therefore \qquad \sinh^{-1}\left(\in^{-1}\right) = \ln_e\left(\in^{-1} + \sqrt{1+\in^{-2}}\right)$$

or $\qquad \beta = e^{\sinh^{-1}\left(\in^{-1}\right)} = \in^{-1} + \sqrt{1+\in^{-2}} \qquad .....(4.3.19)$

From equation (4.3.17), we can write

$$\sinh\theta = \sinh\left(\frac{1}{N}\sinh^{-1}\left(\in^{-1}\right)\right)$$

$$\because \qquad \sinh x = \frac{e^x - e^{-x}}{2}$$

$$= \frac{e^{(1/N)\sinh^{-1}(\epsilon^{-1})} - e^{-(1/N)\sinh^{-1}(\epsilon^{-1})}}{2}$$

$$= \frac{\left[e^{\sinh^{-1}(\epsilon^{-1})}\right]^{1/N} - \left[e^{-\sinh^{-1}(\epsilon^{-1})}\right]^{1/N}}{2}$$

From equation (4.3.19)

$$\sinh\theta = \frac{\beta^{1/N} - \beta^{-1/N}}{2} \qquad \qquad \text{.....(4.3.20a)}$$

Similarly

$$\cosh\theta = \frac{\beta^{1/N} + \beta^{-1/N}}{2} \qquad \qquad \text{.....(4.3.20b)}$$

Now from equation (4.3.18)

$$S_k = \Omega_p\left[-\sin\phi\left[\frac{\beta^{1/N} - \beta^{-1/N}}{2}\right] + j\cos\phi\left[\frac{\beta^{1/N} + \beta^{-1/N}}{2}\right]\right] \qquad \text{.....(4.3.21)}$$

$$= -A\sin\left(\frac{(2k-1)\pi}{2N}\right) + jB\cos\left(\frac{(2k-1)\pi}{2N}\right)$$

$$= A\cos\left(\frac{\pi}{2} + \frac{(2k-1)\pi}{2N}\right) + jB\sin\left(\frac{\pi}{2} + \frac{(2k-1)\pi}{2N}\right)$$

$$= A\cos\phi_k + jB\sin\phi_k \qquad \qquad \text{.....(4.3.22)}$$

$$= \sigma_k + j\Omega_k, \quad k = 1, 2, \dots N \qquad \qquad \text{.....(4.3.23)}$$

The poles of a Chebyshev filter can be obtained by using equation (4.3.22)

where

$$A = \Omega_p\left[\frac{\beta^{1/N} - \beta^{-1/N}}{2}\right] \qquad \qquad \text{.....(4.3.24a)}$$

$$B = \Omega_p\left[\frac{\beta^{1/N} + \beta^{-1/N}}{2}\right] \qquad \qquad \text{.....(4.3.24b)}$$

$$\phi_k = \frac{\pi}{2} + \frac{(2k-1)\pi}{2N}, \quad k = 1, 2, \dots N \qquad \qquad \text{.....(4.3.24c)}$$

The poles of the Chebyshev filter are located on an ellipse in the S-plane as shown in Fig 4.10. The equation of the ellipse is given by

$$\frac{\sigma_k^2}{A^2} + \frac{\Omega_k^2}{B^2} = 1 \qquad \qquad \text{.....(4.3.25)}$$

where   A is minor axis of the ellipse

     B is major axis of the ellipse

S_1^1, S_2^1, S_3^1 and S_4^1 are corresponding poles of S_1, S_2, S_3 and S_4 on the right side of S-plane.

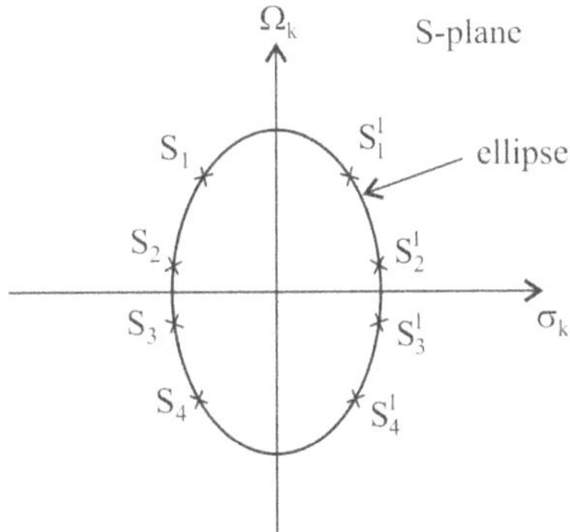

Fig. 4.10 Poles of Chebyshev type I filter.

4.3.3 Type II Chebyshev Filter (or) Inverse Chebyshev Filter

The magnitude–squared response expression is given by

$$\left|H_a\left(j\Omega\right)\right|^2 = \cfrac{1}{1+\epsilon^2 \left[\cfrac{T_N\left(\Omega_s/\Omega_p\right)}{T_N\left(\Omega_s/\Omega\right)}\right]^2} \quad\quad .....(4.3.26)$$

The transfer function has both poles and zeros as follows

$$H_a\left(s\right) = H_0 \frac{\displaystyle\pi_{k=1}^{N}\left(S-Z_k\right)}{\displaystyle\pi_{k=1}^{N}\left(S-P_k\right)} \quad\quad .....(4.3.27)$$

The zeros are on the $j\Omega$ – axis and are given by

$$z_k = j\cfrac{\Omega_S}{\cos\left[\cfrac{\left(2k-1\right)\pi}{2N}\right]}, \quad k = 1, 2, \dots N \quad\quad .....(4.3.28)$$

and $H_0 = \displaystyle\pi_{k=1}^{N} \frac{Z_k}{P_k}$

If N is odd, then $K = (N+1)/2$, the zero is at $S = \infty$

The poles are located at

$$P_k = \sigma_k + j\Omega_k, \quad k = 1, 2, \ldots N \qquad \ldots(4.3.29)$$

$$\text{where} \quad \sigma_k = \frac{\Omega_s \alpha_k}{\alpha_k^2 + \beta_k^2}, \quad \Omega_k = -\frac{\Omega_s \beta_k}{\alpha_k^2 + \beta_k^2} \qquad \ldots(4.3.30a)$$

$$\alpha_k = -\Omega_p C_1 \sin\left[\frac{(2k-1)\pi}{2N}\right], \quad \beta_k = \Omega_p C_2 \cos\left[\frac{(2k-1)\pi}{2N}\right] \qquad \ldots(4.3.30b)$$

$$C_1 = \frac{Y^2 - 1}{2Y}, \quad C_2 = \frac{Y^2 + 1}{2Y}, \quad Y = \left(A + \sqrt{A^2 - 1}\right)^{1/N} \qquad \ldots(4.3.30c)$$

where A is parameter that defines stopband ripple.

The order N of the filter can be found by using equation (4.3.4).

4.3.4 Steps to Design an Analog Butterworth Lowpass Filter

1. From the given specification find the order of the filter N using $N = \log_{10}(k_1) / \log_{10}(k)$

2. Round off it to the next higher integer.

3. Find the normalized (i.e., $\Omega_C = 1$ rad/sec) transfer function $H_a(s)$ for the value N.

4. Calculate the value of cutoff frequency Ω_C.

5. Find the transfer function $H_a(s)$ for the above value of Ω_C by replacing S with $\dfrac{S}{\Omega_C}$ in $H_a(s)$

4.3.5 Steps to Design an Analog Chebyshev Lowpass Filter

1. From the given specifications find the order of the filter N using

$$N = \frac{\cosh^{-1}(1/k_1)}{\cosh^{-1}(1/k)}$$

2. Round off it to the next higher integer.

3. Find the normalized (i.e., $\Omega_p = 1$ rad/sec) transfer function $H_a(s)$ for the value of N using

$$H_a(s) = \frac{H_0}{\prod\limits_{k=1}^{N}(S - P_k)}$$

where $P_k = \sinh(\alpha)\cos(\beta_k) + j\cosh(\alpha)\sin(\beta_k)$

$$\alpha = \frac{1}{N}\sinh^{-1}\left(\frac{1}{\epsilon}\right)$$

$$\beta_k = \frac{(2k + N - 1)\pi}{2N}, \quad k = 1, 2, \ldots, N$$

$$H_0 = \begin{cases} \dfrac{1}{\sqrt{1+\epsilon^2}}\prod\limits_{k=1}^{N}(-P_k) & \text{for even N} \\[2ex] \prod\limits_{k=1}^{N}(-P_k) & \text{for odd N} \end{cases}$$

4. Find the transfer function $H_a(s)$ for the given passband or cutoff frequency Ω_p by replacing S with S/Ω_p in $H_a(s)$.

Example 4.2: Consider an analog filter with transfer function

$$|H_a(s)| = \frac{1}{(s+1)(s^2 + s + 1)}$$

Is this a Butterworth or Chebyshev filter?

Solution: Given

$$|H_a(s)| = \frac{1}{(s+1)(s^2 + s + 1)}$$

$$= \frac{1}{(s+1)\left(s + \dfrac{1}{2} + j\dfrac{\sqrt{3}}{2}\right)\left(s + \dfrac{1}{2} - j\dfrac{\sqrt{3}}{2}\right)}$$

$$S_1 = -1, \quad S_2 = -\frac{1}{2} + j\frac{\sqrt{3}}{2}, \quad S_3 = -\frac{1}{2} - j\frac{\sqrt{3}}{2}$$

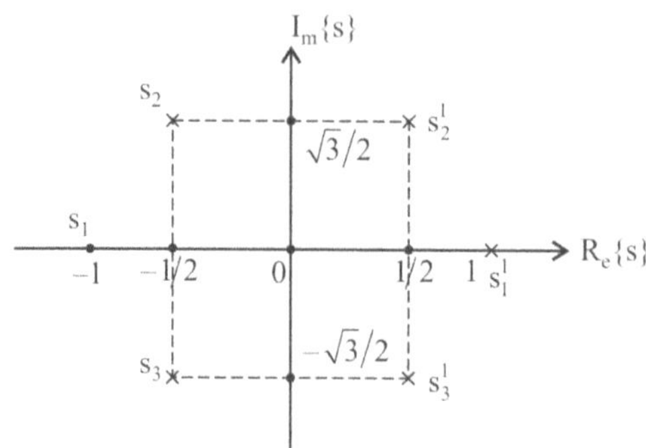

Distance from origin to each pole is unity. Since all the poles lie at the same distance from origin, these form a circle with the corresponding poles $\left(S_1^1,\ S_2^1\ \text{and}\ S_3^1\right)$ on the right side of s-plane.

Since poles lie on the circle, the given transfer function is for Butterworth filter.

Example 4.3: Determine the order of a transfer function $H_a(s)$ having Type I Chebyshev Lowpass filter characteristic with a 1-dB cutoff frequency at 1 kHz and a minimum attenuation of 40 dB at 5 kHz

Solution: From example 4.1, we have $1/k_1 = 196.51334$ and $1/k = 5$

∴ Order N of the Type I Chebyshev Lowpass filter is

$$N = \frac{\cosh^{-1}(196.51334)}{\cosh^{-1}(5)}$$

$$= 2.60591$$

Roundup to next higher integer. So N = 3.

Observe that, for the given same specifications, Chebyshev lowpass filter order is lower than the Butterworth Lowpass filter.

Example 4.4: Design an analog Butterworth filter that has a –2 dB passband attenuation at a frequency of 20 rad/sec, and at least –10 dB stopband attenuation at 30 rad/sec.

Solution: Given $\quad\quad \alpha_p = 2 \text{ dB}; \Omega_p = 20 \text{ rad/sec}$

$$\alpha_s = 10 \text{ dB}; \Omega_s = 30 \text{ rad/sec}$$

$$\alpha_p = -20\log_{10}\frac{1}{\sqrt{1+\epsilon^2}} = 2 \text{ dB}$$

$$= -10\log_{10}\frac{1}{1+\epsilon^2} = 2 \text{ dB}$$

$$= 10\log_{10}\left(1+\epsilon^2\right) = 2 \text{ dB}$$

$$\epsilon = 0.76478$$

$$\alpha_s = -20\log_{10}\frac{1}{A} = 10 \text{ dB}$$

$$= 20\log_{10} A = 10 \text{ dB}$$

$$A = 3.162$$

$$N = \frac{\log_{10}\dfrac{\epsilon}{\sqrt{A^2-1}}}{\log_{10}\left(\Omega_p/\Omega_s\right)}$$

$$N = \frac{\log_{10}\left(0.25495\right)}{\log_{10}\left(0.6666\right)}$$

$$= 3.37$$

So the order of Butterworth filter is 4.

The normalized lowpass Butterworth filter transfer function is

$$H_a(s) = \frac{1}{\prod\limits_{k=1}^{4}\left(S-P_k\right)}$$

$$P_k = e^{j\pi(2k+N-1)/2N}, \quad k = 1, 2, \ldots N$$

$$P_1 = e^{j\pi(2+4-1)/8} = -0.383 + j\,0.934$$

$$P_2 = e^{j\pi(4+4-1)/8} = -0.924 + j\,0.383$$

$$P_3 = e^{j\pi(6+4-1)/8} = -0.924 - j\,0.383$$

$$P_4 = e^{j\pi(8+4-1)/8} = -0.383 - j\,0.934$$

$$H_a(s) = \frac{1}{\left(S+0.383-j\,0.934\right)\left(S+0.383+j\,0.934\right)\left(S+0.924-j\,0.383\right)\left(S+0.924+j\,0.383\right)}$$

$$H_a(s) = \frac{1}{\left(S^2+0.766\,S+1\right)\left(S^2+1.848\,S+1\right)}$$

From equation (4.2.6) (or) (4.2.5)

$$\Omega_C = \frac{\Omega_S}{\left[\sqrt{A^2-1}\right]^{1/N}} = 22.7956 \text{ rad/sec.}$$

(or)

$$\Omega_C = \frac{\Omega_P}{\in^{1/N}} = 21.3868 \text{ rad/sec}$$

We can use either 22.7956 or 21.3868.

Let $\Omega_C = 21.3868$

∴ Unnormalized or the transfer function for $\Omega_C = 21.3868$ can be obtained by substituting

$$S \to \frac{S}{21.3868} \text{ in } H_a(s)$$

i.e.,

$$H_a(s) = \frac{1}{\left[\left(\dfrac{S}{21.3868}\right)^2 + 0.766\left(\dfrac{S}{21.3868}\right)+1\right]\left[\left(\dfrac{S}{21.3868}\right)^2 + 1.848\left(\dfrac{S}{21.3868}\right)+1\right]}$$

$$H_a(s) = \frac{0.20921 \times 10^6}{\left(S^2 + 16.3686\,S + 457.394\right)\left(S^2 + 39.5176\,S + 457.394\right)}$$

Example 4.5 Design a Chebyshev filter with a maximum passband attenuation of 2.5 dB at $\Omega_P = 20$ rad/sec and the stopband attenuation of 30 dB at $\Omega_S = 50$ rad/sec.

Solution: Given

$$\Omega_P = 20 \text{ rad/sec}; \ \alpha_P = 2.5 \text{ dB}$$

$$\Omega_S = 50 \text{ rad/sec}; \ \alpha_S = 30 \text{ dB}$$

We know

$$N = \frac{\cosh^{-1}\left(\dfrac{1}{k_1}\right)}{\cosh^{-1}\left(\dfrac{1}{k}\right)}$$

$$K_1 = \frac{\in}{\sqrt{A^2-1}}; \ K = \frac{\Omega_P}{\Omega_S}$$

and
$$\alpha_p = -\ 20\ \log\frac{1}{\sqrt{1+\epsilon^2}} = 10\ \log_{10} 1 + \epsilon^2$$

$$\epsilon = \sqrt{10^{0.1\alpha_p} - 1} = 0.882$$

$$\alpha_s = -\ 20\ \log_{10}\frac{1}{A} = 20\ \log_{10} A$$

$$A = 10^{0.05\alpha_s} = 10^{0.05\times30} = 31.6$$

$$\therefore\qquad K_1 = \frac{0.882}{\sqrt{(31.6)^2 - 1}} = 0.0279$$

$$K = \frac{\Omega_p}{\Omega_s} = \frac{20}{50} = 0.4$$

$$\therefore\qquad N = \frac{\cosh^{-1}\left(\dfrac{1}{0.0279}\right)}{\cosh^{-1}(1/0.4)} = 2.726$$

Round off $N = 3$

$\therefore$ The normalized transfer function is

$$H_a(s) = \frac{H_o}{\displaystyle\prod_{k=1}^{3}(S - P_k)}$$

where
$$P_k = \sinh(\alpha)\ \cos(\beta_k) + j\cosh(\alpha)\ \sin(\beta_k)$$

$$\alpha = \frac{1}{N}\sinh^{-1}\left(\frac{1}{\epsilon}\right) = 0.3243$$

$$\beta_k = \frac{(2k + N - 1)\pi}{2N},\quad k = 1, 2, 3$$

$$\beta_1 = \frac{(2 + 3 - 1)\pi}{6} = \frac{2}{3}\pi = 120^\circ$$

$$\beta_2 = \frac{(4 + 3 - 1)\pi}{6} = \pi = 180^\circ\ ;\ \beta_3 = \frac{(6 + 3 - 1)\pi}{6} = \frac{4}{3}\pi = 240^\circ$$

$$P_1 = \sinh(0.3243)\cos(120°) + j\cosh(0.3243)\sin(120°)$$

$$= 0.33(-0.5) + j(1.053)(0.866)$$

$$P_1 = -0.165 + j\,0.9119$$

$$P_2 = \sinh(0.3243)\cos(180°) + j\cosh(0.3243)\sin(180°)$$

$$= -0.33$$

$$P_3 = \sinh(0.3243)\cos(240°) + j\cosh(0.3243)\sin(240°)$$

$$= -0.165 - j0.9119$$

So

$$H_a(s) = \frac{H_o}{(S+0.33)\,(S+0.165+j\,0.9119)\,(S+0.165-j\,0.9119)}$$

$$= \frac{H_o}{(S+0.33)\left[(S+0.165)^2 + (0.9119)^2\right]}$$

$$H_a(s) = \frac{H_o}{(S+0.33)\,(S^2 + 0.33\,S + 0.8588)}$$

So here N is odd

$$H_o = \prod_{k=1}^{3}(-P_k) = (0.33)\,(0.165 - j\,0.9119)\,(0.165 + j\,0.9119)$$

$$= 0.28339$$

$\therefore$ Normalized transfer function

$$H_a(s) = \frac{0.28339}{(S+0.33)\,(S^2 + 0.33\,S + 0.8588)}$$

The unnormalized or transfer function with cutoff frequency Ω_p = 20 rad/sec can be obtained by replacing S with $\dfrac{S}{\Omega_p}$ in the above transfer function

$\therefore$

$$H_a(s) = \frac{0.28339}{\left(\dfrac{S}{20}+0.33\right)\left[\left(\dfrac{S}{20}\right)^2 + 0.33\dfrac{S}{20} + 0.8588\right]}$$

$$= \frac{0.28339(20)^3}{(S+6.6)(S^2+6.6\,S+343.52)}$$

$$= \frac{2267.12}{(S+6.6)(S^2+6.6\,S+343.52)}$$

4.4 Analog Spectral Transformations

So far we have discussed designing a lowpass filter for the given specifications. In this section we discuss the frequency transformations (spectral transformations) that can be used to design lowpass filters with different cutoff frequency, highpass filters, bandpass filters and band reject (bandstop) filters from a normalized ($\Omega_c = 1$ rad/sec) lowpass analog filter.

4.4.1   Lowpass to Lowpass Filter

Given a normalized lowpass filter, we need to design a lowpass filter with a different cutoff frequency Ω_c (or passband frequency Ω_p). This can be done by the transformation given in equation (4.4.1).

$$S \to \frac{S}{\Omega_C} \qquad\qquad .....(4.4.1)$$

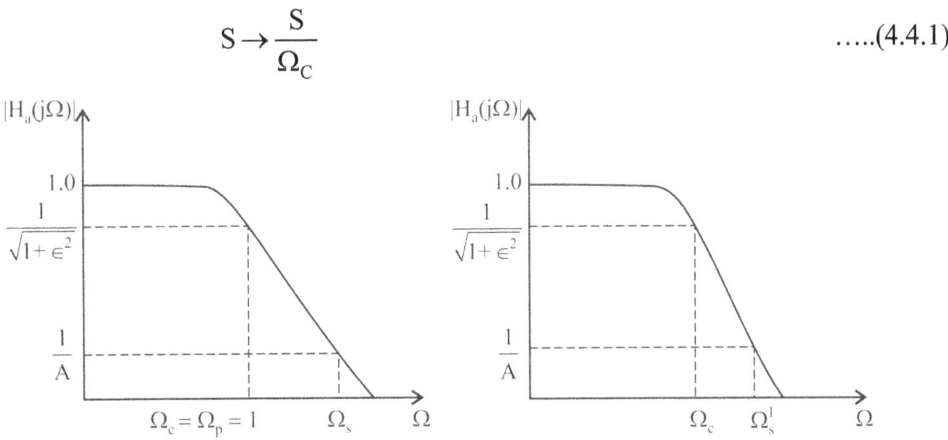

Fig. **4.11** Lowpass to Lowpass transformation.

4.4.2 Lowpass to Highpass Filter

Given a normalized lowpass filter, we need to design a highpass filter with cutoff frequency Ω_c. Then the transformation is given in equation (4.4.2).

$$S \rightarrow \frac{\Omega_C}{S}$$

.....(4.4.2)

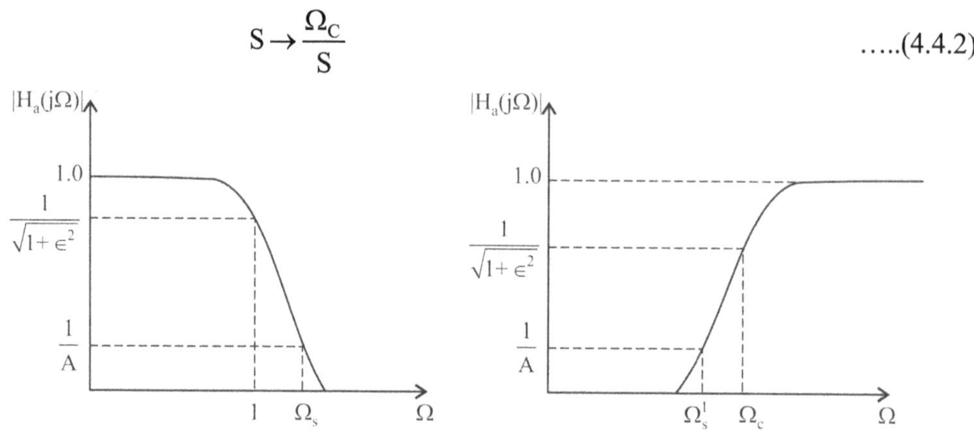

Fig. 4.12 Lowpass to Highpass transformation.

4.4.3 Lowpass to Bandpass Filter

Given a normalized lowpass filter, we need to design a bandpass filter with lower cutoff frequency Ω_l and upper cutoff frequency Ω_u. This can be done by equation (4.4.3).

$$S \rightarrow \frac{S^2 + \Omega_l \Omega_u}{S(\Omega_u - \Omega_l)}$$

.....(4.4.3)

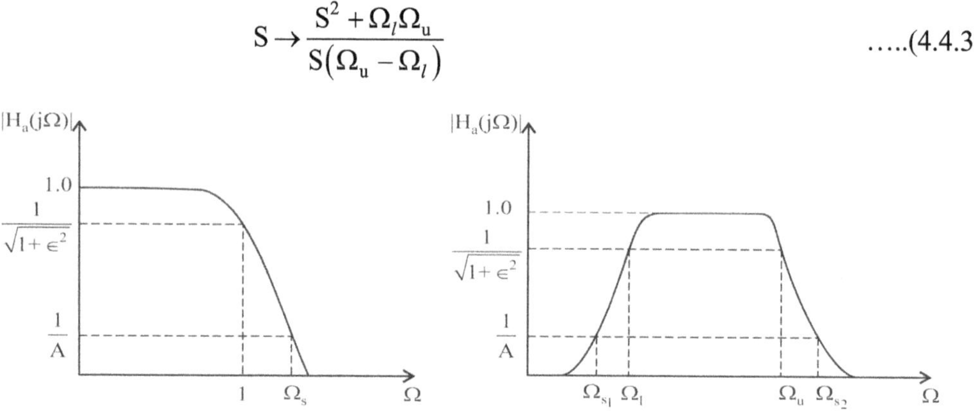

Fig. 4.13 Lowpass to Bandpass transformation.

The given bandpass filter specifications are to be converted into the specifications of normalized lowpass filter. In normalized lowpass filter $\Omega_p = 1$ rad/sec and Ω_s can be calculated by using equations 4.4.4 a, b, and c.

$$\Omega_S = \min\{|\Omega_1|, \ |\Omega_2|\}$$

.....(4.4.4a)

where
$$\Omega_1 = \frac{-\Omega_{s_1}^2 + \Omega_l\Omega_u}{\Omega_{s_1}\left(\Omega_u - \Omega_l\right)} \qquad \ldots..(4.4.4b)$$

$$\Omega_2 = \frac{\Omega_{s_2}^2 - \Omega_l\Omega_u}{\Omega_{s_2}\left(\Omega_u - \Omega_l\right)} \qquad \ldots..(4.4.4c)$$

Now design the normalized lowpass filter with the above specifications. Then transform to Bandpass filter with the transformation given in equation (4.4.3).

4.4.4 Lowpass to Bandstop Filter

The transformation to convert a normalized lowpass filter to a bandstop filter is

$$S \rightarrow \frac{S\left(\Omega_u - \Omega_l\right)}{S^2 + \Omega_l\Omega_u} \qquad \ldots..(4.4.5)$$

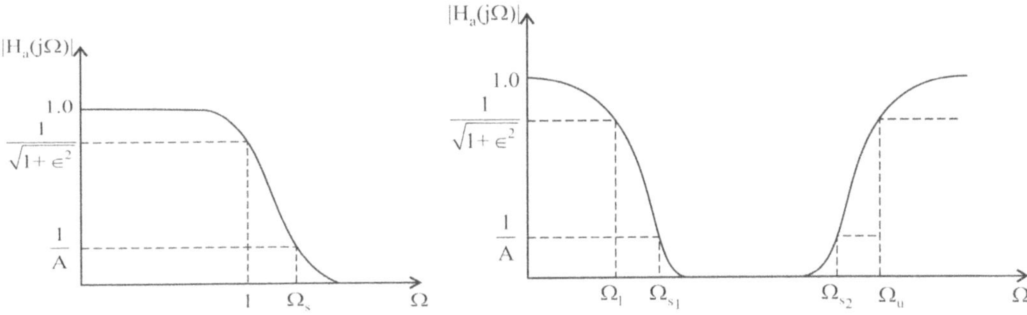

Fig. 4.14 Lowpass to Bandstop transformation.

The given bandstop filter specifications are to be converted into the specifications of lowpass filter. In normalized lowpass filter $\Omega_p = 1$ rad/sec and Ω_s can be calculated by using equations 4.4.6a, b and c.

$$\Omega_s = \min\left\{|\Omega_1|, \ |\Omega_2|\right\} \qquad \ldots..(4.4.6a)$$

where
$$\Omega_1 = \frac{\Omega_{s_1}\left(\Omega_u - \Omega_l\right)}{-\Omega_{s_1}^2 + \Omega_l\Omega_u} \qquad \ldots..(4.4.6b)$$

$$\Omega_2 = \frac{\Omega_{s_2}\left(\Omega_u - \Omega_l\right)}{\Omega_{s_2}^2 - \Omega_l\Omega_u} \qquad \ldots..(4.4.6c)$$

Now design the normalized lowpass filter with the above specifications. Then transform to Bandstop filter with the transformation given in equation (4.4.5).

4.5 Design of IIR Digital Filters from Analog Filters

So far we have discussed designing of IIR analog filters. In this section we discuss different techniques that describe converting an analog filter into digital filter. An effective conversion technique should have the following properties.

1. There should be direct relationship between analog (S-domain) and digital (Z-domain) domain such that the $j\Omega$-axis in the S-plane should map onto the unit circle in the Z-plane.

2. A stable analog filter must be converted to a stable digital filter by mapping the left-half of the S-plane into the inside of the unit circle in the Z-plane.

Following are the methods for digitizing the analog filter.

1. Impulse Invariance (or) Impulse response Invariance method.

2. Step Invariance method.

3. Bilinear Transformation method.

4.5.1 Impulse Invariant Transformation

The impulse response of the digital filter is obtained by uniformly sampling the impulse response of the analog filter.

i.e.,
$$h(n) = h_a(t)\big|_{t=nT} \qquad\qquad .....(4.5.1)$$

where $h_a(t)$ is the impulse response of the analog filter.

$h(n)$ is the impulse response of the digital filter.

T is the sampling period.

Let us consider the transfer function of the analog filter which contains distinct poles.

$$H_a(s) = \sum_{k=1}^{N} \frac{A_k}{S - P_k} \qquad\qquad .....(4.5.2)$$

The impulse response of the above eqn. is

$$h_a(t) = \sum_{k=1}^{N} A_k e^{P_k t} u_a(t) \qquad\qquad .....(4.5.3)$$

where $u_a(t)$ is the continuous-time step function.

$h(n)$ can be obtained by using Eqn. 4.57.

$$h(n) = h_a(nT) = \sum_{k=1}^{N} A_k e^{nP_k T} u_a(nT) \qquad\qquad .....(4.5.4)$$

The system function H(Z) of the digital filter is

$$H(z) = Z\{h(n)\} = \sum_{n=0}^{\infty} h(n) Z^{-n}$$

$$\therefore \quad H(z) = Z\left\{ \sum_{k=1}^{N} A_k e^{nP_k T} \right\}$$

$$= \sum_{k=1}^{N} A_k Z\{e^{nP_k T}\}$$

$$= \sum_{k=1}^{N} \frac{A_k}{1 - e^{P_k T} z^{-1}} \qquad \qquad .....(4.5.5)$$

Comparing expressions in eqns (4.5.2) and (4.5.5), we get

$$\frac{1}{S - P_k} \rightarrow \frac{1}{1 - e^{P_k T} z^{-1}} \quad \text{(or)} \qquad .....(4.5.6a)$$

$$\frac{1}{S - P_k} \rightarrow \frac{z}{z - e^{P_k T}} \qquad \qquad .....(4.5.6b)$$

Eqn. 4.5.6b shows that the analog pole at $S = P_k$ is mapped onto a digital pole at $z = e^{P_k T} = e^{ST}$

i.e., $\qquad \qquad z = e^{ST} \qquad \qquad .....(4.5.7)$

we know $S = \sigma + j\Omega$ and $z = re^{jw}$

$$\therefore \qquad re^{jw} = e^{(\sigma + j\Omega)T}$$

From the above eqn., it is clear that

$$r = e^{\sigma T} \qquad \qquad .....(4.5.8a)$$

and $\qquad \qquad w = \Omega T \qquad \qquad .....(4.5.8b)$

from eqn., 4.5.8a

if $\qquad \qquad \sigma < 0, \qquad \qquad$ r lies between 0 and 1 $\qquad .....(4.5.9a)$

$\qquad \qquad \qquad \sigma = 0, \qquad \qquad r = 1 \qquad \qquad .....(4.5.9b)$

$\qquad \qquad \qquad \sigma > 0, \qquad \qquad r > 1 \qquad \qquad .....(4.5.9c)$

Eqn. (4.5.9a) states that the poles of the left of half of the S-plane will be mapped into inside the unit circle in the Z-plane.

Thus the stable digital filer is obtained.

Eqn. (4.5.9b) states that the poles that lie on the $j\Omega$-axis in the S-plane will be mapped onto the unit circle in the Z-plane.

Eqn. (4.5.9c) states that the poles that lie on the right half of the S-plane will be mapped into the outside of the unit circle in the Z-plane.

Fig. 4.15 illustrates the mapping of S-plane to Z-plane.

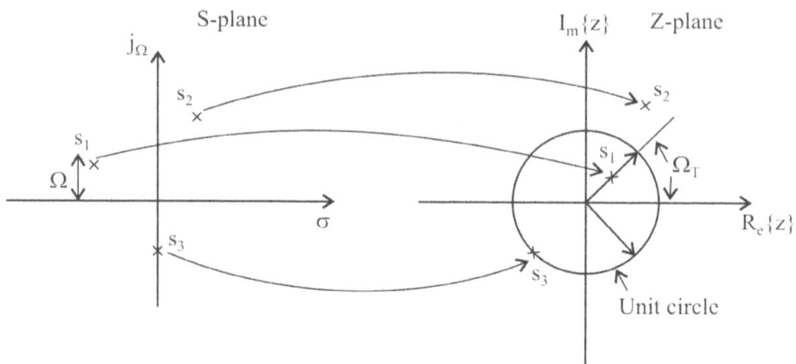

Fig. 4.15 Mapping of S-plane to Z-plane.

This mapping is not one-to-one rather it is many-to-one, where many points in the S-plane are mapped to a single point in the Z-plane. Let us explain this by considering two poles in the S-plane with identical real parts, but with imaginary parts differ by $\dfrac{2\pi}{T}$ as shown in fig. 4.16.

Let the poles are

$$S_1 = \sigma + j\Omega \qquad\qquad\qquad .....(4.5.10)$$

$$S_2 = \sigma + j\left(\Omega + \frac{2\pi}{T}\right)$$

These poles map to Z-plane poles z_1 and z_2

$$z_1 = e^{(\sigma+j\Omega)T} = e^{\sigma T}e^{j\Omega T} \qquad\qquad .....(4.5.11a)$$

$$z_2 = e^{\left(\sigma+j\left(\Omega+\frac{2\pi}{T}\right)\right)T} = e^{\sigma T}e^{j\Omega T}e^{j2\pi}$$

$$z_2 = e^{\sigma T}e^{j\Omega T} \qquad\qquad\qquad .....(4.5.11b)$$

$$\therefore \qquad e^{j2\pi} = 1$$

From Eqs. (4.5.11a) and (4.5.11b), it is clear that these poles map to the same location in the Z-plane. There are infinite number of S-plane poles (which have same real parts and imaginary parts differ by an integer multiple of $\dfrac{2\pi}{T}$) that map to the same location in the Z-plane.

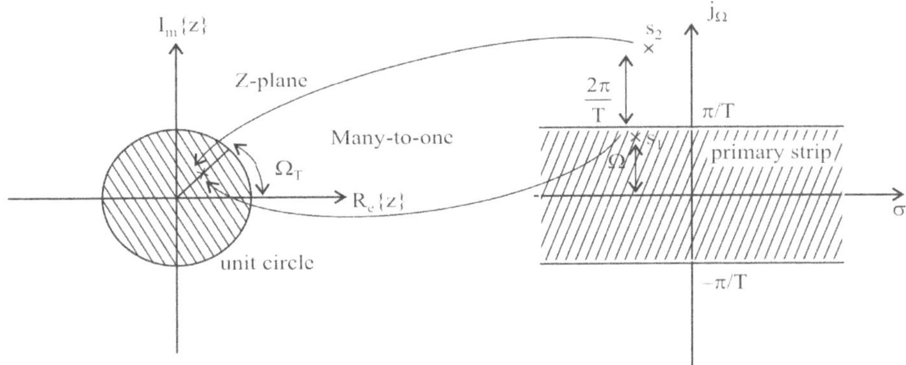

Fig. 4.16 Many-to-one mapping of impulse invariant method.

This is the biggest disadvantage of impulse invariant method. The S-plane poles having imaginary parts greater than $\dfrac{\pi}{T}$ or less than $-\dfrac{\pi}{T}$ cause aliasing, when sampling analog signals.

The analog poles will not be aliased by this method if they are confined to the S-plane's "primary strip" (within π/T) as shown in fig. 4.16.

Due to the presence of aliasing, this method is appropriate for the design of lowpass and bandpass filters only. This method is unsuccessful for designing highpass filters.

Some useful Impulse-Invariant Transformations are

$$\frac{1}{\left(S+p_k\right)^m} \rightarrow \frac{(-1)^{m-1}}{(m-1)!}\frac{d^{m-1}}{dp_k^{m-1}}\frac{1}{1-e^{-p_kT}z^{-1}} \qquad .....(4.5.12)$$

$$\frac{S+a}{\left(S+a\right)^2+b^2} \rightarrow \frac{1-e^{-aT}\left(\cos bT\right)z^{-1}}{1-2e^{-aT}\left(\cos bT\right)z^{-1}+e^{-2aT}z^{-2}} \qquad .....(4.5.13)$$

$$\frac{b}{\left(S+a\right)^2+b^2} \rightarrow \frac{e^{-aT}\left(\sin bT\right)z^{-1}}{1-2e^{-aT}\left(\cos bT\right)z^{-1}+e^{-2aT}z^{-2}} \qquad .....(4.5.14)$$

Example 4.6: Convert the following analog filter with transfer function

$$H_a\left(s\right)=\frac{S+0.2}{\left(S+0.2\right)^2+16}$$ into digital IIR filter using Impulse Response Invariance method.

[JNTU 2002]

Solution:

The given transfer function is of the form

$$H_a(s) = \frac{S+a}{(S+a)^2 + b^2}$$

where $\qquad a = 0.2$ and $b = 4$

Its Impulse Response Invariance transform is

$$H(z) = \frac{1 - e^{-aT}(\cos bT)z^{-1}}{1 - 2e^{-aT}(\cos bT)z^{-1} + e^{-2aT}z^{-2}}$$

Assume $\qquad T = 1$ sec,

$$\therefore \qquad H(z) = \frac{1 - e^{-0.2}(\cos 4)z^{-1}}{1 - 2e^{-0.2}(\cos 4)z^{-1} + e^{-0.4}z^{-2}}$$

$$= \frac{1 - e^{-0.2}\cos\left(4 \times \dfrac{180^\circ}{\pi}\right)z^{-1}}{1 - 2e^{-0.2}\cos\left(4 \times \dfrac{180^\circ}{\pi}\right)z^{-1} + e^{-0.4}z^{-2}}$$

$$= \frac{1 - (0.8187)(-0.654)z^{-1}}{1 - 2(0.8187)(-0.654)z^{-1} + 0.67z^{-2}}$$

$$H(z) = \frac{1 + 0.5\,35\,z^{-1}}{1 + 1.07\,z^{-1} + 0.67\,z^{-2}}$$

Example 4.7: Convert the following analog filter with transfer function $H(s) = \dfrac{1}{(s+1)(s+2)}$ into digital filter using impulse invariance method.

Solution: $\qquad H(s) = \dfrac{1}{(s+1)(s+2)} = \dfrac{c_1}{s+1} + \dfrac{c_2}{s+2}$

$$C_1 = (s+1)\dfrac{1}{(s+1)(s+2)}\bigg|_{s=-1} = 1$$

$$C_2 = (s+2)\frac{1}{(s+1)(s+2)}\bigg|_{s=-2} = -1$$

$$\therefore \qquad H(s) = \frac{1}{s+1} - \frac{1}{s+2}$$

Transfer function of digital filter is, assume T = 1 sec.

$$H(z) = \frac{1}{1-e^{-1}z^{-1}} - \frac{1}{1-e^{-2}z^{-1}}$$

$$= \frac{1-e^{-2}z^{-1} - 1 + e^{-1}z^{-1}}{1-e^{-2}z^{-1} - e^{-1}z^{-1} + e^{-3}z^{-2}}$$

$$= \frac{z^{-1}\left(e^{-1} - e^{-2}\right)}{1 - z^{-1}\left(e^{-2} + e^{-1}\right) + e^{-3}z^{-2}}$$

$$H(z) = \frac{0.2325z^{-1}}{1 - 0.5032z^{-1} + 0.0498z^{-2}}$$

Example 4.8: Convert the following H(s) into H(z) using impulse invariance method

$$H(s) = \frac{1}{(s+0.5)\left(s^2 + 0.5s + 2\right)}$$

Solution:
$$H(s) = \frac{1}{(s+0.5)\left(s^2 + 0.5s + 2\right)} = \frac{C_1}{s+0.5} + \frac{C_2 s + C_3}{s^2 + 0.5s + 2}$$

$$C_1 = (s+0.5)\frac{1}{(s+0.5)\left(s^2 + 0.5s + 2\right)}\bigg|_{s=-0.5} = 0.5$$

$$C_1\left(s^2 + 0.5s + 2\right) + (s+0.5)\left(C_2 s + C_3\right) = 1$$

Comparing s^2 coefficients on both sides

$$C_1 + C_2 = 0 \rightarrow C_2 = -0.5$$

Comparing s coefficient on both sides

$$0.5C_1 + 0.5C_2 + C_3 = 0$$

$$H(s) = \frac{0.5}{s+0.5} - \frac{0.5s}{s^2 + 0.5s + 2}$$

$$= \frac{0.5}{s+0.5} - 0.5 \left(\frac{s}{(s+0.25)^2 + (1.3919)^2} \right)$$

$$= \frac{0.5}{s+0.5} - 0.5 \left(\frac{s+0.25}{(s+0.25)^2 + (1.3919)^2} - \frac{0.25}{(s+0.25)^2 + (1.3919)^2} \right)$$

$$= \frac{0.5}{s+0.5} - 0.5 \left[\frac{s+0.25}{(s+0.25)^2 + (1.3919)^2} \right] + 0.0898 \left[\frac{1.3919}{(s+0.25)^2 + (1.3919)^2} \right]$$

Assume T = 1 sec

$$\therefore \quad H(z) = \frac{0.5}{1 - e^{-0.5}z^{-1}} - 0.5 \left[\frac{1 - e^{-0.25}\cos(1.3919)z^{-1}}{1 - 2\,e^{-0.25}\cos(1.3919)z^{-1} + e^{-0.5}z^{-2}} \right]$$

$$+ 0.0898 \left[\frac{e^{-0.25}(\sin(1.3919))z^{-1}}{1 - 2e^{-0.25}\cos(1.3919)z^{-1} + e^{-0.5}z^{-2}} \right]$$

$$H(z) = \frac{0.5}{1 - 0.6065z^{-1}} - 0.5 \left[\frac{1 - 0.1385z^{-1}}{1 + 0.277z^{-1} + 0.606z^{-2}} \right]$$

$$+ 0.0898 \left[\frac{0.7663z^{-1}}{1 - 0.277z^{-1} + 0.606z^{-2}} \right]$$

It gives

$$H(z) = \frac{0.30325z^{-1} + 0.2194z^{-2}}{1 - 0.8835z^{-1} + 0.774z^{-2} - 0.3675z^{-3}}$$

4.5.2 Step Invariance Transformation

The response of the digital filter to a sampled unit step function is the sampled version of the step response of the analog prototype. If the analog filter has good step response characteristics such as small rise time and low peak overshoot, these characteristics can be preserved in the digital filter.

Let u(t) is the continuous-time step signal, its corresponding response in the step response r(t) of analog filter.

Let u(n) is the discrete-time step signal, its corresponding step response is r(n) of digital filer.

We can obtain r(n) by sampling r(t)

i.e.,
$$r(n) = r(t)\big|_{t=nT} \qquad\qquad .....(4.5.15)$$

Laplace transform of u(t) is $\dfrac{1}{s}$ and let Laplace transform of r(t) as R(s).

Analog transfer function of the filter is

$$H_a(s) = \frac{L\{r(t)\}}{L\{u(t)\}}$$

$$= \frac{R(s)}{1/s}$$

$$R(s) = \frac{H_a(s)}{s} \qquad\qquad .....(4.5.16)$$

Digital transfer function of the filter is

$$H(z) = \frac{z\{r(n)\}}{z\{u(n)\}}$$

$$= \frac{R(z)}{\left(\dfrac{1}{1-z^{-1}}\right)}$$

$$R(z) = \frac{H(z)}{1-z^{-1}} \qquad\qquad .....(4.5.17)$$

Let
$$H_a(s) = \sum_{k=1}^{N} \frac{A_k}{s+P_k} \qquad\qquad .....(4.5.18)$$

From eqn. (4.5.16)

$$R(s) = \frac{1}{s} H_a(s) = \sum_{k=1}^{N} \frac{A_k}{s(s+P_k)}$$

$$R(s) = \sum_{k=1}^{N} \frac{A_k}{P_k}\left(\frac{1}{s} - \frac{1}{s+P_k}\right)$$

so
$$r(t) = L^{-1}\{R(s)\} = \sum_{k=1}^{N} \frac{A_k}{P_k}\left(1 - e^{-P_k t}\right)u(t)$$

from eqn. (4.5.15)

$$r(n) = r(t)\big|_{t=nT}$$

$$= \sum_{k=1}^{N} \frac{A_k}{P_k}\left(1 - e^{-P_k nT}\right)u(nT)$$

$$\therefore \qquad R(z) = z\{r(n)\} = \sum_{k=1}^{N} \frac{A_k}{P_k} z\left\{\left(1 - e^{-P_k nT}\right)u(n)\right\}$$

$$R(z) = \sum_{k=1}^{N} \frac{A_k}{P_k}\left[\frac{1}{1-z^{-1}} - \frac{1}{1-e^{-P_k T}z^{-1}}\right]$$

From eqn. (4.5.17)

$$H(z) = \left(1 - z^{-1}\right)R(z)$$

$$= \sum_{k=1}^{N} \frac{A_k}{P_k}\left[1 - \frac{1-z^{-1}}{1-e^{-P_k T}z^{-1}}\right]$$

$$H(z) = \sum_{k=1}^{N} \frac{A_k}{P_k} \frac{z^{-1}\left(1-e^{-P_k T}\right)}{1-e^{-P_k T}z^{-1}} \qquad\qquad .....(4.5.19)$$

Comparing eqn (4.5.18) and (4.5.19), Transformation is

$$\frac{1}{s+P_k} \to \frac{\left(1-e^{-P_k T}\right)z^{-1}}{P_k\left(1-e^{-P_k T}z^{-1}\right)} \qquad\qquad .....(4.5.20)$$

Similarly

$$\frac{1}{s-P_k} \to \frac{\left(e^{P_k T}-1\right)z^{-1}}{P_k\left(1-e^{P_k T}z^{-1}\right)} \qquad\qquad .....(4.5.21)$$

Example 4.9: Determine H(z) using the step invariant transformation for the analog system function

$$H_a(s) = \frac{1}{(s+1)(s+2)}$$

Solution:

$$H_a(s) = \frac{1}{(s+1)(s+2)} = \frac{C_1}{s+1} + \frac{C_2}{s+2}$$

$$C_1 = (s+1)\frac{1}{(s+1)(s+2)}\Bigg|_{s=-1} = 1$$

$$C_2 = (s+2)\frac{1}{(s+1)(s+2)}\Bigg|_{s=-2} = -1$$

$$H_a(s) = \frac{1}{s+1} - \frac{1}{s+2}$$

By applying step invariant transformation and assuming $T = 1$ sec

$$H(z) = \frac{(1-e^{-1})z^{-1}}{1(1-e^{-1}z^{-1})} - \frac{(1-e^{-2})z^{-1}}{2(1-e^{-2}z^{-1})}$$

$$= \frac{0.632\ z^{-1}}{1-0.368\ z^{-1}} - \frac{0.865\ z^{-1}}{2-0.271\ z^{-1}}$$

$$H(z) = \frac{0.399z^{-1} + 0.147z^{-2}}{2(1-0.503z^{-1}+0.048z^{-2})}$$

4.5.3 Bilinear Transformation (or) Tustin Transformation

The IIR digital filter design using Impulse invariant method is appropriate for designing low pass filter and bandpass filters. This transformation is not suitable for highpass or bandstop filters. This limitation may be resolved by a mapping technique called "Bilinear Transformation". This transformation is a one-to-one mapping from the S-domain to z-domain. That is, the Bilinear Transformation is a conformal mapping that transforms the $j\Omega$-axis onto the unit circle in the z-plane only once, thus avoiding aliasing of frequency components. It also gives stable digital filters by mapping poles of the left half of the s-plane into inside of the unit circle in the z-plane and right half poles of the s-plane into outside of the unit circle in the z-plane.

Consider the first order differential equation which represents the analog system.

$$\frac{dy(t)}{dt} = x(t) \qquad\qquad .....(4.5.22)$$

On integrating eqn. (4.5.22) both sides, we get

$$\int_{(n-1)T}^{nT} \frac{dy(t)}{dt} dt = \int_{(n-1)T}^{nT} x(t) dt \qquad .....(4.5.23)$$

The above integral can be approximated by the trapezoidal rule as given below

$$y(nT) - y((n-1)T) = \left(\frac{T}{2}\right)\left[x(nT) + x(n-1)T\right]$$

For discrete-time system, the above eqn can be written as

$$y(n) - y(n-1) = \left(\frac{T}{2}\right)(x(n) + x(n-1))$$

Convert the above equations into z-domain

$$Y(z) - z^{-1}Y(z) = \frac{T}{2}\left(X(z) + z^{-1}X(z)\right)$$

$$\left[1 - z^{-1}\right]Y(z) = \frac{T}{2}\left[1 + z^{1}\right]X(z)$$

$$\frac{2}{T}\left[\frac{1 - z^{-1}}{1 + z^{-1}}\right]Y(z) = X(z) \qquad .....(4.5.24)$$

Take Laplace transform of 4.78, we get

$$SY(s) = X(s) \qquad .....(4.5.25)$$

On comparing eqn (4.5.24) and (4.5.25), we get relation between S and Z.

$$S \rightarrow \frac{2}{T}\left(\frac{1 - z^{-1}}{1 + z^{-1}}\right) \qquad .....(4.5.26)$$

4.5.3.1 Relation between Analog and Digital Filter Poles

The relation between S and Z is

$$S = \frac{2}{T}\left(\frac{1 - z^{-1}}{1 + z^{-1}}\right)$$

Rearrange the above eqn.

$$\frac{T}{2}S = \frac{1 - Z^{-1}}{1 + Z^{-1}}$$

$$= \frac{z - 1}{z + 1}$$

$$\frac{T}{2}S(z+1) = z-1$$

$$\frac{T}{2}Sz + \frac{T}{2}S = z-1$$

$$\frac{T}{2}Sz - z = -1 - \frac{T}{2}S$$

$$-z\left(1 - \frac{T}{2}S\right) = -\left(1 + \frac{T}{2}S\right)$$

$$z = \frac{1 + \frac{T}{2}S}{1 - \frac{T}{2}S} \qquad\qquad .....(4.5.27)$$

Let $s = \sigma + j\Omega$

$$z = \frac{1 + \frac{T}{2}(\sigma + j\Omega)}{1 - \frac{T}{2}(\sigma + j\Omega)} = \frac{1 + \frac{T}{2}\sigma + j\frac{T}{2}\Omega}{1 - \frac{T}{2}\sigma + j\frac{T}{2}\Omega} \qquad .....(4.5.28)$$

Take magnitude

$$|z| = \left[\frac{\left(1 + \frac{T}{2}\sigma\right)^2 + \left(\frac{T}{2}\Omega\right)^2}{\left(1 - \frac{T}{2}\sigma\right)^2 + \left(\frac{T}{2}\Omega\right)^2}\right]^{1/2} \qquad .....(4.5.29)$$

From eqn. (4.5.29)

1. If $\sigma < 0 \Rightarrow |z| < 1$ i.e., poles which lie in the left half of the S-plane map to inside of the unit circle in the z-plane.

2. If $\sigma = 0 \Rightarrow |z| = 1$ i.e., poles which lie on the imaginary axis in the s-plane map onto the unit circle in the z-plane.

3. If $\sigma > 0 \Rightarrow |z| > 1$ i.e., poles which lie in the right half of the S-plane map to outside of the unit circle in the z-plane.

4.5.3.2 Relation between Analog and Digital Frequency

Put $S = j\Omega$ and $z = e^{j\omega}$ in eqn. (4.5.26)

$$\therefore \qquad j\Omega = \frac{2}{T}\frac{1 - e^{-j\omega}}{1 + e^{-j\omega}}$$

$$= \frac{2}{T} \frac{e^{j\omega/2}\left(e^{j\omega/2} - e^{-j\omega/2}\right)}{e^{-j\omega/2}\left(e^{j\omega/2} + e^{-j\omega/2}\right)}$$

$$j\Omega = \frac{2}{T} 2j \frac{\left(e^{j\omega/2} - e^{-j\omega/2}\right)}{2j} \bigg/ 2\left(\frac{e^{j\omega/2} + e^{-j\omega/2}}{2}\right)$$

$$\therefore \qquad \Omega = \frac{2}{T} \frac{\sin(\omega/2)}{\cos(\omega/2)}$$

$$\Omega = \frac{2}{T} \tan(\omega/2) \qquad\qquad\qquad .....(4.5.30a)$$

or

$$\omega = 2\tan^{-1}\left(\frac{\Omega T}{2}\right) \qquad\qquad\qquad .....(4.5.30b)$$

which is the relation between analog frequency (Ω) and digital frequency (ω), which is as shown in Fig.4.17.

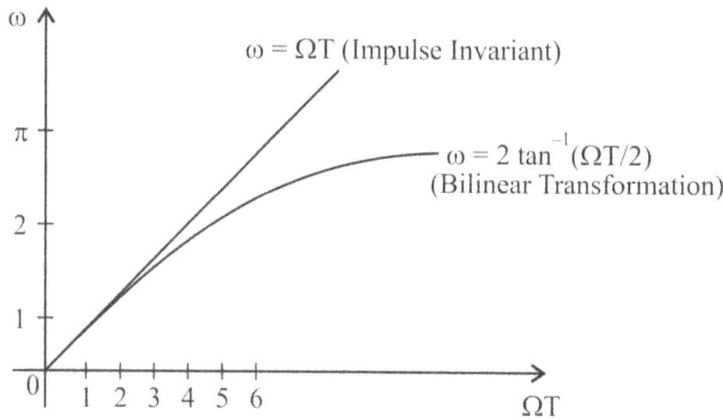

Fig. 4.17 Relation between Ω and ω.

4.5.3.3 Warping Effect

In Bilinear transformation relation between Ω and w is

$$\Omega = \frac{2}{T} \tan\left(\frac{\omega}{2}\right)$$

For small value of ω

$$\Omega = \frac{2}{T} \cdot \frac{\omega}{2} = \frac{\omega}{T}$$

$$\omega = \Omega T \qquad\qquad\qquad .....(4.5.31)$$

For low frequencies the relation between 'Ω' and 'ω' is linear, hence the digital filter will have the same magnitude response as that of analog filter.

For higher frequencies the relation between 'Ω' and 'ω' is non linear, hence the distortion is introduced in the frequency scale of the digital filter to that of the analog filter. This is known as the "Warping effect".

The warping effect can be explained by considering an analog filter with a number of passbands centered at regular intervals. The derived digital filter will have same number of passbands as that of analog filter, but the center frequencies and bandwidths of higher frequency passbands will reduce as shown in Fig. 4.18(a).

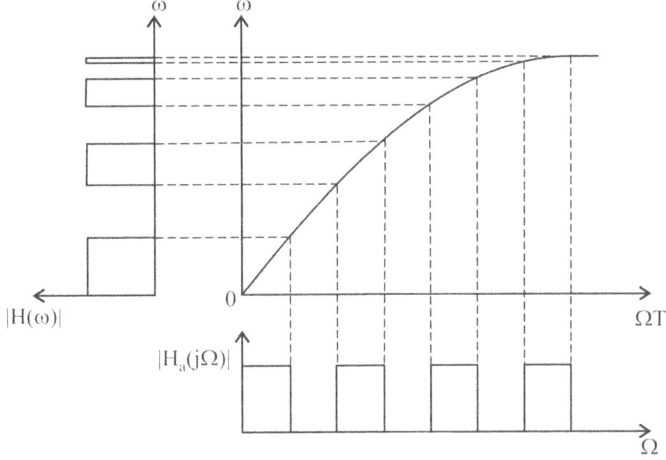

Fig. 4.18 (a) Warping effect on magnitude response.

Consider an analog filter with linear phase response, the phase response of the derived digital filter will be non-linear as shown in fig. 4.18(b).

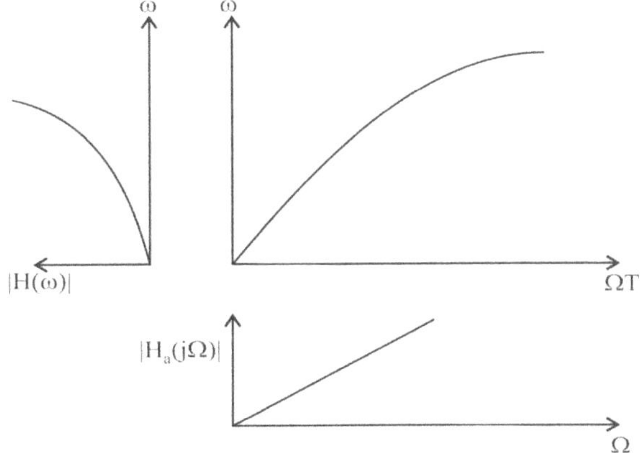

Fig. 4.18 (b) Warping effect on phase response.

The warping effect can be eliminated by prewarping the analog filter. This can be done by finding prewarping analog frequencies using the formula

$$\Omega = \frac{2}{T}\ \tan\left(\frac{\omega}{2}\right)$$

$$\therefore \qquad \Omega_P = \frac{2}{T}\ \tan\left(\frac{\omega_P}{2}\right)\ \text{and}\ \ \Omega_S = \frac{2}{T}\ \tan\left(\frac{\omega_S}{2}\right)$$

Example 4.10: Convert the following analog filter with transfer function

$$H_a(s) = \frac{s+0.2}{(s+0.2)^2 + 16}$$

into digital IIR filter using Bilinear Transformation Method   [JNTU 2002]

Solution: Assume T = 1 sec

Given $\qquad\qquad H_a(S) = \dfrac{s+0.2}{(s+0.2)^2 + 16}$

According to Bilinear Transformation

$$H(z) = H_A(s)\Big|_{S = \frac{2}{T}\left(\frac{1-Z^{-1}}{1+Z^{-1}}\right)}$$

$$\therefore \qquad H(z) = \frac{2\left(\dfrac{1-z^{-1}}{1+z^{-1}}\right)+0.2}{\left[2\left(\dfrac{1-z^{-1}}{1+z^{-1}}\right)+0.2\right]^2 + 16}$$

$$= \frac{\left(2-2z^{-1}+0.2+0.2z^{-1}\right)\left(1+z^{-1}\right)}{4\left(1-z^{-1}\right)^2 + 0.04\left(1+z^{-1}\right)^2 + 0.8\left(1-z^{-1}\right)\left(1+z^{-1}\right) + 16\left(1+z^{-1}\right)^2}$$

$$H(z) = \frac{2.2+0.4z^{-1}-1.8z^{-2}}{20.84+24.08z^{-1}+19.24z^{-2}}$$

4.6 Digital Spectral Transformation

A digital low pass filter can be converted into a digital low pass filter (with new pass band frequency), high pass, band pass and band stop filters. These transformations are given below

Lowpass to Lowpass:

$$z^{-1} \to \frac{z^{-1} - \alpha}{1 - \alpha z^{-1}} \qquad \qquad .....(4.6.1)$$

where $\qquad \alpha = \dfrac{\sin\left[\left(\omega_P - \omega_P^1\right)/2\right]}{\sin\left[\left(\omega_P + \omega_P^1\right)/2\right]} \qquad .....(4.6.2)$

ω_P = pass band frequency of low pass filter

ω_P^1 = Pass band frequency of new filter

Lowpass to Highpass:

$$z^{-1} \to -\left[\frac{z^{-1} + \alpha}{1 + \alpha z^{-1}}\right] \qquad \qquad .....(4.6.3)$$

where $\qquad \alpha = -\dfrac{\cos\left[\left(\omega_P^1 + \omega_P\right)/2\right]}{\cos\left[\left(\omega_P^1 - \omega_P\right)/2\right]} \qquad .....(4.6.4)$

ω_P = pass band frequency of low pass filter

ω_P^1 = Pass band frequency of high pass filter

Lowpass to Bandpass:

$$z^{-1} \to \frac{-\left[z^{-2} - \dfrac{2\alpha K}{1 + K}z^{-1} + \dfrac{K - 1}{K + 1}\right]}{\dfrac{K - 1}{K + 1}z^{-2} - \dfrac{2\alpha K}{K + 1}z^{-1} + 1} \qquad .....(4.6.5)$$

where $\qquad \alpha = \dfrac{\cos\left[\left(\omega_u + \omega_l\right)/2\right]}{\cos\left[\left(\omega_u - \omega_l\right)/2\right]} \qquad .....(4.6.6)$

$$K = \cot\left[\frac{\omega_u - \omega_l}{2}\right]\tan\left(\frac{\omega_P}{2}\right)$$

ω_u = upper cutoff frequency

ω_l = lower cutoff frequency

Lowpass to Bandstop

$$z^{-1} \rightarrow \frac{z^{-2} - \dfrac{2\alpha}{1+K}z^{-1} + \dfrac{1-K}{1+K}}{\dfrac{1-K}{1+K}z^{-2} - \dfrac{2\alpha}{1+K}z^{-1} + 1} \qquad \qquad .....(4.6.7)$$

where

$$\alpha = \frac{\cos\left[(\omega_u + \omega_l)/2\right]}{\cos\left[(\omega_u - \omega_l)/2\right]} \qquad \qquad .....(4.6.8)$$

$$K = \tan\left[(\omega_u - \omega_l)/2\right]\tan\left(\frac{\omega_P}{2}\right)$$

Either analog spectral transformation or digital spectral transformation can be used. But caution must be taken to which transformation to use. For example, the impulse invariant transformation is not suitable for high pass or band reject filters. In such a case, suppose a low pass prototype filter is converted into a high pass filter using "Analog spectral transformation" and transformed later to a digital filter using impulse invariant method, that results aliasing problems. However, if the same prototype low pass filter is first transformed into a digital filter using impulse invariant method and later converted into a high pass filter using "Digital Spectral Transformation" , that will not have any aliasing problem.

When we use bilinear transformation, both analog spectral transformation and digital spectral transformation give same result.

Example 4.11: Using the impulse invariant method obtain the transfer function, H(Z), of the digital filter, assuming a 3 dB cutoff frequency of 150 Hz and a sampling frequency of 1.28 KHz for the following normalized analog transfer function.

$$H(s) = \frac{1}{s^2 + \sqrt{2}\,s + 1}$$

Solution: Given $\qquad \Omega_C = 150$ Hz

$$\omega_C = 2\pi\,\Omega_C = 2\pi \times 150 \text{ rad/sec}$$

$$= 2\pi \times 150 \times \frac{1}{1.28 \times 10^3} \text{ rad}$$

$$= 0.736 \text{ rad}$$

Given transfer function is normalized one, Demoralize H(s) by replacing S with $\dfrac{s}{\omega_C}$

$$\therefore \quad H(s) = \frac{1}{\left(\dfrac{s}{0.736}\right)^2 + \sqrt{2}\left(\dfrac{s}{0.736}\right) + 1} = \frac{0.542}{s^2 + 1.04s + 0.542}$$

$$H(s) = \frac{0.542}{(s + 0.52 + j0.52)(s + 0.52 - j0.52)}$$

$$= \frac{C_1}{s + 0.52 + j0.52} + \frac{C_1^*}{s + 0.52 - j0.52}$$

$$C_1 = (s + 0.52 + j0.52)\frac{0.542}{(s + 0.52 + j0.52)(s + 0.52 - j0.52)}\bigg|_{s = -0.52 - j0.52}$$

$$= j0.52$$

$$\therefore \quad C_1^* = -j0.52$$

$$H(s) = \frac{j0.52}{s + 0.52 + j0.52} - \frac{j0.52}{s + 0.52 - j0.52}$$

by using impulse invariant method

$$H(z) = j0.52\frac{1}{1 - e^{(-0.52 - j0.52)T}z^{-1}} - j0.52\frac{1}{1 - e^{(-0.52 + j0.52)T}z^{-1}}$$

Assume $T = 1$ sec

$$H(z) = \frac{j0.52 - j0.52\, e^{-0.52 + j0.52}\, z^{-1} - j0.52 + e^{-0.52 - j0.52}\, z^{-1}j0.52}{1 - e^{-0.52 - j0.52}\, z^{-1} - e^{-0.52 + j0.52}\, z^{-1} + e^{-1.04}\, z^{-2}}$$

$$= \frac{0.52\, e^{-0.52}\left(2\sin(0.52)\right)z^{-1}}{1 - e^{-0.52}2\cos(0.52)z^{-1} + 0.353z^{-2}}$$

$$= \frac{0.307z^{-1}}{1 - 1.03z^{-1} + 0.353z^{-2}}$$

Example 4.12: It is required to design a digital low pass filter to approximate the following transfer function

$$H(s) = \frac{1}{s^2 + \sqrt{2}\, s + 1}$$

using Bilinear transform method obtain the transfer function, $H(z)$ of the digital filter, assuming a 3 dB cutoff frequency of 150 Hz and a sampling frequency of 1.28 KHz.

Solution: We known relation between Ω and Ω

$$\Omega = \frac{2}{T}\tan\left(\frac{\omega}{2}\right)$$

and relation between s and z is

$$s = \frac{2}{T}\left(\frac{1-z^{-1}}{1+z^{-1}}\right)$$

When we design digital filter using Bilinear transform method, we need to prewarp all the frequencies using $\Omega = \frac{2}{T}\tan\left(\frac{\omega}{2}\right)$ and then transform from S domain to Z domain using $s = \frac{2}{T}\left(\frac{1-z^{-1}}{1+z^{-1}}\right).$ As both formula contain $\frac{2}{T},$ it can be made 1 and we can use

$$\Omega = \tan\left(\frac{\omega}{2}\right) \text{ and } s = \frac{1-z^{-1}}{1+z^{-1}}$$

Given

$$\Omega_C = 150 \text{ Hz}$$

$$\omega_C = 2\pi \times 150 \times \frac{1}{1.28 \times 10^3} = 0.7363 \text{ rad}$$

$\therefore$ Prewarped 3 dB frequency is

$$\omega_C^1 = \tan\left(\frac{\omega_C}{2}\right) = 0.386 \text{ rad}$$

Given transfer function is normalized one,

Demoralized transfer function is

$$H^1(s) = H(s)\Big|_{s=\frac{s}{\omega_C^1}} = \frac{1}{\left(\dfrac{s}{0.386}\right)^2 + \sqrt{2}\left(\dfrac{s}{0.386}\right) + 1}$$

$$H^1(s) = \frac{0.149}{s^2 + 0.546s + 0.149}$$

By using Bilinear transform method

$$H(z) = H'(s)\bigg|_{s=\frac{1-z^{-1}}{1+z^{-1}}}$$

$$= \frac{0.149}{\left(\dfrac{1-z^{-1}}{1+z^{-1}}\right)^2 + 0.546\left(\dfrac{1-z^{-1}}{1+z^{-1}}\right) + 0.149}$$

$$= \frac{0.149 + 0.149\,z^{-2} + 0.298\,z^{-1}}{1+z^{-2}-2\,z^{-1}+0.546-0.546\,z^{-2}+0.149+0.149\,z^{-2}+0.298\,z^{-1}}$$

$$= \frac{0.149\left(1+2\,z^{-1}+z^{-2}\right)}{1.695-1.702\,z^{-1}+0.603\,z^{-2}}$$

$$= \frac{0.149\left(1+2\,z^{-1}+z^{-2}\right)}{1.695\left(1-1.007\,z^{-1}+0.356\,z^{-2}\right)}$$

$$H(z) = \frac{0.088\left(1+2\,z^{-1}+z^{-2}\right)}{1-1.007\,z^{-1}+0.356\,z^{-2}}$$

Example 4.13: The normalized transfer function of a simple, analog lowpass filter is given by

$$H(s) = \frac{1}{s+1}$$

Using Bilinear transform method, determine the transfer function of an equivalent discrete-time highpass filter. Assume a sampling frequency of 150 Hz and a cutoff frequency of 30 Hz.

Solution: Given $\Omega_C = 30$ Hz; sampling frequency $f_s = 150$ Hz, $T = \dfrac{1}{150}$ sec

Digital frequency $\omega_C = 2\pi \times 30 \times \dfrac{1}{150}$

$$= 1.257 \text{ rad}$$

Prewarped digital frequency $\omega_C^1 = \tan\left(\dfrac{\omega_C}{2}\right)$

$$= \tan\left(\frac{1.257}{2} \times \frac{180^\circ}{\pi}\right) = 0.7265$$

Using lowpass to highpass transformation, the denormalized analog transfer function is

$$H'(s) = H(s)\Big|_{s=\frac{W_c^1}{s}}$$

$$= \frac{1}{\dfrac{0.7265}{s}+1}$$

$$= \frac{s}{s+0.7265}$$

Using Bilinear transform method

$$H(z) = H'(s)\Big|_{s=\frac{1-z^{-1}}{1+z^{-1}}}$$

$$H(z) = \frac{\left(\dfrac{1-z^{-1}}{1+z^{-1}}\right)}{\left(\dfrac{1-z^{-1}}{1+z^{-1}}\right)+0.7265}$$

$$= \frac{1-z^{-1}}{1-z^{-1}+0.7265+0.7265\,z^{-1}}$$

$$= \frac{1-z^{-1}}{1.7265-0.2735\,z^{-1}}$$

$$H(z) = 0.5792\,\frac{1-z^{-1}}{1-0.1584\,z^{-1}}$$

Example 4.14: A discrete-time Bandpass filter with Butterworth characteristics meeting the specifications given below is required. Obtain H(Z) using Bilinear transform method:

Passband	: 200-300 Hz
Sampling frequency	: 2 kHz
Filter order, N	: 2

Solution: A first order, normalized analog lowpass filter is required (since the frequency band transformation for bandpass filters – equation 4.53 – will double the filter order).

Thus

$$H(s) = \frac{1}{s+1}$$

Given $\Omega_l = 200$ Hz and $\Omega_u = 300$ Hz, and $f_s = 2$ kHz

Digital frequencies are $\omega_l = 2\pi \times 200 \times \dfrac{1}{2 \times 10^3} = 0.628$ rad

$$\omega_u = 2\pi \times 300 \times \frac{1}{2 \times 10^3} = 0.943 \text{ rad}$$

$\therefore$ Prewarped digital frequencies are

$$\omega_l^1 = \tan\left(\frac{\omega_l}{2}\right) = \tan\left(\frac{0.628 \times 180°}{2 \times \pi}\right) = 0.325 \text{ rad}$$

$$\omega_u^1 = \tan\left(\frac{\omega_u}{2}\right) = \tan\left(\frac{0.943 \times 180°}{2 \times \pi}\right) = 0.5098 \text{ rad}$$

Using lowpass to bandpass transformation (equation 4.53)

$$H'(s) = H(s)\Bigg|_{s = \frac{s^2 + \omega_l^1 \omega_u^1}{s\left(\omega_u^1 - \omega_l^1\right)} = \frac{s^2 + 0.166}{s(0.185)}}$$

$\therefore$

$$H'(s) = \frac{1}{\dfrac{s^2 + 0.166}{s(0.185)}} + 1$$

$$= \frac{s(0.185)}{s^2 + 0.185\,S + 0.166}$$

Using Bilinear transformation method

$$H(z) = H'(s)\Bigg|_{s = \frac{1 - Z^{-1}}{1 + Z^{-1}}}$$

$$= \frac{\left(\dfrac{1 - z^{-1}}{1 + z^{-1}}\right) 0.185}{\left(\dfrac{1 - z^{-1}}{1 + z^{-1}}\right)^2 + 0.185\left(\dfrac{1 - z^{-1}}{1 + z^{-1}}\right) + 0.166}$$

$$= \frac{0.185\left(1 - z^{-2}\right)}{1 + z^{-2} - 2\,z^{-1} + 0.185 - 0.185\,z^{-2} + 0.166 + 0.166\,z^{-2} + 0.332\,z^{-1}}$$

$$H(z) = \frac{0.1369\left(1 - z^{-2}\right)}{1 - 1.235\,z^{-1} + 0.726\,z^{-2}}$$

Example 4.15: A lowpass digital filter meeting the following specifications is required

Passband	:	0-500 Hz
Stopband	:	2-4 kHz
Passband ripple	:	3 dB
Stopband attenuation	:	20 dB
Sampling frequency	:	8 kHz

Obtain H(Z) using

   (i)   Impulse Invariant method, Butterworth characteristics.

   (ii)   Impulse Invariant method, Chebyshev characteristics.

   (iii)   Bilinear Transformation method, Butterworth characteristics.

   (iv)   Bilinear Transformation method, Chebyshev characteristics.

Solution: Given

$$\alpha_p = 3 \text{ dB}$$
$$\alpha_s = 20 \text{ dB}$$
$$f_s = 8 \text{ kHz}$$
$$\Omega_p = 500 \text{ Hz}$$
$$\Omega_s = 2 \text{ kHz}$$

We know

$$\alpha_p = -20\log\frac{1}{\sqrt{1 + \epsilon^2}} = 10\log_{10}\left(1 + \epsilon^2\right)$$

$$3 = 10\log_{10}\left(1 + \epsilon^2\right)$$

$$\epsilon = 0.998$$

$$\alpha_s = -20\log_{10}\frac{1}{A}$$

$$20 = 20\log_{10} A$$

$$A = 10$$

(i)   Order of Butterworth filter is

$$N = \frac{\log_{10} K_1}{\log_{10} K} = \frac{\log_{10}\left(\dfrac{\in}{\sqrt{A^2-1}}\right)}{\log_{10}\left(\Omega_p/\Omega_s\right)}$$

$$N = 1.66$$

$\therefore$ Order $N = 2$

$\therefore$ Normalized Butterworth lowpass filter is

$$H(s) = \frac{1}{\displaystyle\prod_{k=1}^{2}\left(S - P_k\right)}$$

$$P_k = e^{j\pi(2+2k-1)/4}, \qquad\qquad k = 1, 2, 3\ldots$$

$$P_1 = \cos\left(\frac{\pi 3}{4}\right) + j\,\sin\left(\frac{3\pi}{4}\right) = -0.707 + j\,0.707$$

$$P_2 = \cos\left(\frac{5\pi}{4}\right) + j\,\sin\left(\frac{5\pi}{4}\right) = -0.707 - j\,0.707$$

$$H(s) = \frac{1}{\left(s + 0.707 + j\,0.707\right)\left(s + 0.707 - j\,0.707\right)}$$

$$H(s) = \frac{1}{s^2 + 1.414\,s + 1}$$

Cutoff frequency

$$\Omega_c = \frac{\Omega_p}{\in^{1/N}}$$

$$= \frac{500}{(0.998)^{1/2}} = 500 \text{ Hz}$$

Digital frequency $\omega_c = 2\pi \times 500 \times \dfrac{1}{8 \times 10^3} = 0.393 \text{ rad}$

$\therefore$ Denormalized (unnormalized) H(s) is

$$H'(s) = H(s)\Big|_{s = \frac{s}{\omega_c}}$$

$$H'(s) = \cfrac{1}{\left(\dfrac{s}{0.393}\right)^2 + 1.414\left(\dfrac{s}{0.393}\right) + 1}$$

$$= \frac{0.154}{s^2 + 0.55\,s + 0.154}$$

$$= \frac{0.154}{(s + 0.27 + j\,0.27)\,(s + 0.27 - j\,0.27)}$$

$$H'(s) = \frac{C_1}{s + 0.27 + j\,0.27} + \frac{C_1^*}{s + 0.27 - j\,0.27}$$

$$C_1 = (s + 0.27 + j\,0.27)\left.\frac{0.154}{(s + 0.27 + j\,0.27)(s + 0.27 - j\,0.27)}\right|_{s = -0.27 - j\,0.27}$$

$$C_1 = j\,0.285$$

$$C_1^* = -j\,0.285$$

$$H'(s) = \frac{j\,0.285}{s + 0.27 + j\,0.27} - \frac{j\,0.285}{s + 0.27 - j\,0.27}$$

Using Impulse Invariant method

$$H(z) = j\,0.285\,\frac{1}{1 - e^{(-0.27 - j\,0.27)T}\,z^{-1}} - j\,0.285\,\frac{1}{1 - e^{(-0.27 + j\,0.27)T}\,z^{-1}}$$

Assume $T = 1$ sec

$$H(z) = \frac{0.285\,e^{-0.285}\left(2\,\sin(0.285)\right)z^{-1}}{1 - e^{-0.285}\left(2\,\cos(0.285)\right)z^{-1} + e^{-0.57}\,z^{-2}}$$

$$H(z) = \frac{0.12\,z^{-1}}{1 - 1.44\,z^{-1} + 0.56\,z^{-2}}$$

(ii) Order of Chebyshev filter is

$$N = \frac{\cosh^{-1}\left(\dfrac{1}{k_1}\right)}{\cosh^{-1}\left(\dfrac{1}{k}\right)} = \frac{\cosh^{-1}\left(\dfrac{\sqrt{A^2 - 1}}{\in}\right)}{\cosh^{-1}\left(\dfrac{\Omega_S}{\Omega_P}\right)}$$

$$N = 1.45$$

$$\therefore \quad N = 2$$

$\therefore$ Normalized Chebyshev filter is

$$H(s) = \frac{H_0}{\prod\limits_{k=1}^{2}\left(S - P_K\right)}$$

$$P_K = \sinh(\alpha)\cos(\beta_K) + j\cosh(\alpha)\sin(\beta_K)$$

$$\alpha = \frac{1}{N}\sinh^{-1}\left(\frac{1}{\epsilon}\right) = \frac{1}{2}\sinh^{-1}\left(\frac{1}{0.998}\right) = 0.44$$

$$\beta_K = \frac{(2K + N - 1)\pi}{2N}, \quad k = 1, 2, \ldots N$$

$$\beta_1 = \frac{(2 + 2 - 1)\pi}{4} = \frac{3\pi}{4} = 135^\circ$$

$$\beta_2 = \frac{(4 + 2 - 1)\pi}{4} = \frac{5\pi}{4} = 225^\circ$$

$$P_1 = \sinh(0.44)\cos(135^\circ) + j\cosh(0.44)\sin(135^\circ)$$

$$P_1 = 0.454(-0.707) + j\,1.098(0.707) = -0.321 + j\,0.776$$

$$P_2 = 0.454\cos(225^\circ) + j\,1.098\sin(2.25^\circ)$$

$$P_2 = -0.321 - j\,0.776$$

$$H_o = \frac{1}{\sqrt{1 + \epsilon^2}}\prod_{k=1}^{N}(-p_k) \quad \text{for N even}$$

$$= \frac{1}{\sqrt{1 + (0.998)^2}}(-0.321 + j0.776)(-0.321 - j0.776)$$

$$= 0.5$$

$$\therefore \quad H(s) = \frac{0.5}{(s + 0.32 - j0.776)(s + 0.32 + j\,0.776)}$$

$$H(s) = \frac{0.5}{s^2 + 0.64s + 0.705}$$

Denormalized Chebyshev Lowpass filter is

$$H'(s) = H(s)\Big|_{s=\frac{s}{\omega_p}}$$

where $\qquad\qquad \Omega_p = 500 \text{ Hz}$

Digital frequency $\qquad \omega_p = 2\pi \times 500 \times \dfrac{1}{8 \times 10^3}$

$$= 0.393 \text{ rad}$$

$$H'(s) = \frac{0.5}{\left(\dfrac{s}{0.393}\right)^2 + 0.64\left(\dfrac{s}{0.393}\right) + 0.705}$$

$$= \frac{0.077}{s^2 + 0.252s + 0.11}$$

$$= \frac{0.077}{(s + 0.126 - j0.307)(s + 0.126 + j0.307)}$$

$$H'(s) = \frac{C_1}{s + 0.126 - j0.307} + \frac{C_1^*}{s + 0.126 + j0.307}$$

$$C_1 = (s + 0.126 - j0.307)\frac{0.077}{(s + 0.126 - j0.307)s + 0.126 + j\,0.307}\Bigg|_{s=-0.126+j0.307}$$

$$= -j\,0.125$$

$\therefore \qquad C_1^* = j0.125$

$\therefore \qquad H'(s) = \dfrac{j\,0.125}{s + 0.126 + j0.307} - \dfrac{j\,0.125}{s + 0.126 - j0.307}$

Using Impulse Invariant method

$$H(z) = j0.125\frac{1}{1 - e^{(-0.126-j0.307)T}z^{-1}} - j0.125\frac{1}{1 - e^{(-0.126+j0.307)T}z^{-1}}$$

Assume $T = 1$ sec

$$H(z) = \frac{0.125\,e^{-0.126}\big(2\sin(0.307)\big)z^{-1}}{1 - e^{-0.126}\big(2\cos(0.307)\big)z^{-1} + e^{-0.252}z^{-2}}$$

$$H(z) = \frac{0.067z^{-1}}{1 - 1.68z^{-1} + 0.78z^{-2}}$$

(iii) Analog frequencies $\Omega_p = 500$ Hz

$$\Omega_s = 2 \text{ kHz}$$

Digital frequencies $\omega_p = 2\pi \times 500 \times \dfrac{1}{8 \times 10^3} = 0.393 \, \text{rad}.$

$$\omega_s = 2\pi \times 2 \times 10^3 \times \frac{1}{8 \times 10^3} = 1.57 \, \text{rad}$$

Prewarped frequencies are

$$\omega_p^1 = \tan\left(\frac{\omega_p}{2}\right) = \tan\left(\frac{0.393 \times 180^\circ}{2\pi}\right) = 0.199 \, \text{rad}$$

$$\omega_s^1 = \tan\left(\frac{\omega_s}{2}\right) = \tan\left(\frac{1.57 \times 180^\circ}{2\pi}\right) = 0.999 \, \text{rad}$$

Order of Butterworth filter is

$$N = \frac{\log(k_1)}{\log_{10}(k)} = \frac{\log_{10}\dfrac{\in}{\sqrt{A^2 - 1}}}{\log_{10}\left(\omega'_p / \omega'_s\right)}$$

$$= 1.42$$

$\therefore \qquad\qquad N = 2$

Normalized Butterworth filter is

$$H(s) = \frac{1}{\displaystyle\prod_{k=1}^{2}(s - p_k)}$$

$$H(s) = \frac{1}{s^2 + 1.414s + 1}$$

Digital frequency $\omega_c = 0.393$ rad

Prewarped digital frequency $\omega_c^1 = \tan\left(\dfrac{0.393}{2}\right) = \tan\left(\dfrac{0.393 \times 180^\circ}{2\pi}\right) = 0.199 \, \text{rad}$

$\therefore$ Denormalized H(s) is

$$H'(s) = H(s)\Big|_{s=\frac{s}{\omega'_c}}$$

$$H'(s) = \frac{1}{\left(\dfrac{s}{0.199}\right)^2 + 1.414\left(\dfrac{s}{0.199}\right) + 1}$$

$$H'(s) = \frac{0.0396}{s^2 + 0.281s + 0.0396}$$

Using Bilinear Transform method

$$H(z) = H'(s)\Big|_{s=\frac{1-z^{-1}}{1+z^{-1}}}$$

$$H(z) = \frac{0.0396}{\left(\dfrac{1-z^{-1}}{1+z^{-1}}\right)^2 + 0.281\left(\dfrac{1-z^{-1}}{1+z^{-1}}\right) + 0.0396}$$

$$= \frac{0.0396\left(1 + z^{-2} + 2z^{-1}\right)}{1 + z^{-2} - 2z^{-1} + 0.281 - 0.281z^{-2} + 0.0396 + 0.0396z^{-2} + 0.0792z^{-1}}$$

$$H(z) = \frac{0.02998\left(1 + 2z^{-1} + z^{-2}\right)}{1 - 1.4545z^{-1} + 0.5744z^{-2}}$$

(iv) Order of Chebyshev filter is

$$N = \frac{\cosh^{-1}\left(\dfrac{1}{k_1}\right)}{\cosh^{-1}\left(\dfrac{1}{k}\right)} = \frac{\cosh^{-1}\dfrac{\sqrt{A^2 - 1}}{\in}}{\cosh^{-1}\left(w_s^1 / w_p^1\right)}$$

$$= \frac{\cosh^{-1}\left(\dfrac{\sqrt{10^2 - 1}}{0.998}\right)}{\cosh^{-1}\left(0.999/0.199\right)} = 1.3$$

$\therefore \qquad N = 2$

$\therefore$ Normalized Chebyshev filter is

$$H(s) = \frac{H_o}{\prod\limits_{k=1}^{2}(s - p_k)}$$

$\therefore \qquad H(s) = \dfrac{0.5}{s^2 + 0.64s + 0.705}$

Digital frequency $w_p = 0.393$ rad

Prewarped frequency $w_p^1 = \tan\left(\dfrac{0.393 \times 180^o}{2\pi}\right) = 0.199 \, \text{rad.}$

Denormalized Chebyshev Lowpass filter is

$$H'(s) = H(s)\Big|_{s=\frac{s}{w_p^1}}$$

$$H'(s) = \frac{0.5}{\left(\dfrac{s}{0.199}\right)^2 + 0.64\left(\dfrac{s}{0.199}\right) + 0.705}$$

$$H'(s) = \frac{0.019}{s^2 + 0.127s + 0.279}$$

Using Bilinear transformation method

$$H(z) = H'(s)\Big|_{s=\frac{1-z^{-1}}{1+z^{-1}}}$$

$$= \frac{0.019}{\left(\dfrac{1-z^{-1}}{1+z^{-1}}\right)^2 + 0.127\left(\dfrac{1-z^{-1}}{1+z^{-1}}\right) + 0.279}$$

$$= \frac{0.019\left(1 + z^{-2} + 2z^{-1}\right)}{1 + z^{-2} - 2z^{-1} + 0.127 - 0.127z^{-2} + 0.279 + 0.279z^{-2} + 0.558z^{-1}}$$

$$H(z) = \frac{0.0135\left(1 + 2z^{-1} + z^{-2}\right)}{1 - 1.026z^{-1} + 0.189z^{-2}}$$

Example 4.16: A high pass digital filter meeting the following specifications is required.

Passband	2-4 kHz
Stopband	0-500 Hz
Passband ripple	3 dB
Stopband attenuation	20 dB
Sampling frequency	8 kHz

Determine H(z) using Impulse invariant method, assume Butterworth filter characteristics.

Solution:

Given

$$\Omega_p = 2 \text{ kHz}$$
$$\Omega_s = 500 \text{ Hz}$$
$$\alpha_p = 3 \text{ dB} \rightarrow \in = 0.998$$
$$\alpha_s = 20 \text{ dB} \rightarrow A = 10$$

Digital frequencies of high pass filter are

$$\omega_{ph} = 2\pi \times 2 \times 10^3 \times \frac{1}{8 \times 10^3} = 1.57 \text{ rad}$$

$$\omega_{sh} = 2\pi \times 500 \times \frac{1}{8 \times 10^3} = 0.393 \text{ rad}$$

We need to convert highpass filter specifications into lowpass filter specifications.

Assume
$$\omega_{pl} = 1 \text{ rad}$$

We know L.P.F to H.P.F transformation

$$S = \frac{\omega_p}{S}$$

put $S = j\omega_{sl}$ in L.H.S and $S = j\omega_{sh}$ in R.H.S and $\omega_p = \omega_{ph}$ above equation

$$|\omega_{sl}| = \left|\frac{\omega_{ph}}{\omega_{sh}}\right| = \frac{1.57}{0.393} = 3.99 \approx 4 \text{ rad}$$

$$\therefore \qquad \omega_{sl} = 4 \text{rad}$$

$\therefore$ Lowpass filter specification are $\omega_{pl} = 1$ rad $\omega_{sl} = 4$ rad.

$\therefore$ Order of Butterworth Lowpass filter is

$$N = \frac{\log_{10} k_1}{\log_{10}(k)} = \frac{\log_{10}\left(\frac{\in}{\sqrt{A^2-1}}\right)}{\log_{10}\left(w_{pl}/w_{sl}\right)} = 1.65$$

$$\therefore \quad N = 2$$

$\therefore$ Normalized Butterworth Lowpass filter is

$$H(s) = \frac{1}{\prod\limits_{k=1}^{2}(S-P_k)}$$

where $P_k = e^{j\pi(2k+N-1)/2N}$, $\qquad k = 1, 2, 3,\ldots.$

$$H(s) = \frac{1}{s^2+1.414s+1} = \frac{1}{(s+0.707-j0.707)(s+0.707+j0.707)}$$

Since Impulse Invariant method is not applicable for highpass filters analog frequency transformation should not be used to transform from L.P to H.P. Find H(z) from H(s) using Impulse Invariant method, then use digital frequency transformation to transform from L.P to H.P.

$$H(s) = \frac{C_1}{s+0.707+j0.707} + \frac{C_1^*}{s+0.707-j0.707}$$

$$C_1 = (s+0.707+j\,0.707)\frac{1}{(s+0.707+j\,0.707)(s+0.707-j\,0.707)}\Bigg|_{s=-0.707-j0.707}$$

$$= \ j\,0.707$$

$$C_1^* = -j\,0.707$$

$$H(s) = \frac{j\,0.707}{s+0.707+j\,0.707} - \frac{j\,0.707}{s+0.707-j\,0.707}$$

$\therefore$ Normalized digital Butterworth Lowpass filter using Impulse Invariant method is

$$H(z)_L = j\,0.707\frac{1}{1-e^{(-0.707-j0.707)T}z^{-1}} - j\,0.707\frac{1}{1-e^{(-0.707+j0.707)T}z^{-1}}$$

Assume T = 1 sec.

$$H(z)_L = \frac{0.707e^{-0.707}\,2\sin(0.707)z^{-1}}{1-e^{-0.707}(2\cos(0.707))z^{-1}+e^{-1.414}z^{-2}}$$

$$H(z)_L = \frac{0.453z^{-1}}{1 - 0.75z^{-1} + 0.243z^{-2}}$$

$\therefore$ Digital Butterworth Highpass filter is

$$H(z)_H = H(z)_L \Big|_{z^{-1} \to -\frac{z^{-1} + \alpha}{1 + \alpha z^{-1}}}$$

where

$$\alpha = -\frac{\cos\left[(\omega_{ph} + \omega_{pl})/2\right]}{\cos\left[(\omega_{ph} - \omega_{pl})/2\right]}$$

$$\alpha = -\frac{\cos\left[(1.57 + 1)/2\right]}{\cos\left[(1.57 - 1)/2\right]}$$

$$\alpha = -0.294$$

$$\therefore \quad H(z)_H = \frac{0.453\left(-\dfrac{z^{-1} - 0.294}{1 - 0.294z^{-1}}\right)}{1 - 0.75\left(-\dfrac{z^{-1} - 0.294}{1 - 0.294z^{-1}}\right) + 0.243\left(-\dfrac{z^{-1} - 0.294}{1 - 0.294z^{-1}}\right)^2}$$

$$H(z)_H = \frac{0.166\left(1 - 3.694z^{-1} + z^{-2}\right)}{1 + 0.104z^{-1} + 0.135z^{-2}}$$

Example 4.17: Repeat example 4.16 using Bilinear transformation method and assume Butterworth filter characteristics.

Solution:

Digital frequencies of highpass filter are

$$\omega_{ph} = 2\pi \times 2 \times 10^3 \times \frac{1}{8 \times 10^3} = 1.57 \, \text{rad}$$

$$\omega_{sh} = 2\pi \times 500 \times \frac{1}{8 \times 10^3} = 0.393 \, \text{rad}$$

Prewarped frequencies are

$$\omega_{ph}^1 = \tan\left(\frac{1.57 \times 180^\circ}{2\pi}\right) = 0.999 \, \text{rad.}$$

$$\omega_{sh}^1 = \tan\left(\frac{0.393 \times 180^\circ}{2\pi}\right) = 0.199 \, \text{rad}$$

We need to convert highpass filter specifications into lowpass filter specifications.

Assume
$$\omega_{pl}^1 = 1 \, \text{rad}$$

and
$$\left|\omega_{sl}^1\right| = \left|\frac{\omega_{ph}^1}{\omega_{sh}^1}\right| = \frac{0.999}{0.199} = 5.018 \, \text{rad}$$

$\therefore$ Lowpass filter specifications are $\omega_{pl}^1 = 1 \, \text{rad}$ and $\omega_{sl}^1 = 5.018 \, \text{rad}$

$\therefore$ Order of Butterworth lowpass filter is

$$N = \frac{\log_{10}(k_1)}{\log_{10}(k)} = \frac{\log_{10}\left(\dfrac{\in}{\sqrt{A^2 - 1}}\right)}{\log\left(\omega_{pl}^1 / \omega_{sl}^1\right)}$$

$$= \log_{10}\left(\frac{0.998}{\sqrt{10^2 - 1}}\right) \Big/ \log_{10}\left(1/5.018\right)$$

$$= 1.426$$

$\therefore$
$$N = 2$$

$\therefore$ Normalized Butterworth Lowpass filter is

$$H(s) = \frac{1}{\displaystyle\prod_{k=1}^{2}(s - p_k)}$$

$$P_k = e^{j\pi(2k+N-1)/2N}, \quad k = 1, 2, 3, \ldots$$

$$H(s) = \frac{1}{s^2 + 1.414s + 1}$$

Since we are using Bilinear transformation, we can use analog frequency transformation.

$\therefore$ Denormalized Butterworth Highpass filter is

$$H'(s) = H(s)\Big|_{s = \frac{\omega_{ph}^1}{s}}$$

where
$$\omega_{ph}^1 = 0.999 = 1 \, \text{rad}.$$

$$\therefore \quad H'(s) = \frac{1}{\left(\dfrac{1}{s}\right)^2 + 1.414\left(\dfrac{1}{s}\right) + 1}$$

$$= \frac{s^2}{s^2 + 1.414s + 1}$$

Digital Butterworth Highpass filter can be obtained by using Bilinear transformation

$$H(z) = H'(s)\Big|_{s=\frac{1-z^{-1}}{1+z^{-1}}}$$

$$H(z) = \frac{\left(\dfrac{1-z^{-1}}{1+z^{-1}}\right)^2}{\left(\dfrac{1-z^{-1}}{1+z^{-1}}\right)^2 + 1.414\left(\dfrac{1-z^{-1}}{1+z^{-1}}\right) + 1}$$

$$H(Z) = \frac{1 + z^{-2} - 2\,z^{-1}}{1 + z^{-2} - 2\,z^{-1} + 1.414 - 1.414\,z^{-2} + 1 + z^{-2} + 2\,z^{-1}}$$

$$= \frac{1 + z^{-2} - 2\,z^{-1}}{3.414 + 0.586\,z^{-2}}$$

$$H(z) = \frac{0.2929\left(1 - 2\,z^{-1} + z^{-2}\right)}{1 + 0.1716\,z^{-2}}$$

Example 4.18: A requirement exists for a bandpass filter, with a Butterworth magnitude-frequency response, that satisfies the following specifications:

Lower passband edge frequency	200 Hz
Upper passband edge frequency	300 Hz
Lower stopband edge frequency	50 Hz
Upper stopband edge frequency	450 Hz
Passband ripple	3 dB
Stopband attenuation	20 dB
Sampling frequency	1 kHz

Determine the transfer function of the discrete-time filter using Bilinear transformation method.

Solution: Given

$$\Omega_l = 200 \text{ Hz} \qquad \alpha_p = 3 \text{ dB} \rightarrow \in = 0.998$$

$$\Omega_u = 300 \text{ Hz} \qquad \alpha_s = 20 \text{ dB} \rightarrow A = 10$$

$$\Omega_{S1} = 50 \text{ Hz} \qquad f_s = 1 \text{ kHz} \rightarrow T = \frac{1}{f_s} = \frac{1}{1k} \text{sec}$$

$$\Omega_{S2} = 450 \text{ Hz}$$

Digital frequencies are

$$\omega_l = 2\pi \times 200 \times \frac{1}{1 \times 10^3} = 1.25 \text{ rad}$$

$$\omega_u = 2\pi \times 300 \times \frac{1}{1 \times 10^3} = 1.885 \text{ rad}$$

$$\omega_{s1} = 2\pi \times 50 \times \frac{1}{1 \times 10^3} = 0.314 \text{ rad}$$

$$\omega_{s2} = 2\pi \times 450 \times \frac{1}{1 \times 10^3} = 2.83 \text{ rad}$$

Prewarped frequencies are

$$\omega_l^1 = \tan\left(\frac{1.25}{2}\right) = 0.72 \text{ rad}$$

$$\omega_u^1 = \tan\left(\frac{1.885}{2}\right) = 1.376 \text{ rad}$$

$$\omega_{s1}^1 = \tan\left(\frac{0.314}{2}\right) = 0.158 \text{ rad}$$

$$\omega_{s2}^1 = \tan\left(\frac{2.83}{2}\right) = 6.37 \text{ rad}$$

We need to convert Bandpass filter specifications into Lowpass filter specifications

Assume $\omega_p^1 = 1 \text{ rad}$

and

$$\omega_s^1 = \min\left\{\left|\omega_1^1\right|, \left|\omega_2^1\right|\right\}$$

$$\omega_1^1 = \frac{-\left(\omega_{s1}^1\right)^2 + \omega_l^1 \omega_u^1}{\omega_{s1}^1 \left(\omega_u^1 - \omega_l^1\right)}$$

$$= \frac{-(0.158)^2 + (0.72)(1.376)}{(0.158)(1.376 - 0.72)}$$

$$= 9.318 \text{ rad}$$

$$\omega_2^1 = \frac{-\left(\omega_{s2}^1\right)^2 + \omega_l^1 \omega_u^1}{\omega_{s2}^1 \left(\omega_u^1 - \omega_l^1\right)}$$

$$= \frac{-(6.37)^2 + (0.72)(1.376)}{6.37(1.376 - 0.72)}$$

$$= -9.473 \text{ rad}$$

$$\therefore \quad \omega_s^1 = \min\left\{|9.318|, |-9.473|\right\}$$

$$= 9.318 \text{ rad}$$

$\therefore$ Lowpass filter specifications are $\omega_p^1 = 1 \text{ rad}$ and $\omega_s^1 = 9.318 \text{ rad}$

Order of Butterworth lowpass filter is

$$N = \frac{\log_{10}\left(\dfrac{\in}{\sqrt{A^2 - 1}}\right)}{\log_{10}\left(\omega_p^1 / \omega_s^1\right)}$$

$$= \frac{\log_{10}\left(\dfrac{0.998}{\sqrt{10^2 - 1}}\right)}{\log_{10}(1/9.318)}$$

$$= 1.03$$

For simplicity we can take $N = 1$

$\therefore$ Normalized Butterworth Lowpass filter is

$$H(s) = \frac{1}{s - p_1}, \quad P_1 = e^{j\pi(2+1-1)/2} = -1$$

Using Lowpass to Bandpass analog frequency transformation, we get Bandpass filter transfer function

$$H'(s) = H(s)\Big|_{s = \frac{s^2 + \omega_l^1 \omega_u^1}{s\left(\omega_u^1 - \omega_l^1\right)}}$$

$$s = \frac{s^2 + (0.72)(1.376)}{s(1.376 - 0.72)} = \frac{s^2 + 1}{s(0.656)}$$

$$\therefore \quad H'(s) = \frac{1}{\dfrac{s^2 + 1}{s(0.656)} + 1}$$

$$= \frac{0.656\, s}{s^2 + 0.656\, s + 1}$$

By using Bilinear transformation, we obtain H(z) of bandpass filter.

$$H(z) = H'(s)\Big|_{s = \frac{1-z^{-1}}{1+z^{-1}}}$$

$$= \frac{0.656\left(\dfrac{1-z^{-1}}{1+z^{-1}}\right)}{\left(\dfrac{1-z^{-1}}{1+z^{-1}}\right)^2 + 0.656\left(\dfrac{1-z^{-1}}{1+z^{-1}}\right) + 1}$$

$$= \frac{0.656\left(1 - z^{-2}\right)}{1 + z^{-2} + 0.656 - 0.656\, z^{-2} - 2\, z^{-1} + 1 + z^{-2} + 2\, z^{-1}}$$

$$H(z) = \frac{0.246\left(1 - z^{-2}\right)}{1 + 0.506\, z^{-2}}$$

Example 4.19: A requirement exists for a Bandstop digital I I R filter, with a Butterworth magnitude-frequency response, that meets the following specifications.

Lower passband	0-50 Hz
Upper passband	450-500 Hz
Stopband	200-300 Hz
Passband ripple	3 dB
Stopband attenuation	20 dB
Sampling frequency	1 kHz

Determine the transfer function of Bandstop filter using Bilinear transformation method.

Solution: Given

$$\alpha_p = 3 \text{ dB} \rightarrow \in = 0.998$$

$$\alpha_s = 20 \text{ dB} \rightarrow A = 10$$

Digital frequencies are

$$\omega_l = 2\pi \times 50 \times \frac{1}{10^3} = 0.314 \text{ rad}$$

$$\omega_u = 2\pi \times 450 \times \frac{1}{10^3} = 2.827 \text{ rad}$$

$$\omega_{s1} = 2\pi \times 200 \times \frac{1}{10^3} = 1.2566 \text{ rad}$$

$$\omega_{s2} = 2\pi \times 300 \times \frac{1}{10^3} = 1.885 \text{ rad}$$

Prewarped frequencies are

$$\omega_l^1 = \tan\left(\frac{0.314}{2}\right) = 0.1583 \text{ rad}$$

$$\omega_u^1 = \tan\left(\frac{2.827}{2}\right) = 6.305 \text{ rad}$$

$$\omega_{s1}^1 = \tan\left(\frac{1.2566}{2}\right) = 0.7265 \text{ rad}$$

$$\omega_{s2}^1 = \tan\left(\frac{1.885}{2}\right) = 1.3765 \text{ rad}$$

We need to convert Bandstop filter specifications into Lowpass filter specifications.

Assume $\omega_p^1 = 1 \text{ rad}$

and

$$\omega_s^1 = \min\left\{\left|\omega_1^1\right|, \left|\omega_2^1\right|\right\}$$

$$\omega_1^1 = \frac{\omega_{s1}^1\left(\omega_u^1 - \omega_l^1\right)}{-\left(\omega_{s1}^1\right)^2 + \omega_l^1 \ \omega_u^1}$$

$$= \frac{(0.7265)(6.305 - 0.1583)}{-(0.7265)^2 + (0.1583)(6.305)}$$

$$= 9.495 \text{ rad}$$

$$\omega_2^1 = \frac{\omega_{s2}^1 \left(\omega_u^1 - \omega_l^1 \right)}{-\left(\omega_{s2}^1 \right)^2 + \omega_l^1 \, \omega_u^1}$$

$$= \frac{1.3765(6.305 - 0.1583)}{-(1.3765)^2 + (0.1583)(6.305)}$$

$$= -9.436 \text{ rad}$$

$$\omega_s^1 = \min \left\{ |9.495|, |-9.436| \right\}$$

$$= 9.436 \text{ rad}$$

$\therefore$ Lowpass filter specifications are $\omega_p^1 = 1 \text{ rad}$ and $\omega_s^1 = 9.436 \text{ rad}$

$\therefore$ Order of Butterworth Lowpass filter is

$$N = \frac{\log_{10} \left(\dfrac{\in}{\sqrt{A^2 - 1}} \right)}{\log_{10} \left(\omega_p^1 \middle| \omega_s^1 \right)}$$

$$= \frac{\log_{10} \left(\dfrac{0.998}{\sqrt{10^2 - 1}} \right)}{\log_{10} \left(1 \middle| 9.436 \right)}$$

$$= 1.02$$

For simplicity consider $N = 1$

$\therefore$ Normalized Butterworth Lowpass filter is

$$H(s) = \frac{1}{s + 1}$$

Butterworth Bandstop filter is

$$H'(s) = H(s) \Big|_{s = \frac{\left(\omega_u^1 - \omega_l^1 \right)}{s^2 + \omega_l^1 \, \omega_u^1}}$$

$$s = \frac{s\left(6.305 - 0.1583\right)}{s^2 + \left(0.1583\right)\left(6.305\right)}$$

$$= \frac{s\ 6.147}{s^2 + 1}$$

$$\therefore \qquad H'(s) = \frac{1}{\left(\dfrac{6.147\ s}{s^2 + 1}\right) + 1}$$

$$H'(s) = \frac{s^2 + 1}{s^2 + 6.147\ s + 1}$$

Digital Butterworth Bandstop filter is

$$H(z) = H'(s)\Big|_{s = \frac{1 - z^{-1}}{1 + z^{-1}}}$$

$$= \frac{\left(\dfrac{1 - z^{-1}}{1 + z^{-1}}\right)^2 + 1}{\left(\dfrac{1 - z^{-1}}{1 + z^{-1}}\right)^2 + 6.147\left(\dfrac{1 - z^{-1}}{1 + z^{-1}}\right) + 1}$$

$$H(z) = \frac{0.2454\ \left(1 + z^{-2}\right)}{1 - 0.509\ z^{-2}}$$

Example 4.20: Obtain the transfer function of a lowpass digital filter meeting the following specifications.

Passband	0-60 Hz
Stopband	> 85 Hz
Stopband attenuation	> 15 dB

Assume a sampling frequency 256 Hz and a Butterworth characteristics.

Solution: Digital frequencies $\omega_p = 2\pi \times 60 \times \dfrac{1}{256} = 1.473$ rad $= \omega_c$

which is 3 dB cutoff frequency

$$\therefore \qquad \alpha_p = 3 \text{ dB} \rightarrow \in = 0.998$$

$$\omega_s = 2\pi \times 85 \times \frac{1}{256} = 2.0862 \text{ rad}$$

$$\alpha_{\backslash s} = 15 \text{ dB}$$

$$15 = 20\log_{10} A \rightarrow A = 5.623$$

Prewarped frequencies are

$$\omega_c^1 = \omega_p^1 = \tan\left(\frac{1.473}{2}\right) = 0.9067 \text{ rad}$$

$$\omega_s^1 = \tan\left(\frac{2.0862}{2}\right) = 1.7158 \text{ rad}$$

$\therefore$ Order of Butterworth Lowpass filter is

$$N = \frac{\log_{10}\left(\dfrac{0.998}{\sqrt{(5.623)^2 - 1}}\right)}{\log_{10}\left(\dfrac{0.9067}{1.7158}\right)}$$

$$= 2.68$$

$\therefore \qquad N = 3$

Normalized Butterworth lowpass filter is

$$H(s) = \frac{1}{\prod\limits_{k=1}^{3}(S - P_k)}$$

$$P_1 = e^{j\pi(2+3-1)/6} = -0.5 + j\,0.866$$

$$P_2 = e^{j\pi(4+3-1)/6} = -1$$

$$P_3 = e^{j\pi(6+3-1)/6} = -0.5 - j\,0.866$$

$\therefore$

$$H(s) = \frac{1}{(s+1)(s+0.5+j\,0.866)(s+0.5-j\,0.866)}$$

$$= \frac{1}{(s+1)(s^2+s+1)}$$

Denormalized H(s) is

$$H'(s) = H(s)\Big|_{s=\frac{s}{\omega_c^1}}$$

$$= \frac{1}{\left[\left(\dfrac{s}{0.9067}\right)+1\right]\left[\left(\dfrac{s}{0.9067}\right)^2+\left(\dfrac{s}{0.9067}\right)+1\right]}$$

$$H'(s) = \frac{0.745}{(s+0.9067)(s^2+0.9067s+0.822)}$$

Digital Butterworth Lowpass filter is

$$H(z) = H'(s)\Big|_{s=\frac{1-z^{-1}}{1+z^{-1}}}$$

$$H(z) = \frac{0.745}{\left[\dfrac{1-z^{-1}}{1+z^{-1}}+0.9067\right]\left[\left(\dfrac{1-z^{-1}}{1+z^{-1}}\right)^2+0.9067\left(\dfrac{1-z^{-1}}{1+z^{-1}}\right)+0.822\right]}$$

$$H(z) = 0.143\,\frac{1+3z^{-1}+3z^{-2}+z^{-3}}{1-0.1801z^{-1}+0.342z^{-2}-0.0165z^{-3}}$$

Example 4.21: The specifications of the desired lowpass filter is

$$\frac{1}{\sqrt{2}} \le |H(\omega)| \le 1.0; \quad 0 \le \omega \le 0.2\pi$$

$$|H(\omega)| \le 0.08; \quad 0.4\pi \le \omega \le \pi$$

Design a Butterworth digital filter using Bilinear transformation.

Solution:

Given

$$\frac{1}{\sqrt{1+\epsilon^2}} = \frac{1}{\sqrt{2}} \qquad \text{and} \qquad \frac{1}{A} = 0.08$$

$$1+\epsilon^2 = 2 \qquad\qquad A = 12.5$$

$$\epsilon = 1$$

$$\omega_p = 0.2\pi \qquad \text{and} \qquad \omega_s = 0.4\pi$$

Since we are using Bilinear transformation we need to find prewarped frequencies.

Prewarped frequencies $\quad \omega_p^1 = \tan\left(\dfrac{0.2\pi}{2}\right) = 0.325\,\text{rad}$

$$\omega_s^1 = \tan\left(\frac{0.4\pi}{2}\right) = 0.7265\,\text{rad}$$

$\therefore$ Order of Butterworth filter is

$$N = \frac{\log_{10} k_1}{\log_{10} k} = \frac{\log_{10}\left(\frac{\in}{\sqrt{A^2-1}}\right)}{\log_{10}\left(w_p^1/w_s^1\right)} = \frac{\log_{10}\frac{1}{\sqrt{(12-5)^2-1}}}{\log_{10}\left(\frac{0.325}{0.7265}\right)} = 3.14$$

$\therefore \qquad\qquad N = 4$

$\therefore$ Normalized Butterworth Lowpass filter is

$$H(s) = \frac{1}{\prod\limits_{k=1}^{4}(s-p_k)}$$

$$P_1 = e^{j\pi(2+4-1)/8} = -0.383 + j0.924$$

$$P_2 = e^{j\pi(4+4-1)/8} = -0.924 + j0.383$$

$$P_3 = e^{j\pi(6+4-1)/8} = -0.924 - j0.383$$

$$P_4 = e^{j\pi(8+4-1)/8} = -0.383 - j0.924.$$

$$\therefore H(s) = \frac{1}{(s+0.385-j0.924)(s+0.383-j0.924)(s+0.924-j0.383)(s+0.924+j0.383)}$$

$$H(s) = \frac{1}{(s^2+0.766s+1)(s^2+1.848s+1)}$$

3 dB cutoff frequency (Prewarped)

$$\omega_c^1 = \frac{\omega_p^1}{(\in)^{1/N}} = \frac{0.325}{(1)^{1/4}} = 0.325\,\text{rad}.$$

$\therefore$ Denormalized Butterworth Lowpass filter is

$$H'(s) = H(s)\Big|_{s=\frac{s}{\omega_c^1}}$$

$$= \frac{1}{\left[\left(\frac{s}{0.325}\right)^2 + 0.766\left(\frac{s}{0.325}\right) + 1\right]\left[\left(\frac{s}{0.325}\right)^2 + 1.848\left(\frac{s}{0.325}\right) + 1\right]}$$

$$H'(s) = \frac{0.0112}{\left(s^2 + 0.249s + 0.106\right)\left(s^2 + 0.6s + 0.106\right)}$$

Digital Butterworth Lowpass filter can be obtained by using Bilinear Transformation.

$$H(z) = H'(s)\Big|_{s=\frac{1-z^{-1}}{1+z^{-1}}}$$

$$H(z) = \frac{0.0112}{\left[\left(\frac{1-z^{-1}}{1+z^{-1}}\right)^2 + 0.249\left(\frac{1-z^{-1}}{1+z^{-1}}\right) + 0.106\right]\left[\left(\frac{1-z^{-1}}{1+z^{-1}}\right)^2 + 0.6\left(\frac{1-z^{-1}}{1+z^{-1}}\right) + 0.106\right]}$$

$$H(z) = \frac{0.0112\left(1+z^{-1}\right)^2\left(1+z^{-1}\right)^2}{\left[\left(1-z^{-1}\right)^2 + 0.249\left(1-z^{-1}\right)\left(1+z^{-1}\right) + 0.106\left(1+z^{-1}\right)^2\right]\left[\left(1-z^{-1}\right)^2 + 0.6\left(1-z^{-1}\right)\left(1+z^{-1}\right) + 0.106\left(1+z^{-1}\right)^2\right]}$$

$$H(z) = \frac{0.0112\left(1 + 4z^{-1} + 6z^{-2} + 4z^{-3} + z^{-4}\right)}{1.355\left(1 - 1.32z^{-1} + 0.633z^{-2}\right)1.706\left(1 - 1.048z^{-1} + 0.297z^{-2}\right)}$$

$$H(z) = \frac{0.005\left(1 + 4z^{-1} + 6z^{-2} + 4z^{-3} + z^{-4}\right)}{\left(1 - 1.32z^{-1} + 0.633z^{-2}\right)\left(1 - 1.048z^{-1} + 0.297z^{-2}\right)}$$

Example 4.22: The specifications of the desired lowpass digital filter is

$$0.9 \le \left|H(\omega)\right| \le 1.0; \quad 0 \le \omega \le 0.25\pi$$

$$\left|H(\omega)\right| \le 0.24; \qquad 0.5\pi \le \omega \le \pi$$

Design a Chebyshev digital filter using Bilinear transformation method.

Solution: Given $\dfrac{1}{\sqrt{1+\epsilon^2}} = 0.9$ and $\dfrac{1}{A} = 0.24$

$$\epsilon = 0.484 \qquad\qquad\qquad A = 4.167$$

Digital frequencies are

$$\omega_p = 0.25\pi$$

$$\omega_s = 0.5\pi$$

Since we are using Bilinear transformation we need to find prewarped frequencies prewarped frequencies are

$$\omega_P^1 = \tan\left(\frac{0.25\pi}{2}\right) = 0.414 \,\text{rad.}$$

$$\omega_s^1 = \tan\left(\frac{0.5\pi}{2}\right) = 1 \,\text{rad.}$$

Order of Chebyshev filter is

$$N = \frac{\cosh^{-1}\left(\dfrac{\sqrt{A^2-1}}{\in}\right)}{\cosh^{-1}\left(\omega_s^1/\omega_p^1\right)}$$

$$= \frac{\cosh^{-1}\left(\dfrac{\sqrt{(4.167)}-1}{0.484}\right)}{\cosh^{-1}\left(1/0.414\right)}$$

$$= 1.84$$

$$\therefore \qquad\qquad N = 2.$$

$\therefore$ Normalized Chebyshev Lowpass filter is

$$H(s) = \frac{H_o}{\prod\limits_{k=1}^{2}(s-p_k)}$$

$$p_k = \sinh(\alpha)\cos(\beta_k) + j\cosh(\alpha)\sin(\beta_k)$$

$$\alpha = \frac{1}{N}\sinh^{-1}\left(\frac{1}{\in}\right) = \frac{1}{2}\sinh^{-1}\left(\frac{1}{0.484}\right) = 0.7364.$$

$$\sinh(0.7364) = 0.81; \quad \cosh(0.7364) = 1.284$$

$$\beta_k = \frac{(2k+N-1)\pi}{2N}, \quad k = 1, 2. \text{ and } N=2$$

$$\beta_1 = \frac{(2+2-1)\pi}{4} = 135^\circ; \quad \beta_2 = \frac{(4+2-1)\pi}{4} = 225^\circ$$

$$\therefore \quad P_1 = 0.81 \cos(135^\circ) + j\,(1.284)\sin(135^\circ)$$

$$= -0.573 + j\,0.91$$

$$P_2 = 0.81 \cos(225^\circ) + j\,(1.284)\sin(225^\circ)$$

$$= -0.573 - j\,0.91$$

$$H_o = \frac{1}{\sqrt{1+\epsilon^2}} \prod_{k=1}^{2}(-p_k) = 0.9(0.573 - j0.91)(0.573 + j0.91)$$

$$= 1.04$$

$$\therefore \quad H(s) = \frac{1.04}{(s+0.573-j0.91)(s+0.573+j0.91)}$$

$$H(s) = \frac{1.04}{s^2 + 1.146s + 1.156}$$

Denormalized Chebyshev Lowpass filter is

$$H'(s) = H(s)\Big|_{s=\frac{s}{\omega_p}}$$

$$= \frac{1.04}{\left(\dfrac{s}{0.414}\right)^2 + 1.146\left(\dfrac{s}{0.414}\right) + 1.156}$$

$$H'(s) = \frac{0.178}{s^2 + 0.474s + 0.198}$$

Digital Chebyshev Lowpass filter is

$$H(z) = H'(s)\Big|_{s=\frac{1-z^{-1}}{1+z^{-1}}}$$

$$H(z) = \frac{0.178}{\left(\dfrac{1-z^{-1}}{1+z^{-1}}\right)^2 + 0.474\left(\dfrac{1-z^{-1}}{1+z^{-1}}\right) + 0.198}$$

$$H(z) = \frac{0.1065\left(1 + 2z^{-1} + z^{-2}\right)}{1 - 0.959z^{-1} + 0.435z^{-2}}$$

Example 4.23: Design a IIR highpass filter with the following specifications with maximally flat approximation:

$$A_p = 0.5 \text{ dB}; A_a = 12 \text{ dB}; \omega_p = 2 \text{ r/s}; \omega_a = 6 \text{ r/s,} \qquad \text{[JNTU: 2002]}$$

Solution: Given specifications are

Passband edge frequency	2 r/s
Stopband edge frequency	6 r/s = $(2\pi - 6) = 0.283$ r/s
Passband attenuation	0.5 dB
Stopband attenuation	12 dB

Assume T = 1 sec and Bilinear Transformation.

maximally flat approximation indicates, it is Butterworth approximation.

$$\therefore \qquad \alpha_p = 0.5 \text{ dB}$$
$$\alpha_S = 12 \text{ dB}$$

$$0.5 = -20\log_{10} \frac{1}{\sqrt{1+\epsilon^2}} = 10\log_{10}\left(1+\epsilon^2\right)$$

$$\epsilon = 0.349$$

$$12 = -20\log_{10}\frac{1}{A} = 20\log_{10} A$$

$$A = 3.98$$

Digital frequencies are

$$\omega_{SH} = (6 \times 1 = 6 \text{ rad}) = 0.283 \text{ rad}$$
$$\omega_{PH} = 2 \times 1 = 2 \text{ rad}$$

Since we are using Bilinear transform, we need to find prewarped frequencies. Prewarped frequencies are

$$\omega_{SH}^1 = \tan\left(\frac{0.283}{2}\right) = 0.143 \text{ rad}$$

$$\omega_{PH}^1 = \tan\left(\frac{2}{2}\right) = 1.557 \text{ rad}$$

Given specifications are for highpass filter, we need to convert them to Lowpass filter specifications.

Assume $\qquad \omega_{PL}^1 = 1 \text{ rad}$

$$\therefore \qquad \omega^1_{SL} = \omega^1_{PH} \Big/ \omega^1_{SH} = \frac{1.557}{0.143} = 10.88 \text{ rad}$$

Order of Butterworth Lowpass filter is

$$N = \frac{\log_{10}\left(\dfrac{\in}{\sqrt{A^2-1}}\right)}{\log_{10}\omega^1_{PL}\big/\omega^1_{SL}} = \frac{\log_{10}\left(\dfrac{0.349}{\sqrt{(3.98)^2-1}}\right)}{\log_{10}\left(\dfrac{1.557}{10.88}\right)} = 1.006$$

For simplicity consider $N = 1$

$\therefore$ Normalized Butterworth Lowpass filter is

$$H(s) = \frac{1}{s+1}$$

Butterworth highpass filter can obtained by replacing s with $\dfrac{\omega'_{PH}}{s}$ in the above H(s).

$$\therefore \qquad H'(s) = H(s)\Big|_{s=\frac{\omega'_{PH}}{s}}$$

$$H'(s) = \frac{1}{\dfrac{1.557}{s}+1} = \frac{s}{s+1.557}$$

Digital Highpass (Butterworth) filter can be obtained by using Bilinear transformation.

$$H(z) = H'(s)\Big|_{s=\frac{1-z^{-1}}{1+z^{-1}}}$$

$$H(z) = \left(\frac{1-z^{-1}}{1+z^{-1}}\right)\Big/\left(\frac{1-z^{-1}}{1+z^{-1}}\right)+1.557$$

$$= \frac{1-z^{-1}}{1-z^{-1}+1.557+1.557\ z^{-1}}$$

$$H(z) = \frac{1-z^{-1}}{2.557\left(1+0.218\ z^{-1}\right)} = 0.39\frac{\left(1-z^{-1}\right)}{1+0.218\ z^{-1}}$$

Example 4.24: A Lowpass filter with the following specifications is required:

(i)   Frequency response within 3 dB : 0 to 1000 Hz

(ii)   Attenuation at 2000 Hz : ≥ 15 dB

(iii) Sampling frequency : 10 kHz

Obtain H(Z) of the corresponding I I R digital filter using Butterworth approximation and Bilinear transformation technique. [JNTU: 2001]

Solution: Given

$$\alpha_p = 3 \text{ dB} \qquad\qquad \Omega_p = 1000 \text{ Hz}$$

$$\alpha_s = 15 \text{ dB} \qquad\qquad \Omega_s = 2000 \text{ Hz}$$

$$f_s = 10 \text{ kHz}$$

$$3 = 10\log_{10}\left(1 + \epsilon^2\right) \qquad 15 = 20\log_{10}A$$

$$\epsilon = 0.997 \qquad\qquad A = 5.623$$

Digital frequencies are

$$\omega_p = 2\pi \times 1000 \times \frac{1}{10 \times 10^3} = 0.628 \text{ rad}$$

$$\omega_s = 2\pi \times 2000 \times \frac{1}{10 \times 10^3} = 1.257 \text{ rad}$$

Since we are using Bilinear transformation, we need to find prewarped frequencies

$\therefore$ Prewarp w_p and w_s by using $\omega' = \tan\left(\dfrac{\omega}{2}\right)$

$\therefore$
$$\omega_p^1 = \tan\left(\frac{\omega_p}{2}\right) = \tan\left(\frac{0.628}{2}\right) = 0.325 \text{ rad}$$

$$\omega_s^1 = \tan\left(\frac{\omega_s}{2}\right) = \tan\left(\frac{1.257}{2}\right) = 0.727 \text{ rad}$$

$\therefore$ Order of Butterworth lowpass filter is

$$N = \log_{10}\left(\frac{\epsilon}{\sqrt{A^2 - 1}}\right) \Bigg/ \log_{10}\left(\frac{\omega_p^1}{\omega_s^1}\right) = 2.13$$

$\therefore$ $\qquad\qquad N = 3$

Normalized Butterworth Lowpass filter is

$$H(s) = \frac{1}{\prod_{k=1}^{3}\left(S - P_k\right)}$$

$$P_k = e^{j\pi(2K+N-1)/2N}, \quad k = 1, 2, 3, \ldots N$$

$$P_1 = e^{j\pi(2+3-1)/6} = -0.5 + j\,0.866$$

$$P_2 = e^{j\pi(4+3-1)/6} = -1$$

$$P_3 = e^{j\pi(6+3-1)/6} = -0.5 - j\,0.866$$

$$H(s) = \frac{1}{(s+1)\left(s^2 + s + 1\right)}$$

Prewarped cutoff frequency $\omega_c^1 = \dfrac{\omega_p^1}{(\in)^{1/N}} = \dfrac{0.325}{(0.997)^{1/3}} = 0.325$ rad

Denormalized Lowpass filter is

$$H'(s) = H(s)\Big|_{s = \frac{s}{\omega_c^1}}$$

$$= \frac{1}{\left[\left(\dfrac{s}{0.325}\right)+1\right]\left[\left(\dfrac{s}{0.325}\right)^2 + \left(\dfrac{s}{0.325}\right)+1\right]}$$

$$H'(s) = \frac{0.0343}{(s+0.325)\left(s^2 + 0.325\,s + 0.106\right)}$$

Digital Butterworth Lowpass filter is

$$H(Z) = H'(s)\Big|_{s = \frac{1-z^{-1}}{1+z^{-1}}}$$

$$= \frac{0.0343}{\left[\left(\dfrac{1-Z^{-1}}{1+Z^{-1}}\right)+0.325\right]\left[\left(\dfrac{1-Z^{-1}}{1+Z^{-1}}\right)^2 + 0.325\left(\dfrac{1-Z^{-1}}{1+Z^{-1}}\right)+0.106\right]}$$

$$= \frac{0.0343\left(1+Z^{-1}\right)^3}{1.325\left(1-0.509\ Z^{-1}\right)\ 1.431\ \left(1-1.25\ Z^{-1}+0.546\ Z^{-2}\right)}$$

$$= \frac{0.018\left(1+Z^{-1}\right)^3}{\left(1-0.509\ Z^{-1}\right)\ \left(1-1.25\ Z^{-1}+0.546\ Z^{-2}\right)}$$

Example 4.25: An analogue Chebyshev filter has the following specifications:

   (i)   cut-off frequency = 500 Hz

   (ii)   order of the filter = 2

   (iii) sampling frequency = 10 kHz

   (iv) ripple in the pass-band ≤ 0.01

Obtain the transfer function H(Z) of the corresponding IIR digital filter using Bilinear Transformation Technique.

[JNTU 99/s]

Solution: Given

$$\Omega_c\ = 500\ \text{Hz}$$

$$N = 2$$

$$f_s = 10\ \text{kHz}$$

$$\in\ = 0.01$$

Digital cutoff frequency $\omega_c = 2\pi \times 500 \times \dfrac{1}{10\times 10^3} = 0.3142\ \text{rad}$

Normalized Chebyshev lowpass filter is

$$H(s) = \frac{H_o}{\displaystyle\prod_{k=1}^{2}\left(S-P_k\right)}$$

where

$$P_k = \sinh(\alpha)\cos(\beta_k) + j\ \cosh(\alpha)\sin(\beta_k)$$

$$\alpha = \frac{1}{N}\sinh^{-1}\left(\frac{1}{\in}\right) = \frac{1}{2}\sinh^{-1}\left(\frac{1}{0.01}\right) = 2.65$$

$$\beta_k = \frac{(2K+N-1)\pi}{2N},\quad k = 1,\ 2,\ ....N$$

$$\beta_1 = \frac{(2+2-1)\pi}{4} = 135°$$

$$\beta_2 = \frac{(4+2-1)\pi}{4} = 225°$$

$$\therefore \quad P_1 = \sinh(2.65)\cos(135°) + j\,\cosh(2.65)\sin(135°)$$

$$= -4.98 + j\,5.03$$

$$P_2 = \sinh(2.65)\cos(225°) + j\,\cosh(2.65)\sin(225°)$$

$$= -4.98 - j\,5.03$$

Since N is even

$$H_o = \frac{1}{\sqrt{1+\epsilon^2}} \prod_{k=1}^{2}(-P_k)$$

$$= \frac{1}{\sqrt{1+(0.01)^2}}(4.98 - j\,5.03)(4.98 + j\,5.03)$$

$$H_o = 50$$

$$\therefore \quad H(s) = \frac{50}{(s+4.98-j\,5.03)(s+4.98+j\,5.03)}$$

$$= \frac{50.1}{s^2 + 9.96\,s + 50.1}$$

Since we are using Bilinear transform,

Prewarped cutoff frequency $\omega_c^1 = \tan\left(\dfrac{\omega_c}{2}\right) = \tan\left(\dfrac{0.3142}{2}\right) = 0.158$ rad

Denormalized H(s) is

$$H'(s) = H(s)\Big|_{s = \frac{s}{\omega_c^1}}$$

$$H'(s) = \frac{50.1}{\left(\dfrac{s}{0.158}\right)^2 + 9.96\left(\dfrac{s}{0.158}\right) + 50.1}$$

$$= \frac{1.25}{s^2 + 1.57\,s + 1.25}$$

Digital Chebyshev Lowpass filter is

$$H(z) = H'(s)\Big|_{s = \frac{1-z^{-1}}{1+z^{-1}}}$$

$$H(z) = \frac{1.25}{\left(\dfrac{1-z^{-1}}{1+z^{-1}}\right)^2 + 1.57\left(\dfrac{1-z^{-1}}{1+z^{-1}}\right) + 1.25}$$

$$= \frac{1.25\left(1+z^{-1}\right)^2}{3.82\left(1+0.131\,z^{-1}+0.196\,z^{-2}\right)}$$

$$H(z) = \frac{0.327\left(1+z^{-1}\right)^2}{\left(1+0.131\,z^{-1}+0.196\,z^{-2}\right)}$$

Example 4.26: Design an IIR digital lowpass filter using Butterworth approximation and Bilinear transformation for the following specifications: 3 dB at 500 Hz, monotonic stopband and response of at least 15 dB at 750 Hz. Sampling frequency 2 kHz. Obtain H(Z) of the desired filter.

[JNTU: 2000]

Solution: Given

$$\alpha_p = 3 \text{ dB} ; \qquad 3 = 10\log_{10}\left(1+\in^2\right)$$

$$\in = 0.997$$

$$\alpha_s = 15 \text{ dB} ; \qquad 15 = 20\log_{10} A$$

$$A = 5.623$$

$$\Omega_p = 500 \text{ Hz}$$

$$\Omega_s = 750 \text{ Hz}$$

$$f_s = 2 \text{ kHz}$$

Digital frequencies are

$$\omega_p = 2\pi \times 500 \times \frac{1}{2\times 10^3} = 1.57 \text{ rad}$$

$$\omega_s = 2\pi \times 750 \times \frac{1}{2\times 10^3} = 2.36 \text{ rad}$$

Since transformation is Bilinear,

Prewarped frequencies are $\omega_p^1 = \tan\left(\dfrac{\omega_p}{2}\right) = \tan\left(\dfrac{1.57}{2}\right) = 0.999 \approx 1\ \text{rad}$

$$\omega_s^1 = \tan\left(\dfrac{\omega_s}{2}\right) = \tan\left(\dfrac{2.36}{2}\right) = 2.43\,\text{rad}$$

Order of Butterworth lowpass filter is

$$N = \frac{\log_{10}\left(\dfrac{\in}{\sqrt{A^2-1}}\right)}{\log_{10}\left(\dfrac{\omega_p^1}{\omega_s^1}\right)} = 1.99$$

$\therefore \qquad\qquad\qquad N = 2$

$\therefore \qquad$ Normalized Butterworth Lowpass filter is

$$H(s) = \frac{1}{\prod\limits_{k=1}^{2}\left(S - P_k\right)}$$

$$P_k = e^{j\pi(2K+N-1)/2N}, \quad k = 1, 2, \ldots N$$

$$P_1 = e^{j\pi(2+2-1)/4} = -0.707 + j\,0.707$$

$$P_2 = e^{j\pi(4+2-1)/4} = -0.707 - j\,0.707$$

$\therefore \qquad$ $$H(s) = \frac{1}{(s + 0.707 - j\,0.707)(s + 0.707 + j\,0.707)}$$

$$= \frac{1}{s^2 + 1.414\,s + 1}$$

Prewarped cutoff frequency $\omega_c^1 = \dfrac{\omega_p^1}{(\in)^{1/N}} = \dfrac{1}{(0.997)^{1/2}} = 1\ \text{rad}$

Denormalized Butterworth Lowpass filter is

$$H'(s) = H(s)\Big|_{s = \frac{s}{\omega_c^1}}$$

$$H'(s) = \cfrac{1}{\left(\dfrac{s}{1}\right)^2 + 1.414\left(\dfrac{s}{1}\right) + 1}$$

$$H'(s) = \frac{1}{s^2 + 1.414\,s + 1}$$

Digital Butterworth Lowpass filter is

$$H(z) = H'(s)\Big|_{s=\frac{1-z^{-1}}{1+z^{-1}}}$$

$$H(z) = \cfrac{1}{\left(\dfrac{1-z^{-1}}{1+z^{-1}}\right)^2 + 1.414\left(\dfrac{1-z^{-1}}{1+z^{-1}}\right) + 1}$$

$$H(z) = 0.293\,\frac{\left(1+z^{-1}\right)^2}{\left(1+0.172\,z^{-2}\right)}$$

Example 4.27: Using bilinear transformation technique obtain H(Z) of a recursive IIR digital filter with a passband extending from 0 to 1000 Hz with 1 dB ripple and magnitude response monotonically falling to atleast 19 dB at 1800 Hz. Use sampling frequency of 5000 Hz. [JNTU 2000/S]

Solution: Given $\qquad \alpha_P = 1$ dB; $\qquad 1 = 10 \log_{10}\left(1 + \epsilon^2\right)$

$$\epsilon = 0.509$$

$$\alpha_S = 19 \text{ dB}; \qquad 19 = 20 \log_{10} A$$

$$A = 8.91$$

$$\Omega_P = 1000 \text{Hz}$$

$$\Omega_S = 1800 \text{ Hz}$$

$$F_S = 5000 \text{ Hz}$$

Digital frequencies are $\omega_P = 2\pi \times 1000 \times \dfrac{1}{5000} = 1.257$ rad

$$\omega_S = 2\pi \times 1800 \times \frac{1}{5000} = 2.262 \text{ rad}$$

Since transformation is Bilinear,

Prewarped frequencies are $\omega_P^1 = \tan\left(\dfrac{\omega_P}{2}\right) = \tan\left(\dfrac{1.257}{2}\right) = 0.727$ rad

$$\omega_S^1 = \tan\left(\dfrac{\omega_S}{2}\right) = \tan\left(\dfrac{2.262}{2}\right) = 2.125 \text{ rad}$$

Since magnitude response is monotonically falling, filter is butter worth

Order of butter worth low pass filter is

$$N = \log_{10}\left(\dfrac{\in}{\sqrt{A^2-1}}\right) \Big/ \log_{10}\left(\omega_P^1/\omega_S^1\right) = 2.66$$

$\therefore \qquad\qquad N = 3$

Normalized Butterworth Low pass filter is

$$H(s) = \dfrac{1}{\displaystyle\prod_{k=1}^{3}(S - P_k)}$$

$$P_k = e^{j\pi(2k + N - 1)/2N}, \quad k = 1, 2, 3.$$

$$P_1 = e^{j\pi(2 + 3 - 1)/6} = -0.5 + j\,0.87$$

$$P_2 = e^{j\pi(4 + 3 - 1)/6} = -1$$

$$P_3 = e^{j\pi(6 + 3 - 1)/6} = -0.5 - j\,0.87$$

$$H(s) = \dfrac{1}{(s+1)(s+0.5 - j\,0.87)(s+0.5 + j\,0.87)}$$

$$H(s) = \dfrac{1}{(s+1)(s^2 + s + 1)}$$

Prewarped cutoff frequency $\omega_c^1 = \dfrac{\omega_P^1}{(\in)^{1/N}} = \dfrac{0.727}{(0.509)^{1/3}} = 0.91$ rad

Denoralized Butterworth Lowpass filter is

$$H'(s) = H_{(s)}\Big|_{s = \frac{s}{\omega_c^1}}$$

$$H'(s) = \cfrac{1}{\left(\cfrac{s}{0.91}+1\right)\left[\left(\cfrac{s}{0.91}\right)^2+\left(\cfrac{s}{0.91}\right)+1\right]}$$

$$H'(s) = \cfrac{0.75}{(s+0.91)(s^2+0.91\,s+0.828)}$$

Digital Butterworth Lowpass filter is

$$H(z) = H'(s)\Big|_{s=\frac{1-z^{-1}}{1+z^{-1}}}$$

$$H(z) = \cfrac{0.75}{\left[\left(\cfrac{1-z^{-1}}{1+z^{-1}}\right)+0.91\right]\left[\left(\cfrac{1-z^{-1}}{1+z^{-1}}\right)^2+0.91\left(\cfrac{1-z^{-1}}{1+z^{-1}}\right)+0.828\right]}$$

$$= \cfrac{0.75\left(1+z^{-1}\right)^3}{1.91\left(1-0.047z^{-1}\right)2.738\left(1-0.125z^{-1}+0.335z^{-2}\right)}$$

$$H(z) = \cfrac{0.143\left(1+z^{-1}\right)^3}{\left(1-0.047z^{-1}\right)\left(1-0.125z^{-1}+0.335z^{-2}\right)}$$

Example 4.28: A discrete-time second-order bandpass Butterworth filter is to be designed to the following specifications, using bilinear transformation method:

(i)   Lower cut-off frequency = 10 Hz

(ii) Upper cut-off frequency = 20 Hz

(iii) Sampling frequency = 100 samples/sec   [JNTU 2000/s]

Solution: Given bandpass filter specifications are

Order $N = 2,$

$\qquad\qquad\qquad\quad \Omega_l = 10$ Hz,

$\qquad\qquad\qquad\quad \Omega_u = 20$ Hz,

$\qquad\qquad\qquad\quad f_s = 100$ samples/sec

Digital frequencies are

$$\omega_l = 2\pi \times 10 \times \frac{1}{100} = 0.628 \text{ rad}$$

$$\omega_u = 2\pi \times 20 \times \frac{1}{100} = 1.257 \text{ rad}$$

Since order of bandpass filter is 2, consider 1st order

$\therefore$ Butterworth low pass filter is

i.e., $$H(s) = \frac{1}{s+1}$$

Since transformation is bilinear,

Prewarped digital frequencies are

$$\omega_l^1 = \tan\left(\frac{\omega_l}{2}\right) = \tan\left(\frac{0.628}{2}\right) = 0.325 \text{ rad}$$

$$\omega_u^1 = \tan\left(\frac{\omega_u}{2}\right) = \tan\left(\frac{1.257}{2}\right) = 0.727 \text{ rad}$$

$\therefore$ Butterworth band pass filter is

$$H'(s) = H(s)\Big|_{s = \frac{s^2 + \omega_u^1 \omega_l^1}{s(\omega_u^1 - \omega_l^1)}}$$

$$s = \frac{s^2 + (0.325)(0.727)}{s(0.727 - 0325)} = \frac{s^2 + 0.236}{s(0.402)}$$

$\therefore$ $$H'(s) = \frac{1}{\dfrac{s^2 + 0.236}{0.402s} + 1}$$

$$H'(s) = \frac{0.402s}{s^2 + 0.402\,s + 0.236}$$

Discrete-time Butterworth band pass filter is

$$H(z) = H'(s)\Big|_{s = \frac{1 - z^{-1}}{1 + z^{-1}}}$$

$$H(z) = \frac{0.402\left(\dfrac{1 - z^{-1}}{1 + z^{-1}}\right)}{\left(\dfrac{1 - z^{-1}}{1 + z^{-1}}\right)^2 + 0.402\left(\dfrac{1 - z^{-1}}{1 + z^{-1}}\right) + 0.236}$$

$$H(z) = \frac{0.245\left(1 - z^{-2}\right)}{\left(1 - 0.934z^{-1} + 0.509z^{-2}\right)}$$

Example 4.29: Find the transfer function of a Chebyshev Lowpass discrete time filter to meet the following specifications

Passband: 0 to 0.5 MHz with 0.2 dB ripple

Stopband edge: 1 MHz with attenuation of atleast 50 dB. [JNTU 2000/s]

Solution: Given
$$\alpha_p = 0.2 \text{ dB}; \qquad 0.2 = 10 \log_{10}\left(1+\epsilon^2\right) \quad \Rightarrow \epsilon = 0.217$$

$$\alpha_s = 50 \text{ dB}; \qquad 50 = 20 \log_{10} A \qquad \Rightarrow A = 316.23$$

$$\Omega_p = 0.5 \text{ MHz}$$

$$\Omega_s = 1 \text{ MHz}$$

Assume $f_s = 2.5$ MHz (if f_s is not given then take $f_s > 2\,\Omega_s$)

Digital frequencies are $\omega_p = 2\pi \times 0.5 \times 10^6 \times \dfrac{1}{2.5 \times 10^6} = 1.257$ rad

$$\omega_S = 2\pi \times 1 \times 10^6 \times \dfrac{1}{2.5 \times 10^6} = 2.513 \text{ rad}$$

Let as use bilinear transformation

So prewarped digital frequencies are

$$\omega_P^1 = \tan\left(\frac{\omega_P}{2}\right) = \tan\left(\frac{1.257}{2}\right) = 0.727 \text{ rad}$$

$$\omega_S^1 = \tan\left(\frac{\omega_S}{2}\right) = \tan\left(\frac{2.513}{2}\right) = 3.076 \text{ rad}$$

Order of Butterworth Lowpass filter is

$$N = \frac{\log_{10}\left(\dfrac{\epsilon}{\sqrt{A^2-1}}\right)}{\log_{10}\left(\dfrac{\omega_P^1}{\omega_S^1}\right)} = 5.05 \qquad \therefore N = 6$$

Normalized Butterworth Lowpass filter is

$$H(s) = \frac{1}{\displaystyle\prod_{k=1}^{6}\left(S - P_k\right)};$$

$$P_k = e^{j\pi(2k+N-1)/2N}, \quad k = 1, 2, 3, 4, 5, 6.$$

$$P_1 = e^{j\pi(2+6-1)/12} = -0.259 + j\,0.966$$

$$P_2 = e^{j\pi(4+6-1)/12} = -0.707 + j\,0.707$$

$$P_3 = e^{j\pi(6+6-1)/12} = -0.966 + j\,0.259$$

$$P_4 = e^{j\pi(8+6-1)/12} = -0.966 - j\,0.259$$

$$P_5 = e^{j\pi(10+6-1)/12} = -0.707 - j\,0.707$$

$$P_6 = e^{j\pi(12+6-1)/12} = -0.259 - j\,0.966$$

$$H(s) = \frac{1}{(s+0.259-j\,0.966)(s+0.259+j\,0.966)(s+0.707-j\,0.707)}$$
$$\times \frac{1}{(s+0.707+j\,0.707)(s+0.966-j\,0.259)(s+0.966+j\,0.259)}$$

$$H(s) = \frac{1}{(s^2+0.518s+1)(s^2+1.414s+1)(s^2+1.932s+1)}$$

Prewarped cut-off frequency $\omega_c^1 = \dfrac{\omega_P^1}{(\in)^{1/N}} = \dfrac{0.727}{(0.217)^{1/6}} = 0.938$ rad

$\therefore$ Denormalized Butterworth Lowpass filter is

$$H'(s) = H(s)\Big|_{s=\frac{s}{\omega_c^1}}$$

$$H'(s) = \frac{1}{\left[\left(\dfrac{s}{0.938}\right)^2 + 0.518\left(\dfrac{s}{0.938}\right) + 1\right]\left[\left(\dfrac{s}{0.938}\right)^2 + 1.414\left(\dfrac{s}{0.938}\right) + 1\right]}$$
$$\times \frac{1}{\left[\left(\dfrac{s}{0.938}\right)^2 + 1.932\left(\dfrac{s}{0.938}\right) + 1\right]}$$

$$H'(s) = \frac{0.681}{(s^2+0.486s+0.879)(s^2+1.326s+0.879)(s^2+1.812s+0.879)}$$

Digital Butterworth Lowpass filter is

$$H(z) = H'(s)\Big|_{s=\frac{1-z^{-1}}{1+z^{-1}}}$$

$$H(z) = \frac{0.0243\left(1+z^{-1}\right)^{6}}{\left(1-0.102z^{-1}+0.589z^{-2}\right)\left(1-0.075z^{-1}+0.172z^{-2}\right)\left(1-0.066z^{-1}+0.0182z^{-2}\right)}$$

Example 4.30: Design an analog band pass filter to satisfy the following specifications

  (i)   3 dB upper and lower cut-off frequencies are 100 Hz and 3.8 kHz

  (ii) Stopband attenuation of 20 dB at 20 Hz and 8 kHz

  (iii) No ripple with both passband and stopband   [JNTU 2000]

Solution: Given $\Omega_l = 100$ Hz, $\Omega_u = 3.8$ kHz

$$\Omega_{S_1} = 20 \text{ Hz}, \quad \Omega_{S_2} = 8 \text{ kHz}$$

$$\alpha_P = 3 \text{ dB}; \quad 3 = 10 \log_{10}\left(1+\epsilon^2\right) \quad \Rightarrow \epsilon = 0.997$$

$$\alpha_S = 20 \text{ dB}; \quad 20 = 20 \log_{10} A \quad \Rightarrow A = 10$$

Analog frequencies are $\Omega_l^1 = 2\pi \times 100 = 200\pi$ rad/sec

$$\Omega_u^1 = 2\pi \times 3.8 \times 10^3 = 7600\pi \text{ rad/sec}$$

$$\Omega_{S_1}^1 = 2\pi \times 20 = 40\pi \text{ rad/sec}$$

$$\Omega_{S_2}^1 = 2\pi \times 8 \times 10^3 = 16000\pi \text{ rad/sec}$$

We need to convert bandpass filter specifications into lowpass filter specifications.

Assume $\Omega_P^1 = 1$ rad/sec

and $\Omega_S^1 = \min\ \left\{|\Omega_1|,|\Omega_2|\right\}$

$$\Omega_1 = \frac{-\left(\Omega_{S_1}^1\right)^2 + \Omega_l^1\,\Omega_u^1}{\Omega_{S_1}^1\left(\Omega_u^1 - \Omega_l^1\right)}$$

$$= \frac{-\left(40\pi\right)^2 + \left(200\pi\right)\left(7600\pi\right)}{40\pi\left(7600\pi - 200\pi\right)} = 5.129 \text{ rad/sec}$$

$$\Omega_2 = \frac{-\left(\Omega_{S_2}^1\right)^2 + \Omega_l^1\,\Omega_u^1}{\Omega_{S_2}^1\left(\Omega_u^1 - \Omega_l^1\right)}$$

$$= -\frac{\left(1600\pi\right)^2 + \left(200\pi\right)\left(7600\pi\right)}{1600\pi\left(7600\pi - 200\pi\right)} = -2.149 \text{ rad/sec}$$

$$\therefore \qquad \Omega_S^1 = 2.149 \text{ rad} / \text{sec}$$

Order of Butterworth Lowpass filter is

$$N = \log_{10}\left(\frac{\in}{\sqrt{A^2-1}}\right)\Big/\log_{10}\left(\frac{\Omega_P^1}{\Omega_S^1}\right) = 3.007 \qquad \therefore N = 4$$

Normalized Butterworth Lowpass filter is

$$H(s) = \frac{1}{\overset{4}{\underset{k=1}{\pi}}\left(S - P_k\right)}$$

$$P_k = e^{j\pi(2k+N-1)/2N}, \quad k = 1, 2, 3, 4.$$

$$P_1 = e^{j\pi(2+4-1)/8} = -0.383 + j\,0.924$$

$$P_2 = e^{j\pi(4+4-1)/8} = -0.924 + j\,0.383$$

$$P_3 = e^{j\pi(6+4-1)/8} = -0.924 - j\,0.383$$

$$P_4 = e^{j\pi(8+4-1)/8} = -0.383 - j\,0.924$$

$$H(s) = \frac{1}{(s+0.383 - j\,0.924)(s+0.383 + j\,0.924)(s+0.924 - j\,0.383)(s+0.924 + j\,0.383)}$$

$$H(s) = \frac{1}{(s^2 + 0.766s + 1)(s^2 + 1.848s + 1)}$$

$\therefore$ Analog Butterworth bandpass filter is

$$H'(s) = H(s)\Big|_{s = \frac{s^2 + \Omega_l^1 \Omega_u^1}{s(\Omega_u^1 - \Omega_l^1)}}$$

$$s = \frac{s^2 + (200\pi)(7600\pi)}{s(7600\pi - 200\pi)} = \frac{s^2 + 1.5 \times 10^7}{s\,2.32 \times 10^4}$$

$$H'(s) = \frac{1}{\left[\left(\dfrac{s^2 + 1.5 \times 10^7}{2.32 \times 10^4 s}\right)^2 + 0.766\left(\dfrac{s^2 + 1.5 \times 10^7}{2.32 \times 10^4 s}\right) + 1\right]\left[\left(\dfrac{s^2 + 1.5 \times 10^7}{2.32 \times 10^4 s}\right)^2 + 1.848\left(\dfrac{s^2 + 1.5 \times 10^7}{2.32 \times 10^4 s}\right) + 1\right]}$$

$$H'(s) = \frac{2.89 \times 10^{17} s^4}{\left(s^4 + 1.77 \times 10^4 s^3 + 56.8 \times 10^7 s^2 + 2.67 \times 10^{11} s + 2.25 \times 10^{14}\right)}$$

$$\times \frac{1}{\left(s^4 + 4.28 \times 10^4 s^3 + 56.8 \times 10^7 s^2 + 2.25 \times 10^{14}\right)}$$

4.7 Filters and Transformations

Here comparison of Digital and Analog filters, advantages and disadvantages of digital filters, comparison of Butterworth and Chebyshev Filters and Comparison of Impulse Invariant and Bilinear Transformations are explained.

4.7.1 Comparison of Digital and Analog Filters

Digital filter	Analog filter
1. A digital filter operates on digital samples and produces digital output	1. An analog filter operates on analog signals and produces analog output
2. Digital filters are represented by difference equations	2. Analog filters are represented by differential equations
3. Digital filter transfer function is represented by Z-transform	3. Analog filter transfer function is represented by Laplace transform
4. It is constructed by adders, multipliers and delay elements	4. It is constructed by active or passive electronic components
5. By changing the filter coefficients, the frequency response can be changed	5. By changing the electronic components, the frequency response can be changed

Advantages of digital filters:
1. The digital filters have high thermal stability than analog filters as digital filters do not have resistor, capacitors and inductors, whose values changes with temperature.
2. Adaptive features can be implemented by changing the filter coefficients as the digital filters are programmable.
3. By using multiplexing technique a single digital filter can be used to process multiple signals.
4. In digital filters the precision of the filter depends on the length (or size) of the registers used to store the filter coefficients. The performance characteristics of the filters like accuracy, dynamic range, stability and frequency response tolerance can be enhanced by increasing the register bits length.

Disadvantages of digital filters:

1. The band width of the discrete signal is limited by the sampling frequency.

2. The quantization error arises due to finite word length in the representation of signals and parameters.

4.7.2 Comparison of Butterworth and Chebyshev Filters

Butterworth	Chebyshev
1. All pole design	1. All pole design
2. The poles lie on a circle in the S-plane	2. The poles lie on an ellipse in the S-plane
3. The magnitude response is maximally flat at the origin and monotonically decreasing function of Ω	3. The magnitude response is equiripple in the pass band and monotonically decreasing in the stop band
4. The normalized magnitude response has a value of $\dfrac{1}{\sqrt{2}}$ at the cutoff frequency $\Omega_C \approx \Omega_p$	4. The normalized magnitude response has a value of $\dfrac{1}{\sqrt{1+\epsilon^2}}$ at the cutoff frequency $\Omega_C \approx \Omega_p$
5. Only few parameters have to be calculated to determine the transfer function	5. Large number of parameters have to be calculated to determine the transfer function
6. Order of the filter is $$N = \frac{\log_{10}(k_1)}{\log_{10}(k)}$$	6. Order of the filter is $$N = \frac{\cosh^{-1}\left(\dfrac{1}{k_1}\right)}{\cosh^{-1}\left(\dfrac{1}{k}\right)}$$
7. The order of the Chebyshev filter to satisfy a set of specifications is lower than that of Butterworth.	
8. For a given order of the filter, the transition band of Chebyshev response is narrower than that of Butterworth.	
9. Butterworth filter gives fairly linear phase response than Chebyshev filter.	

4.7.3 Comparison of Impulse Invariant Transformation and Bilinear Transformation

Impulse Invariant transformation	Bilinear transformation
1. It is many-to-one mapping	1. It is one-to-one mapping
2. The relation between analog and digital frequency is linear	2. The relation between analog and digital frequency is nonlinear
3. To prevent the problem of aliasing the analog filters should be band limited	3. There is no problem of aliasing and so the analog filter need not be band limited
4. The magnitude and phase response of analog filter can be preserved by choosing low sampling time	4. Due to the effect of warping, the phase response of analog filter cannot be preserved. But the magnitude response can be preserved by prewarping

Advantage of Bilinear transformation

1. The bilinear transformation is one-to-one mapping.
2. There is no aliasing and so the analog filter need not have a band limited frequency response.
3. The effect of warping on amplitude response can be limited by prewarping the analog filter.
4. Bilinear transformation can be used to design digital filters with prescribed magnitude response with piecewise constant values.

Disadvantages of Bilinear transformation

1. The nonlinear relationship between analog and digital frequencies introduces frequency distortion which is called frequency warping.

2. Using bilinear transformation, a linear phase analog filter cannot be transformed to a linear phase digital filter.

Review Questions

1. Digital filters are represented by

    (a) Differential equations   (b) Difference equations

    (c) Both a and b   (d) None   *Ans*: [b]

2. Analog filters are represented by

    (a) Differential equations   (b) Difference equations

    (c) Both a and b   (d) None   *Ans*: [a]

3. Digital filter system function is represented by

    (a) Laplace transform   (b) Fourier transform

    (c) Z – transform   (d) Both a and c   *Ans*: [c]

4. The range of frequencies of signal that are passed through the filter is called

    (a) Stopband   (b) Transition band

    (c) Passband   (d) Either b or c   *Ans*: [c]

5. The range of frequencies that are rejected is called

    (a) Stopband   (b) Transition band

    (c) Passband   (d) Either a or b   *Ans*: [a]

6. Any IIR digital filter design starts from designing of

    (a) Analog high pass filter   (b) Analog band pass filter

    (c) Analog band reject filter   (d) Analog low pass filter   *Ans*: [d]

7. The magnitude – squared response of an analog low pass butterworth filter $H_a(s)$ of Nth order is

    (a) $\dfrac{1}{1+\left(\dfrac{\Omega}{\Omega_C}\right)^N}$

    (b) $\dfrac{1}{1+\epsilon^2\left(\dfrac{\Omega}{\Omega_P}\right)^N}$

    (c) $\dfrac{1}{1+\left(\dfrac{\Omega}{\Omega_C}\right)^{2N}}$

    (d) $\dfrac{1}{1+\left(\dfrac{\Omega}{\Omega_S}\right)^{2N}}$ *Ans:* [c]

8. The magnitude – squared response of an analog low pass Chebyshev filter is

    (a) $\dfrac{1}{1+T_N^2\left(\dfrac{\Omega}{\Omega_S}\right)}$

    (b) $\dfrac{1}{1+\epsilon^2\,T_N^2\left(\dfrac{\Omega}{\Omega_P}\right)}$

    (c) $\dfrac{1}{1+\epsilon^2\,T_N\left(\dfrac{\Omega}{\Omega_P}\right)}$

    (d) $\dfrac{1}{1+\epsilon^2\,T_N\left(\dfrac{\Omega}{\Omega_S}\right)}$ *Ans:* [b]

9. The magnitude response of Butterworth lowpass filter will have

    (a) Monotonic passband   (b) Monotonic stopband

    (c) Equiripple passband   (d) Both a and b   *Ans:* [d]

10. The magnitude response of Chebyshev (type I) low pass filter will have

    (a) Monotonic passband   (b) Monotonic stopband

    (c) Equiripple passband   (d) Both b and c   *Ans:* [d]

11. When the order of the Chebyshev filter is even, the magnitude response in the passband starts at

    (a) The maximum value of the magnitude   (b) $\dfrac{1}{\sqrt{1+\epsilon^2}}$

    (c) In between a and b   (d) Can't say   *Ans:* [b]

12. When the order of the Chebyshev filter is odd, the magnitude response in the passband starts at

    (a) The maximum value of the magnitude   (b) $\dfrac{1}{\sqrt{1+\epsilon^2}}$

    (c) In between a and b   (d) Can't say   *Ans:* [a]

13. Order of the Chebyshev filter is

    (a) $\log_{10}(k_1)/\log_{10}(k)$

    (b) $\dfrac{\cosh(k_1)}{\cosh(k)}$

    (c) $\dfrac{\cosh\left(\dfrac{1}{k_1}\right)}{\cosh\left(\dfrac{1}{k}\right)}$

    (d) $\dfrac{\cosh^{-1}\left(\dfrac{1}{k_1}\right)}{\cosh^{-1}\left(\dfrac{1}{k}\right)}$
    $\qquad$ *Ans*: [d]

14. Order of the Butterworth filter is

    (a) $\log_e(k_1)/\log_{10}(k)$

    (b) $\log_{10}(k_1)/\log_{10}(k)$

    (c) $\log_{10}(k)/\log_{10}(k_1)$

    (d) $\log_e(k_1)/\log_e(k)$
    $\qquad$ *Ans*: [b]

15. Poles of Butterworth filter lies on ….. in the S-plane

    (a) Ellipse

    (b) Circle

    (c) Either a or b

    (d) Can't say
    $\qquad$ *Ans*: [b]

16. Poles of Chebyshev filter lies on …. in the S-plane

    (a) Ellipse

    (b) Circle

    (c) Either a or b

    (d) Can't say
    $\qquad$ *Ans*: [a]

17. The poles P_k of Butterworth filter of order N is

    (a) $e^{j\pi(2k-1)/N}$

    (b) $e^{j\pi(N+k-1)/2N}$

    (c) $e^{j\pi(2k+N-1)/2N}$

    (d) $e^{j\pi(2N+k-1)/2N}$
    $\qquad$ *Ans*: [c]

18. Low pass to low pass of cutoff frequency Ω_C, transformation is

    (a) $S \rightarrow \dfrac{\Omega_C}{S}$

    (b) $S \rightarrow \dfrac{S}{\Omega_C}$

    (c) $S \rightarrow (S+\Omega_C)$

    (d) $S \rightarrow (S-\Omega_C)$
    $\qquad$ *Ans*: [b]

19. Lowpass to highpass of cutoff frequency Ω_C, transformation is

    (a) $S \rightarrow \dfrac{\Omega_C}{S}$

    (b) $S \rightarrow \dfrac{S}{\Omega_C}$

    (c) $S \rightarrow (S+\Omega_C)$

    (d) $S \rightarrow (S-\Omega_C)$
    $\qquad$ *Ans*: [a]

20. Low pass to band pass of lower cutoff frequency Ω_l and upper cutoff frequency Ω_u, transformation is

    (a) $S \to \dfrac{S^2 + \Omega_l \Omega_u}{S(\Omega_u - \Omega_l)}$ (b) $S \to \dfrac{S^2 + \Omega_l \Omega_u}{S(\Omega_u + \Omega_l)}$

    (c) $S \to \dfrac{S(\Omega_u - \Omega_l)}{S^2 + \Omega_l \Omega_u}$ (d) $S \to \dfrac{S(\Omega_u + \Omega_l)}{S^2 + \Omega_l \Omega_u}$ *Ans:* [a]

21. Lowpass to band reject of lower cutoff frequency Ω_l and upper cutoff frequency Ω_u, transformation is

    (a) $S \to \dfrac{S^2 + \Omega_l \Omega_u}{S(\Omega_u - \Omega_l)}$ (b) $S \to \dfrac{S^2 + \Omega_l \Omega_u}{S(\Omega_u + \Omega_l)}$

    (c) $S \to \dfrac{S(\Omega_u - \Omega_l)}{S^2 + \Omega_l \Omega_u}$ (d) $S \to \dfrac{S(\Omega_u + \Omega_l)}{S^2 + \Omega_l \Omega_u}$ *Ans:* [c]

22. Transformation(s) which transform from S-domain to Z-domain is (are)

    (a)   Impulse Invariance method   (b)   Step invariance method

    (b)   Bilinear Transformation   (d)   All the above   *Ans:* [d]

23. The impulse invariant transformation is

    (a) $\dfrac{1}{s - P_k} \to \dfrac{1}{1 - e^{-P_k T} Z^{-1}}$ (b) $\dfrac{1}{s - P_k} \to \dfrac{1}{1 - e^{P_k T} Z^{-1}}$

    (c) $\dfrac{1}{s + P_k} \to \dfrac{1}{1 - e^{-P_k T} Z^{-1}}$ (d)   Both b & c   *Ans:* [d]

24. The relation between analog frequency Ω and digital frequency w in impulse invariance method is

    (a) $\Omega = wT$ (b) $\omega = \Omega T$

    (c) $\omega = \Omega / T$ (d) $\omega = \Omega^T$ *Ans:* [b]

25. Impulse invariant transformation is

    (a)   Many-to-one   (b)   One-to-one

    (c)   One-to-many   (d)   Many-to-many   *Ans:* [a]

26. Due to many-to-one transformation of impulse invariant method, it has

    (a)   Aliasing problem   (b)   Warping effect

    (c)   Both a & b   (d)   None   *Ans:* [a]

27. Due to aliasing problem, impulse invariant method is applicable only for
   (a) Lowpass filters
   (b) Band pass filters with low cutoff frequency
   (c) Highpass filter
   (d) Both a & b   *Ans*: [d]

28. Step invariance transformation is

   (a) $\dfrac{1}{s+p_k} \to \dfrac{\left(1-e^{-p_kT}\right)_{z^{-1}}}{p_k\left(1-e^{-p_kT}z^{-1}\right)}$

   (b) $\dfrac{1}{s-p_k} \to \dfrac{\left(e^{p_kT}-1\right)_{z^{-1}}}{p_k\left(1-e^{p_kT}z^{-1}\right)}$

   (c) $\dfrac{1}{s+p_k} \to \dfrac{\left(1-e^{p_kT}\right)_{z^{-1}}}{p_k\left(1-e^{p_kT}z^{-1}\right)}$

   (d) both a & b   *Ans*: [d]

29. Bilinear transformation is

   (a) $S \to \dfrac{2}{T}\dfrac{1-z^{-1}}{1+z^{-1}}$

   (b) $S \to \dfrac{1}{T}\dfrac{1-z^{-1}}{1+z^{-1}}$

   (c) $S \to \dfrac{T}{2}\dfrac{z-1}{z+1}$

   (d) $S \to \dfrac{T}{2}\dfrac{z+1}{z-1}$ *Ans*: [a]

30. Bilinear transformation is

   (a) Many-to-one
   (b) One-to-one
   (c) One-to-many
   (d) Many-to-many   *Ans*: [b]

31. The relation between analog frequency Ω and digital frequency w in Bilinear transformation is

   (a) $\Omega = \dfrac{2}{T}\tan\left(\dfrac{\omega}{2}\right)$

   (b) $\Omega = \dfrac{T}{2}\tan\left(\dfrac{\omega}{2}\right)$

   (b) $\Omega = \dfrac{2}{T}\tan\left(\dfrac{2}{\omega}\right)$

   (d) $\Omega = \tan(\omega)$ *Ans*: [a]

32. The relation between Ω and w in Bilinear transformation is

   (a) Linear
   (b) Nonlinear
   (c) Inverse proportional
   (d) None   *Ans*: [b]

33. Due to nonlinear relation between Ω and w in Bilinear transformation, it has

   (a) Aliasing problem
   (b) Warping effect
   (c) Both a & b
   (d) None   *Ans*: [b]

34. The warping effect in Bilinear transformation can be eliminated by

    (a) Prewarping   (b) Converting analog frequencies into digital frequencies

    (c) Both a & b   (d) None   *Ans*: [a]

35. Bilinear transformation is applicable for

    (a) Lowpass, bandpass   (b) Highpass

    (c) Band reject filters   (d) All the above   *Ans*: [d]

36. By prewarping the frequencies, warping effect in Bilinear transformation can be eliminated for

    (a) Magnitude response   (b) Phase response

    (c) Both a & b   (d) None   *Ans*: [a]

37. If the Impulse invariant transformation is used to design highpass filter, ... spectral transformation is more appropriate to avoid aliasing problem.

    (a) Analog   (b) Digital

    (c) Both a & b   (d) None   *Ans*: [b]

38. The order of the Chebyshev filter to satisfy a set of specifications is .... that of Butterworth

    (a) Greater than   (b) Lower than

    (c) Equal to   (d) Can't say   *Ans*: [b]

39. For a given order of the filter, the transition band of Chebyshev response is ..... than (to) that of Butterworth

    (a) Narrow   (b) Wider

    (c) Equal   (d) Can't say   *Ans*: [a]

40. The filter which gives fairly linear phase response is

    (a) Butterworth   (b) Chebyshev

    (c) Both a & b   (d) None   *Ans*: [a]

Chapter 5
FIR Digital Filters

5.0 Introduction

In the previous unit our discussion was on the design of digital IIR filter and its properties. In many applications, for example, data transmission, biomedicine, digital audio and image processing, FIR filters are preferred over their IIR counterparts.

The following are the main advantages of the FIR filters over IIR filters.

1. FIR filters can be easily designed with exact linear phase response.

2. FIR filters realized nonrecursively, i.e., by direct evaluation of equation (5.1.1) are always stable. FIR filters can also be realized recursively.

3. Nonrecursive FIR filters suffer less from the effects of finite word length than IIR filters.

4. FIR filters are very simple to implement.

5. Excellent methods are available for designing various kinds of FIR filters.

The disadvantages of FIR filters are:

1. FIR requires more coefficients for sharp cutoff filters than IIR.

2. Analog filters can be readily transformed into equivalent IIR digital filters meeting similar specifications. This is not possible with FIR filters as they have no analog counterpart.

3. Narrow transition band FIR filter implementation requires more arithmetic operations and hardware components such as multipliers, adders and delay elements. So its implementation is costly.

4. Memory requirement and execution time are very high.

5.1 Characteristics of FIR Digital Filters

5.1.1 Linear Phase FIR Filters

The basic FIR filter is characterized by the following equation

$$y(n) = \sum_{k=0}^{N-1} h(k)\, x(n-k) \qquad\qquad .....(5.1.1)$$

Its transfer function is

$$H(z) = \sum_{k=0}^{N-1} h(k)\, z^{-k} \qquad\qquad .....(5.1.2)$$

where $h(k)$, $k = 0, 1, ..., N-1$, are the impulse response coefficients of the filter and N is the filter length.

equation (5.1.2) can also be written as

$$H(z) = \sum_{n=0}^{N-1} h(n)\, z^{-n} \qquad\qquad .....(5.1.3)$$

Frequency response can be obtained by replacing Z with $e^{j\omega}$

$$\therefore \qquad H(\omega) = H\left(e^{j\omega}\right) = \sum_{n=0}^{N-1} h(n)\, e^{-j\omega n} \qquad\qquad .....(5.1.4)$$

which is periodic with period 2π

From equation (5.1.4),

$$H(\omega) = \pm \left| H(\omega) \right|\, e^{j\theta(\omega)} \qquad\qquad .....(5.1.5)$$

where $\left| H(\omega) \right|$ is the magnitude response

and $\theta(\omega)$ is the phase response

The phase delay or group delay of the filter provides a useful measure of how the filter modifies the phase characteristics of the signal. If we consider a signal that consists of several frequency components (such as speech waveform or a modulated signal). The phase delay of the filter is "the amount of time delay each frequency component of the signal suffers in going through the filter". The group delay is "the average time delay the composite signal suffers at each frequency".

Mathematically the phase delay (τ_p) and group delay (τ_g) are defined as

$$\tau_p = \frac{-\theta(\omega)}{\omega} \quad \text{and} \quad \tau_g = \frac{-d\theta(\omega)}{\omega} \qquad\qquad .....(5.1.6)$$

For FIR filters to have linear phase, its phase response must be

$$\theta(\omega) = -\alpha\omega, \qquad -\pi \le \omega \le \pi \qquad\qquad .....(5.1.7)$$

where α is a constant.

Substituting equation (5.1.7) in equation (5.1.6), gives

$\tau_p = \tau_g = \alpha$, which means filter will have constant phase delay and group delay.

From equation (5.1.4) and (5.1.5), we can write

$$\sum_{n=0}^{N-1} h(n)\, e^{-j\omega n} = \pm |H(\omega)| e^{j\theta(\omega)} \qquad\qquad .....(5.1.8)$$

which gives

$$\sum_{n=0}^{N-1} h(n)\cos(\omega n) = \pm |H(\omega)| \cos(\theta(\omega)) \qquad\qquad .....(5.1.9)$$

and

$$-\sum_{n=0}^{N-1} h(n)\sin(\omega n) = \pm |H(\omega)| \sin(\theta(\omega)) \qquad\qquad .....(5.1.10)$$

Taking ratio of equation (5.1.10) to (5.1.9) and substituting equation (5.1.7), we get

$$\frac{\displaystyle\sum_{n=0}^{N-1} h(n)\sin(\omega n)}{\displaystyle\sum_{n=0}^{N-1} h(n)\cos(\omega n)} = \frac{\sin(\alpha\omega)}{\cos(\alpha\omega)} \qquad\qquad .....(5.1.11)$$

After simplifying equation (5.1.11), we get

$$\sum_{n=0}^{N-1} h(n)\sin(\alpha - n)\omega = 0 \qquad\qquad .....(5.1.12)$$

The equation (5.1.12) will be zero when

$$h(n) = h(N - 1 - n) \qquad\qquad .....(5.1.13a)$$

and
$$\alpha = \frac{N-1}{2} \qquad\qquad .....(5.1.13b)$$

Therefore, FIR filters will have Linear phase response (constant τ_p and τ_g) when the impulse response is symmetrical about α.

The impulse response satisfying equation (5.1.13a) is shown in Fig. 5.1.

When N = 5 the centre of symmetry of the sequence occurs at 2nd sample and when N = 4 the centre of symmetry occurs at 1.5 samples.

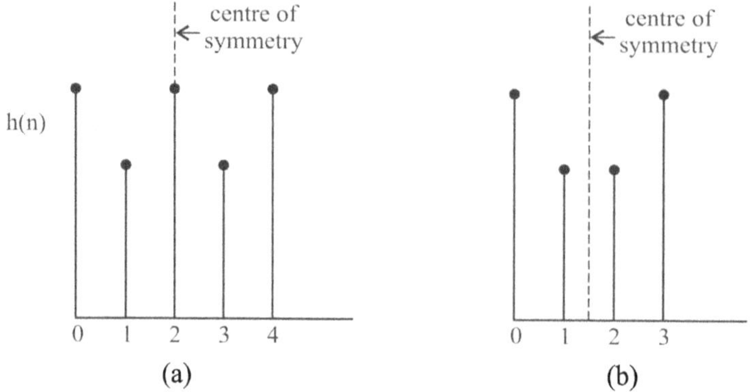

Fig. 5.1 Impulse responses of FIR filter for symmetry condition when (a) N odd (b) N even.

One more case for FIR filter to have Linear phase is

$$\theta(\omega) = \beta - \alpha\omega \qquad\qquad .....(5.1.14)$$

where β and α are constants.

in this case FIR filter will have only constant group delay not the phase delay.

equation (5.1.5) is

$$H(\omega) = \pm|H(\omega)|e^{j(\beta-\alpha\omega)} \qquad\qquad .....(5.1.15)$$

equation (5.1.8) can be written as

$$\sum_{n=0}^{N-1}h(n)\, e^{-j\omega n} = \pm|H(\omega)|e^{j(\beta-\alpha\omega)} \qquad\qquad .....(5.1.16)$$

from equation (5.1.16), we can write

$$\sum_{n=0}^{N-1}h(n)\, \cos(\omega n) = \pm|H(\omega)|\cos(\beta-\alpha\omega) \qquad\qquad .....(5.1.17)$$

and
$$-\sum_{n=0}^{N-1}h(n)\, \sin(\omega n) = \pm|H(\omega)|\sin(\beta-\alpha\omega) \qquad\qquad .....(5.1.18)$$

Taking ratio of equation (5.1.18) to equation (5.1.17), we get

$$\frac{-\displaystyle\sum_{n=0}^{N-1}h(n)\, \sin(\omega n)}{\displaystyle\sum_{n=0}^{N-1}h(n)\, \cos(\omega n)} = \frac{\sin(\beta-\alpha\omega)}{\cos(\beta-\alpha\omega)}$$

which can be written as

$$\sum_{n=0}^{N-1} h(n)\,\sin\left[\beta-(\alpha-n)\omega\right]=0 \qquad\qquad\dots(5.1.19)$$

If $\beta=\pi/2$, equation (5.1.19) becomes

$$\sum_{n=0}^{N-1} h(n)\,\cos(\alpha-n)\omega=0 \qquad\qquad\dots(5.1.20)$$

equation (5.1.20) will be zero when

$$h(n)=-h(N-1-n) \qquad\qquad\dots(5.1.21a)$$

and $$\alpha=\frac{N-1}{2} \qquad\qquad\dots(5.1.21b)$$

Therefore, FIR filters have Linear phase (τ_g constant and τ_p not constant) when the impulse response is antisymmetrical about α.

The impulse response satisfying equation (5.1.21) is shown in Fig (5.2). When N = 5 the centre of antisymmetry occurs at 2nd sample, and when N = 4 the centre of antisymmetry occurs at 1.5 sample.

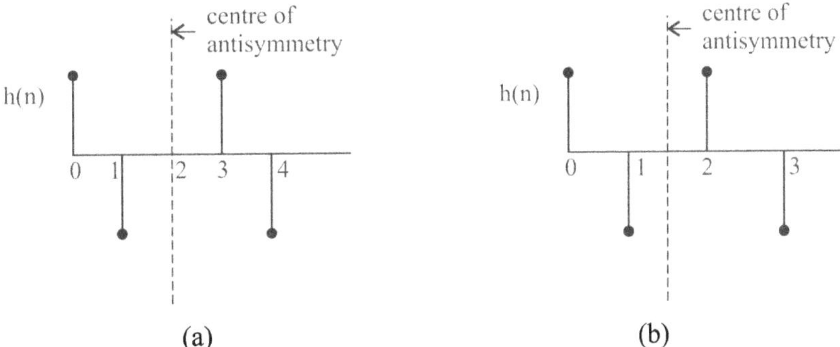

Fig. 5.2 Impulse responses of FIR filter for Antisymmetry condition when (a) N odd (b) N even.

From Fig 5.2(a), $h\left(\dfrac{N-1}{2}\right)=0$ for antisymmetric condition.

5.1.2 Frequency Response

As per the discussion from the previous section, for FIR filters to have linear phase frequency response, it should satisfy two conditions.

Symmetry condition $\qquad h(n)=h(N-1-n)$

Antisymmetry condition $\quad h(n)=-h(N-1-n) \qquad\qquad\dots(5.1.22)$

Let us find frequency response by imposing the above two conditions on the unit sample response.

Case I: Frequency response by imposing symmetry condition $[h(n) = h(N{-}1{-}n)]$:

(a) When N is odd

Let N = 5

$\therefore$ $h(0) = h(5{-}1{-}0) = h(4)$

$h(1) = h(5{-}1{-}1) = h(3)$

$h(2) = h(5{-}1{-}2) = h(2)$

$h(3) = h(5{-}1{-}3) = h(1)$

$h(4) = h(5{-}1{-}4) = h(0)$

Observe $h(2)$ does not have matching term. Thus the filter impulse response is symmetric about the point $h(2)$.

The corresponding frequency response from equation 5.1.4 is

$$H(\omega) = h(0) + h(1)\,e^{-j\omega} + h(2)\,e^{-j2\omega} + h(3)\,e^{-j3\omega} + h(4)\,e^{-j4\omega}$$

$$= e^{-j2\omega}\left[h(2) + h(0)\,e^{j2\omega} + h(4)\,e^{-j2\omega} + h(1)\,e^{j\omega} + h(3)\,e^{-j\omega} \right]$$

Since $h(0) = h(4)$ and $h(1) = h(3)$, $H(\omega)$ can be expressed as

$$H(\omega) = e^{-j2\omega}\left[h(2) + 2h(0)\cos2\omega + 2h(1)\cos\omega \right] \qquad \text{.....(5.1.23)}$$

Comparing equation (5.1.5) with equation (5.1.23), let $|H(\omega)| = |H_r(\omega)|$

$$H_r(\omega) = h(2) + 2h(0)\cos 2\omega + 2h(1)\cos\omega \qquad \text{.....(5.1.24)}$$

The phase characteristic of the filter is

$$\theta(\omega) = \begin{cases} -2\omega, & \text{if } H_r(\omega) > 0 \\ -2\omega + \pi, & \text{if } H_r(\omega) < 0 \end{cases} \qquad \text{.....(5.1.25)}$$

We observe that $\theta(\omega)$ is a linear function of ω

(b) When N is even

Let N = 4

$\therefore$ $h(0) = h(4{-}1{-}0) = h(3)$

$h(1) = h(4{-}1{-}1) = h(2)$

$h(2) = h(4{-}1{-}2) = h(1)$

$h(3) = h(4{-}1{-}3) = h(0)$

In this case each term has matching term.

The corresponding frequency response is

$$H(\omega) = h(0) + h(1)\, e^{-j\omega} + h(2)\, e^{-j2\omega} + h(3)\, e^{-j3\omega}$$

$$H(\omega) = e^{-j3\omega/2}\left[h(0)\, e^{j3\omega/2} + h(3)\, e^{-j3\omega/2} + h(1)\, e^{j\omega/2} + h(2)\, e^{-j\omega/2} \right]$$

Since $h(0) = h(3)$ and $h(1) = h(2)$

$$\therefore \qquad H(\omega) = e^{-j3\omega/2}\left[2h(0)\cos(3\omega/2) + 2h(1)\cos(\omega/2) \right] \qquad\qquad .....(5.1.26)$$

Let $\qquad |H(\omega)| = |H_r(\omega)|$

$$H_r(\omega) = 2h(0)\cos(3\omega/2) + 2h(1)\cos(\omega/2) \qquad\qquad .....(5.1.27)$$

The phase characteristic of the filter is

$$\theta(\omega) = \begin{cases} \dfrac{-3\omega}{2} & \text{if } H_r(\omega) > 0 \\[2ex] \dfrac{-3\omega}{2} + \pi & \text{if } H_r(\omega) < 0 \end{cases} \qquad\qquad .....(5.1.28)$$

We observe that $\theta(\omega)$ is a linear function of ω

From the above discussions, a generalized frequency response of filter length N, having impulse response that satisfies symmetry condition may be expressed as

$$H(\omega) = H_r(\omega)\, e^{-j\omega(N-1)/2} \qquad\qquad .....(5.1.29)$$

where

$$H_r(\omega) = h\left(\frac{N-1}{2}\right) + 2\sum_{n=0}^{(N-3)/2} h(n)\cos\left[\omega\left(\frac{N-1}{2} - n\right)\right], \quad N \text{ odd} \quad .....(5.1.30a)$$

$$H_r(\omega) = 2\sum_{n=0}^{\frac{N}{2}-1} h(n)\cos\left[\omega\left(\frac{N-1}{2} - n\right)\right], \qquad\qquad N \text{ even} \quad .....(5.1.30b)$$

The phase characteristic of the filter for both N odd and N even is

$$\theta(\omega) = \begin{cases} -\omega\left(\dfrac{N-1}{2}\right), & \text{if } H_r(\omega) > 0 \\[2ex] -\omega\left(\dfrac{N-1}{2}\right) + \pi, & \text{if } H_r(\omega) < 0 \end{cases} \qquad\qquad .....(5.1.31)$$

***Case* II:** Frequency response by imposing antisymmetry condition $[h(n) = -h(N-1-n)]$:

    (a)   When N is odd

Let N = 5

$\therefore$　$h(0) = -h(5-1-0) = -h(4)$

　　$h(1) = -h(5-1-1) = -h(3)$

　　$h(2) = -h(5-1-2) = -h(2) \implies 2h(2) = 0, \quad \text{i.e.,} h(2) = 0$

　　$h(3) = -h(5-1-3) = -h(1)$

　　$h(4) = -h(5-1-4) = -h(0)$

We observe that $h\left(\dfrac{N-1}{2}\right) = h(2) = 0$

$\therefore$　Antisymmetry occurs at h(2)

The corresponding frequency response is

$$H(\omega) = h(0) + h(1)\, e^{-j\omega} + h(3)\, e^{-j3\omega} + h(4)\, e^{-j4\omega}$$

$$= e^{-j2\omega}\left[h(0)\, e^{j2\omega} + h(4)\, e^{-j2\omega} + h(1)\, e^{j\omega} + h(3)\, e^{-j\omega} \right]$$

Since h(0) = -h(4) and h(1) = -h(3), $H(\omega)$ can be expressed as

$$H(\omega) = e^{-j2\omega} \cdot j \left\{ 2h(0)\left[\frac{e^{j2\omega} - e^{-j2\omega}}{2j}\right] + 2h(1)\left[\frac{e^{j\omega} - e^{-j\omega}}{2j}\right] \right\}$$

$$= e^{-j2\omega} e^{j\frac{\pi}{2}} \left[2h(0)\sin 2\omega + 2h(1)\sin \omega \right]$$

$$H(\omega) = e^{j\left(\frac{\pi}{2} - 2\omega\right)} \left[2h(0)\sin 2\omega + 2h(1)\sin \omega \right] \qquad\qquad \text{.....(5.1.32)}$$

$$H_r(\omega) = 2h(0)\sin 2\omega + 2h(1)\sin \omega \qquad\qquad \text{.....(5.1.33)}$$

The phase characteristic of the filter is

$$\theta(\omega) = \begin{cases} \dfrac{\pi}{2} - 2\omega, & \text{if } H_r(\omega) > 0 \\[2mm] \dfrac{\pi}{2} - 2\omega + \pi, & \text{if } H_r(\omega) < 0 \end{cases} \qquad\qquad \text{.....(5.1.34)}$$

We observe that $\theta(\omega)$ is a linear function of ω

(b)　When N is even

　　Let N = 4

$\therefore$　$h(0) = -h(4-1-0) = -h(3)$

$$h(1) = -h(4-1-1) = -h(2)$$

$$h(2) = -h(4-1-2) = -h(1)$$

$$h(3) = -h(4-1-3) = -h(0)$$

The corresponding frequency response is

$$H(\omega) = h(0) + h(1)\,e^{-j\omega} + h(2)\,e^{-j2\omega} + h(3)\,e^{-j3\omega}$$

$$= e^{-j3\omega/2}\left[h(0)\,e^{j3\omega/2} + h(3)\,e^{-j3\omega/2} + h(1)\,e^{j\omega/2} + h(4)\,e^{-j\omega/2} \right]$$

Since $h(0) = -h(3)$ and $h(1) = -h(2)$

$$H(\omega) = e^{-j3\omega/2}\cdot j\left\{ 2h(0)\left[\frac{e^{j3\omega/2} - e^{-j3\omega/2}}{2j} \right] + 2h(1)\left[\frac{e^{j\omega/2} - e^{-j\omega/2}}{2j} \right] \right\}$$

$$H(\omega) = e^{j\left(\frac{\pi}{2} - \frac{3\omega}{2}\right)}\left[2h(0)\sin\left(\frac{3\omega}{2}\right) + 2h(1)\sin\left(\frac{\omega}{2}\right) \right] \qquad .....(5.1.35)$$

$$H_r(\omega) = 2h(0)\sin\left(\frac{3\omega}{2}\right) + 2h(1)\sin\left(\frac{\omega}{2}\right) \qquad .....(5.1.36)$$

The phase characteristic of the filter is

$$\theta(\omega) = \begin{cases} \dfrac{\pi}{2} - \dfrac{3\omega}{2}, & \text{if } H_r(\omega) > 0 \\[2ex] \dfrac{\pi}{2} - \dfrac{3\omega}{2} + \pi, & \text{if } H_r(\omega) < 0 \end{cases} \qquad .....(5.1.37)$$

A generalized frequency response of filter length N, having impulse response that satisfies antisymmetry condition may be expressed as

$$H(\omega) = H_r(\omega)\,e^{j\left[-\omega\left(\frac{N-1}{2}\right) + \frac{\pi}{2} \right]} \qquad .....(5.1.38)$$

where

$$H_r(\omega) = 2\sum_{n=0}^{(N-3)/2} h(n)\sin\left[\omega\left(\frac{N-1}{2} - n\right) \right], \quad N \text{ odd} \qquad .....(5.1.39)$$

$$H_r(\omega) = 2\sum_{n=0}^{\frac{N}{2}-1} h(n)\sin\left[\omega\left(\frac{N-1}{2} - n\right) \right], \qquad N \text{ even} \qquad .....(5.1.40)$$

The phase characteristic of the filter for both N odd and N even is

$$\theta(\omega) = \begin{cases} \dfrac{\pi}{2} - \omega\left(\dfrac{N-1}{2}\right), & \text{if } H_r(\omega) > 0 \\[2mm] \dfrac{3\pi}{2} - \omega\left(\dfrac{N-1}{2}\right), & \text{if } H_r(\omega) < 0 \end{cases} \qquad .....(5.1.41)$$

5.1.3 Location of Zeros of Linear Phase FIR Filter

The positions of the zeros of linear phase FIR filters are highly constrained by the symmetry conditions on the impulse response.

The transfer function of a linear phase FIR filter is given by

$$H(z) = \sum_{n=0}^{N-1} h(n)\, z^{-n} \qquad .....(5.1.42)$$

which can be written as

$$H(z) = \sum_{n=0}^{N-1} h(n)\, z^{-n} = h(0) + h(1)\, z^{-1} + h(2)\, z^{-2} + ...$$

$$\pm h(2)\, z^{-(N-3)} \pm h(1)\, z^{-(N-2)} \pm h(0)\, z^{-(N-1)} \quad .....(5.1.43)$$

where the + sign corresponds to symmetry in the impulse response and − sign corresponds to anti-symmetry in the impulse response.

equation (5.1.43) can be written as

$$H(z) = z^{-(N-1)/2} \left\{ h(0)\left[z^{(N-1)/2} \pm z^{-(N-1)/2} \right] + h(1)\left[z^{(N-3)/2} \pm z^{-(N-3)/2} \right] \right.$$

$$\left. + h(2)\left[z^{(N-5)/2} \pm z^{-(N-5)/2} \right] + ... \right\} \qquad .....(5.1.44)$$

If z is replaced by z^{-1} in equation (5.1.44), the result is

$$H(z^{-1}) = z^{(N-1)/2} \left\{ h(0)\left[z^{-(N-1)/2} \pm z^{(N-1)/2} \right] + h(1)\left[z^{-(N-3)/2} \pm z^{(N-3)/2} \right] \right.$$

$$\left. + h(2)\left[z^{-(N-5)/2} \pm z^{(N-5)/2} \right] + ... \right\} \qquad .....(5.1.45)$$

relating equation (5.1.45) to equation (5.1.44), gives

$$H(z^{-1}) = \pm z^{(N-1)}\, H(z) \qquad .....(5.1.46)$$

Equation (5.1.46) shows that H(z) and $H(z^{-1})$ are identical to within a delay of (N–1) samples and a multiplier of ±1. Thus the zeros of $H(z^{-1})$ are identical to the zeros of H(z).

Let us assume, H(z) has a complex zero

at $\qquad z_i = r_i e^{j\theta_i}$, with $r_i \neq 1$, $\theta_i \neq 0, \pi$

equation (5.1.46) says that H(z) must also have a mirror-image zero

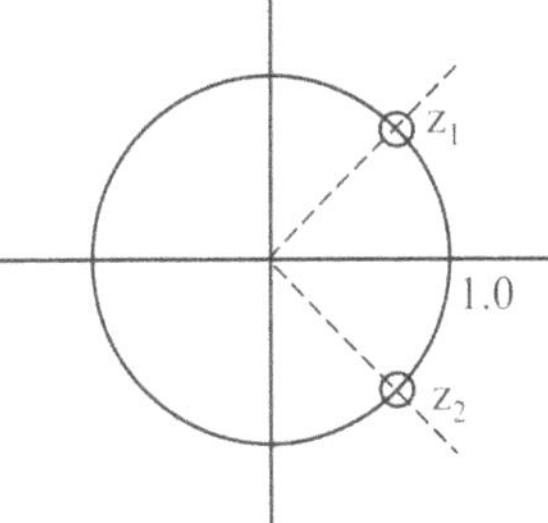

at $\qquad z_i^{-1} = \left(\dfrac{1}{r_i}\right) e^{-j\theta_i}$

(i) So for every complex zero of H(z)

If $r_i \neq 1$ and $\theta_i \neq 0, \pi$

H(z) consists of atleast the elemental factor $H_i(z)$ of the form

$$H_i(z) = \left(1 - z^{-1} r_i e^{j\theta_i}\right)\left(1 - z^{-1} r_i e^{-j\theta_i}\right)\left(1 - z^{-1}\frac{1}{r_i}e^{j\theta_i}\right)\left(1 - z^{-1}\frac{1}{r_i}e^{-j\theta_i}\right) \quad .....(5.1.47)$$

From equation (5.1.47), we can say for every complex zero of H(z), it contains extra three more zeros as shown in fig 5.3

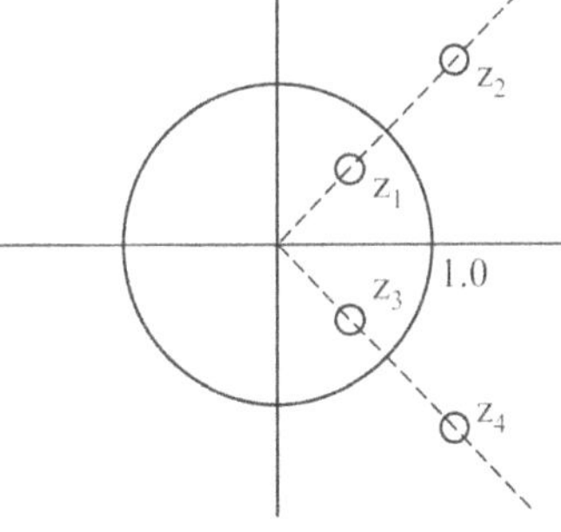

z_1 coordinates are (r, θ)

z_2 coordinates are $\left(\dfrac{1}{r}, \theta\right)$

z_3 coordinates are $(r, -\theta)$

z_4 coordinates are $\left(\dfrac{1}{r}, -\theta\right)$

Fig. 5.3 Zeros locations of FIR filter.

(ii) If $r_i = 1$ and $\theta_i \neq 0, \pi$, the zeros lie on the unit circle and its complex conjugate is its own mirror image;

elemental factor is

$$H_i(z) = \left(1 - z^{-1} e^{j\theta_i}\right)\left(1 - z^{-1} e^{-j\theta_i}\right) \qquad .....(5.1.48)$$

If a complex pole lies on the unit circle, there will be one more pole of it's mirror image as shown in Fig 5.4

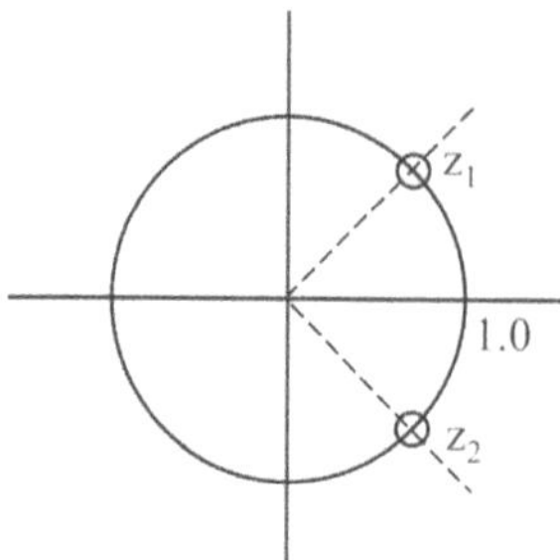

z_1 coordinates are $(1, \theta)$
z_2 coordinates are $(1, -\theta)$

Fig. 5.4 Zeros locations of FIR filter.

(iii) If $r_i \neq 1$ but $\theta_i = 0$ or π, the zeros are real;

the elemental factor is

$$H_i(z) = \left(1 \pm r_i\, z^{-1}\right)\left(1 \pm \frac{1}{r_i}\, z^{-1}\right) \qquad \qquad .....(5.1.49)$$

where the $+$ sign corresponds to $\theta_i = \pi$

$-$ sign corresponds to $\theta_i = 0$

In this case there will be two zeros.

(iv) If $r_i = 1$ and $\theta_i = 0$ or π, the zeros lie at either $z = +1$ or $z = -1$

the elemental factor is

$$H_i(z) = (1 \pm z^{-1}) \qquad \qquad .....(5.1.50)$$

Where $+$ sign corresponds to $\theta_i = \pi$

$-$ sign corresponds to $\theta_i = 0$

in this case there will be simple zero.

5.2 Design of FIR Filters using Window Techniques

We begin with the desired frequency response specification $H_d(\omega)$ and determine the corresponding unit sample response $h_d(n)$.

Causal filter desired frequency response can be related with $h_d(n)$ as

$$H_d(\omega) = \sum_{n=0}^{\infty} h_d(n)\, e^{-j\omega n} \qquad \qquad .....(5.2.1)$$

where
$$h_d(n) = \frac{1}{2\pi} \int_{-\pi}^{\pi} H_d(\omega)\, e^{j\omega n}\, d\omega \qquad \ldots\ldots(5.2.2)$$

In general, the impulse response $h_d(n)$ obtained from equation (5.2.2) is infinite in duration. To yield an FIR filter of length N, $h_d(n)$ must be truncated at some point, say at $n = N - 1$.

Truncation of $h_d(n)$ to a length N is equivalent to multiplying $h_d(n)$ by a "window" $w(n)$ of length N.

$\therefore$ Impulse response of FIR filter is $h(n) = h_d(n)\, \omega(n)$ $\qquad \ldots\ldots(5.2.3)$

We have different types of windows that are used namely

1. Rectangular window
2. Triangular (or) Barlett window
3. Raised cosine window
4. Hanning window
5. Hamming window
6. Blackman window

5.2.1 Rectangular Window

Causal rectangular window is defined as

$$W_R(n) = \begin{cases} 1, & n = 0,1,\ldots,N-1 \\ 0, & \text{otherwise} \end{cases} \qquad \ldots\ldots(5.2.4a)$$

Non-causal rectangular window is

$$W_R(n) = \begin{cases} 1, & -(N-1)/2 \le n \le (N-1)/2 \\ 0, & \text{otherwise} \end{cases} \qquad \ldots\ldots(5.2.4b)$$

equation (5.2.4a) is shown in Fig 5.5

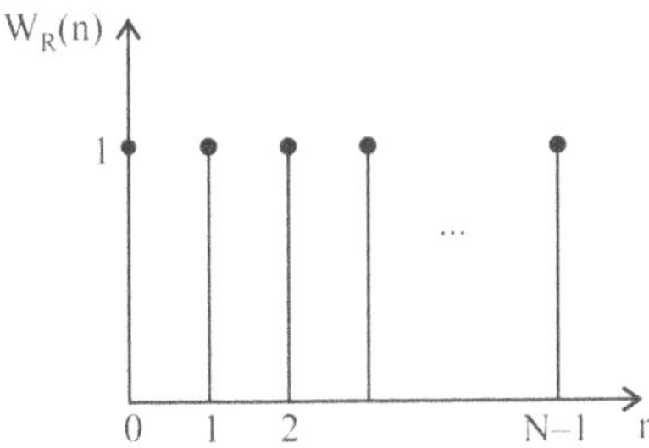

Fig. 5.5 Rectangular window.

Spectrum of rectangular window is given by

$$W_R(\omega) = \sum_{n=0}^{N-1} \omega_R(n)\, e^{-j\omega n}$$

$$= \sum_{n=0}^{N-1} e^{-j\omega n} = \frac{1 - e^{-j\omega n}}{1 - e^{-j\omega}} = e^{-j\omega(N-1)/2}\, \frac{\sin(\omega N/2)}{\sin(\omega/2)}$$

$$W_R(\omega) = e^{-j\omega\alpha}\, \frac{\sin(\omega N/2)}{\sin(\omega/2)}, \qquad \text{where } \alpha = \frac{N-1}{2}$$

Magnitude response

$$\left|W_R(\omega)\right| = \frac{\left|\sin(\omega N/2)\right|}{\left|\sin(\omega/2)\right|}; \qquad -\pi \le \omega \le \pi \qquad\qquad \dots\dots(5.2.5)$$

Phase response

$$\theta(\omega) = \begin{cases} -\omega\alpha, & \text{if } \sin(\omega N/2) \ge 0 \\ -\omega\alpha + \pi, & \text{if } \sin(\omega N/2) < 0 \end{cases} \qquad\qquad \dots\dots(5.2.6)$$

equation (5.2.5) is shown in Fig 5.6 and it's Log magnitude response is shown in Fig 5.7.

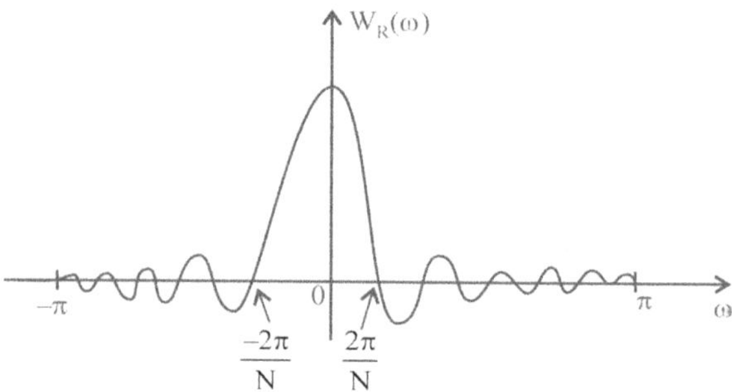

Fig. 5.6 Frequency response of rectangular window.

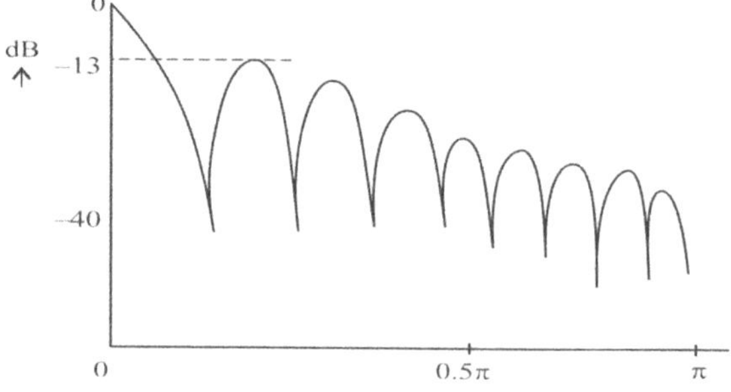

Fig. 5.7 Log magnitude response of rectangular window.

The unit sample response of the FIR filter becomes

$$h(n) = h_d(n)W_R(n)$$

where $h_d(n)$ is the desired unit sample response and

$w_R(n)$ is the Rectangular window

Frequency response of the truncated filter can be obtained by convolving $H_d(W)$ with $W_R(W)$.

$$H(\omega) = \frac{1}{2\pi} \int_{-\pi}^{\pi} H_d(v)\omega_R(\omega - v)\,dv \qquad \ldots\ldots(5.2.7)$$

$$= H_d(\omega) * W_R(\omega)$$

(a) Lowpass FIR Filter

Let us design a symmetric lowpass linear-phase FIR filter having a desired frequency response

$$H_d(\omega) = \begin{cases} 1\,e^{-j\omega\alpha}, & 0 \le |\omega| \le \omega_c \\ 0, & \text{otherwise} \end{cases} \qquad \ldots\ldots(5.2.8)$$

where $\alpha = \dfrac{N-1}{2}$

equation (5.2.8) is shown in Fig 5.8

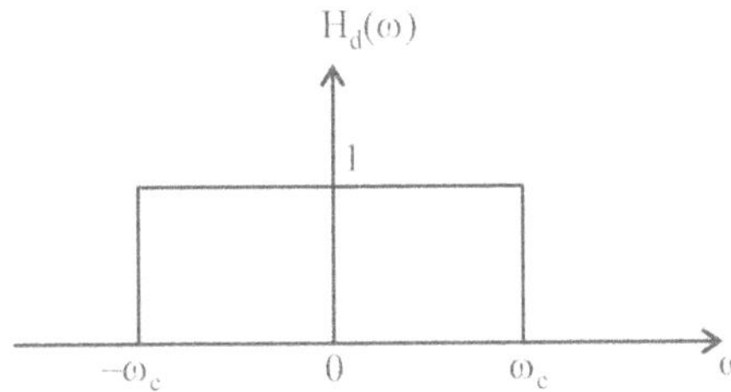

Fig. 5.8 Desired frequency response of lowpass filter.

The corresponding impulse response, obtained by evaluating the integral in equation (5.2.2)

$$\therefore \qquad h_d(n) = \frac{1}{2\pi} \int_{-\omega_c}^{\omega_c} e^{-j\omega\alpha}e^{j\omega n}\,d\omega$$

$$= \frac{1}{2\pi} \int_{-\omega_c}^{\omega_c} e^{j\omega(n-\alpha)}\,d\omega$$

$$= \frac{1}{2\pi \, j(n-\alpha)} \left[e^{j\omega(n-\alpha)} \right]_{-\omega_c}^{\omega_c}$$

$$= \frac{1}{\pi \, (n-\alpha)} \left[\frac{e^{j\omega_c(n-\alpha)} - e^{-j\omega_c(n-\alpha)}}{2 \, j} \right]$$

$$h_d(n) = \frac{\sin\left[\omega_c(n-\alpha)\right]}{\pi \, (n-\alpha)}, \; n \neq \alpha \qquad\qquad \ldots..(5.2.9a)$$

$$= \frac{\omega_c}{\pi}, \qquad\qquad n = \alpha \qquad\qquad \ldots..(5.2.9b)$$

Clearly $h_d(n)$ is noncausal and infinite in duration.

equations (5.2.29) is shown in Fig 5.9

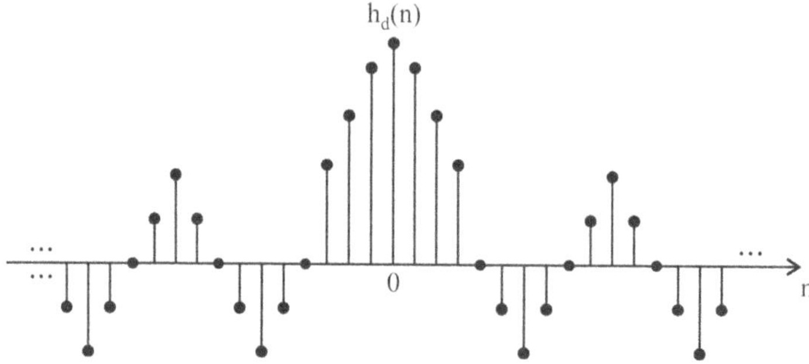

Fig. 5.9 Impulse response of lowpass filter.

If we multiply $h_d(n)$ by the rectangular window in equation (5.2.4a) we obtain an FIR filter of length N having the impulse response.

$$h(n) = \frac{\sin\left[\omega_c(n-\alpha)\right]}{\pi \, (n-\alpha)}, \quad 0 \leq n \leq N-1, \; n \neq \alpha \qquad \ldots..(5.2.10a)$$

$$= \frac{\omega_c}{\pi}, \quad n = \alpha \qquad\qquad \ldots..(5.2.10b)$$

Frequency response of Linear phase Lowpass FIR filter can be obtained by using equation (5.2.7)

i.e., $\qquad\qquad H(\omega) = H_d(\omega) * W_R(\omega)$

and it is shown in Fig 5.8.

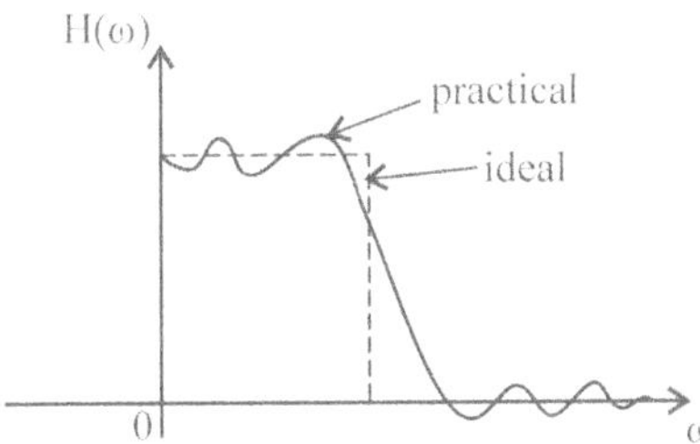

Fig. 5.10 Frequency response of Lowpass FIR filter.

Gibb's Phenomenon:

The magnitude of the frequency response $H(\omega)$ of lowpass filter is shown in Fig 5.10. We observe that relatively large oscillations or ripples occur near the band edge of the filter. The oscillations increase in frequency as N increases, but they do not diminish in amplitude. These large oscillations are the direct result of the large side lobes that exist in the frequency response of rectangular window $W_R(\omega)$ shown in fig 5.7. As this window function is convolved with the desired frequency response $H_d(\omega)$ shown in fig 5.8, the oscillations occur as the large constant area side lobes of $W_R(w)$ move across the discontinuity that exist in $H_d(\omega)$. Since equation (5.2.1) is basically a Fourier series representation of $H_d(\omega)$, the multiplication of $h_d(n)$ with a rectangular window is identical to truncating the Fourier series representation of the desired filter $H_d(\omega)$. The truncation of the Fourier series is known to introduce ripples in the frequency response characteristic $H(\omega)$ due to the nonuniform convergence of the Fourier series at a discontinuity. These ripples or oscillatory behavior near the band edge of the filter is called the Gibb's phenomenon as shown in Fig 5.10.

The large side lobes in the frequency response of rectangular window are due to abrupt discontinuities in the time-domain characteristic of rectangular window shown in Fig 5.5. Due to these large side lobes there exist ripples or oscillations, which is also called Gibb's phenomenon, at the band edges of the lowpass FIR filter. This can be reduced by using windows which do not have abrupt discontinuities in the time domain characteristics, and, correspondingly, low side lobes in their frequency–domain characteristics. Examples for these type of windows are Triangular, Hanning, Hamming, Blackman windows etc.

(b) Highpass FIR Filter

Let us design a symmetric highpass linear–phase FIR filter having a desired frequency response.

$$H_d(\omega) = \begin{cases} 1\, e^{-j\omega\alpha}, & \omega_c \le |\omega| \le \pi \\ 0, & \text{otherwise} \end{cases} \qquad \ldots\ldots(5.2.11)$$

where $\qquad \alpha = \dfrac{N-1}{2}$

equation (5.2.2) is shown in Fig 5.11

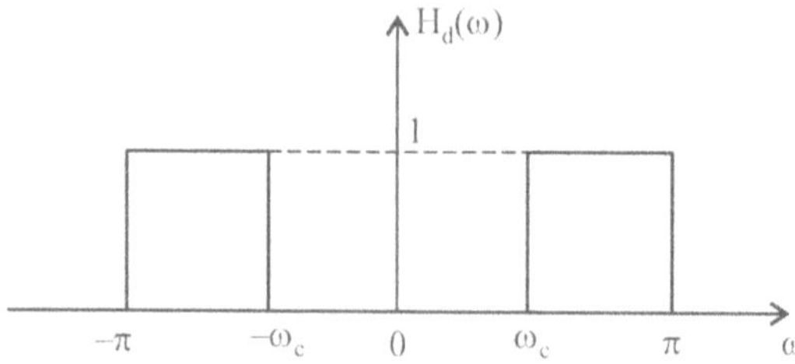

Fig. 5.11 Desired frequency response of highpass filter.

The corresponding impulse response is

$$h_d(n) = \frac{1}{2\pi}\left[\int_{-\pi}^{-\omega_c} e^{j\omega(n-\alpha)}\,d\omega + \int_{\omega_c}^{\pi} e^{j\omega(n-\alpha)}\,d\omega\right]$$

$$= \frac{1}{2\pi(n-\alpha)j}\left[e^{-j\omega_c(n-\alpha)} - e^{-j\pi(n-\alpha)} + e^{j\pi(n-\alpha)} - e^{j\omega_c(n-\alpha)}\right]$$

$$= \frac{1}{\pi(n-\alpha)}\left[\frac{e^{j\pi(n-\alpha)} - e^{-j\pi(n-\alpha)}}{2j} - \frac{e^{j\omega_c(n-\alpha)} - e^{-j\omega_c(n-\alpha)}}{2j}\right]$$

$$h_d(n) = \frac{1}{\pi(n-\alpha)}\left[\sin\left[\pi(n-\alpha)\right] - \sin\left[\omega_c(n-\alpha)\right]\right],\ n \neq \alpha \qquad \ldots(5.2.12a)$$

$$= \frac{\pi - \omega_c}{\pi},\qquad\qquad\qquad\qquad n = \alpha \qquad \ldots(5.2.12b)$$

(c)　Bandpass FIR Filter

Let us design a symmetric Bandpass linear–phase FIR filter having a desired frequency response

$$H_d(\omega) = \begin{cases} 1\,e^{-j\omega\alpha}, & \omega_l \leq |\omega| \leq \omega_u \\ 0, & \text{otherwise} \end{cases} \qquad \ldots(5.2.13)$$

where $\qquad \alpha = \dfrac{N-1}{2}$

equation (5.2.13) is shown in Fig 5.12

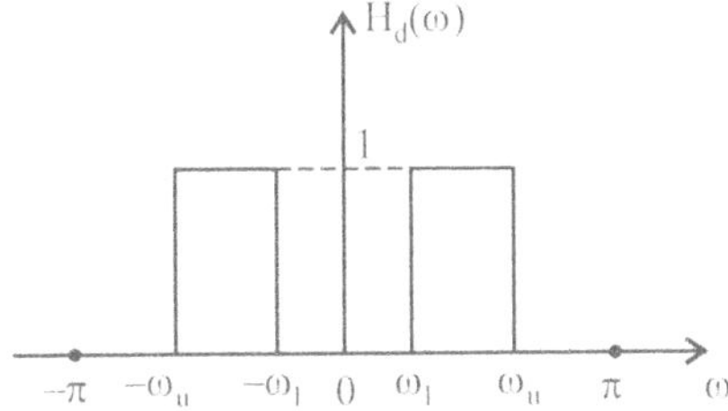

Fig. 5.12 Desired frequency response of Bandpass filter.

The corresponding impulse response is

$$h_d(n) = \frac{1}{2\pi}\left[\int_{-\omega_u}^{-\omega_l} e^{j\omega(n-\alpha)} \, d\omega + \int_{\omega_l}^{\omega_u} e^{j\omega(n-\alpha)} \, d\omega\right]$$

$$= \frac{1}{\pi(n-\alpha)}\left[\sin\left[\omega_u(n-\alpha)\right] - \sin\left[\omega_l(n-\alpha)\right]\right], n \neq \alpha \qquad \ldots\ldots(5.2.14a)$$

$$= \frac{\omega_u - \omega_l}{\pi}, \ n = \alpha \qquad \ldots\ldots(5.2.14b)$$

(d)   Band Reject FIR Filter

Let us design a symmetric band reject linear–phase FIR filter having a desired frequency response

$$H_d(\omega) = \begin{cases} 1 \ e^{-j\omega\alpha}, & \text{otherwise} \\ 0, & \omega_l \leq |\omega| \leq \omega_u \end{cases} \qquad \ldots\ldots(5.2.15)$$

equation (5.2.15) is shown in Fig 5.13

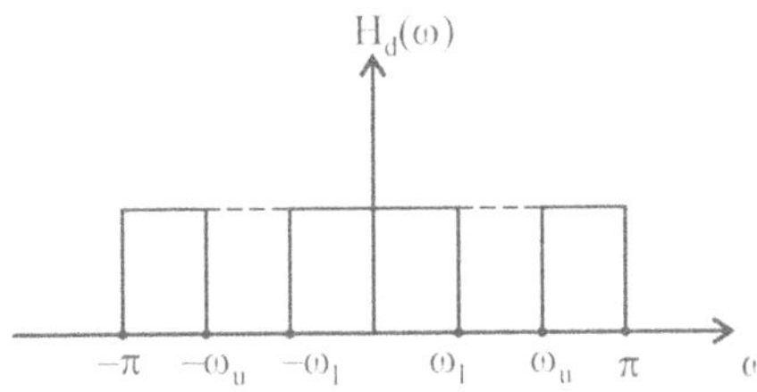

Fig. 5.13 Desired frequency response of band reject filter.

The corresponding impulse response is

$$h_d(n) = \frac{1}{2\pi}\left[\int_{-\pi}^{-\omega_u} e^{j\omega(n-\alpha)}\, d\omega + \int_{-\omega_l}^{\omega_l} e^{j\omega(n-\alpha)}\, d\omega + \int_{\omega_u}^{\pi} e^{j\omega(n-\alpha)}\, d\omega\right]$$

$$h_d(n) = \frac{1}{2\pi j(n-\alpha)}\left[e^{-j\omega_u(n-\alpha)} - e^{-j\pi(n-\alpha)} + e^{j\omega_l(n-\alpha)} - e^{-j\omega_l(n-\alpha)} + e^{j\pi(n-\alpha)} - e^{j\omega_u(n-\alpha)}\right]$$

$$= \frac{1}{\pi(n-\alpha)}\left[\sin\left[\omega_l(n-\alpha)\right] - \sin\left[\omega_u(n-\alpha)\right] + \sin\left[\pi(n-\alpha)\right]\right], \quad n \neq \alpha \quad .....(5.2.16a)$$

$$= \frac{\omega_l - \omega_u + \pi}{\pi}, \quad n = \alpha \qquad\qquad .....(5.2.16b)$$

5.2.2 Triangular (or) Barlett Window

Causal triangular window is defined as

$$W_T(n) = 1 - \frac{|n-\alpha|}{\alpha}, \quad 0 \leq n \leq N-1$$

$$= 0, \qquad\qquad \text{otherwise} \qquad\qquad .....(5.2.17a)$$

Non-causal triangular window is defined as

$$W_T(n) = 1 - \frac{|n|}{\alpha}, \quad -\alpha \leq n \leq \alpha ,$$

$$= 0, \qquad \text{otherwise} \qquad\qquad .....(5.2.17b)$$

$$\text{Where } \alpha = \frac{N-1}{2}$$

Causal time domain window can be obtained by shifting non-causal time domain window towards right i.e., replacing n with $n - \alpha$ in non-causal window.

Spectrum of the triangular window is

$$W_T(\omega) = \left[\frac{\sin\left(\dfrac{\alpha\omega}{2}\right)}{\sin\left(\dfrac{\omega}{2}\right)}\right]^2 \qquad\qquad .....(5.2.18)$$

Equation (5.2.17a) is shown in Fig 5.14 for N = 25 and it's frequency response is shown in Fig 5.15.

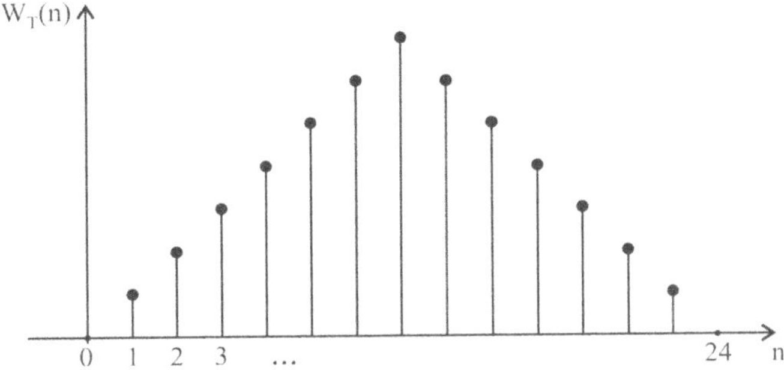

Fig. 5.14 Triangular window.

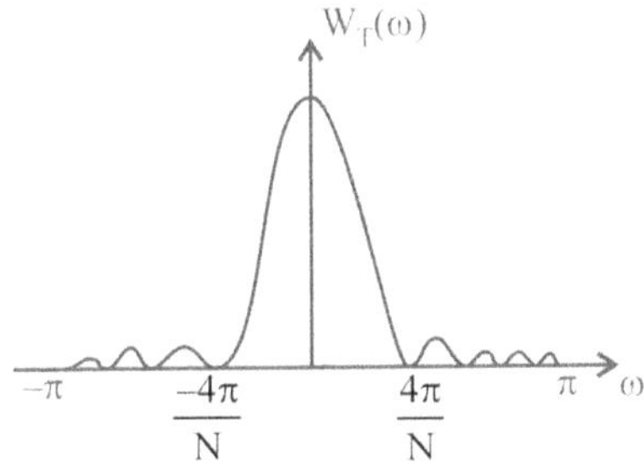

Fig. 5.15 Frequency response of triangular window.

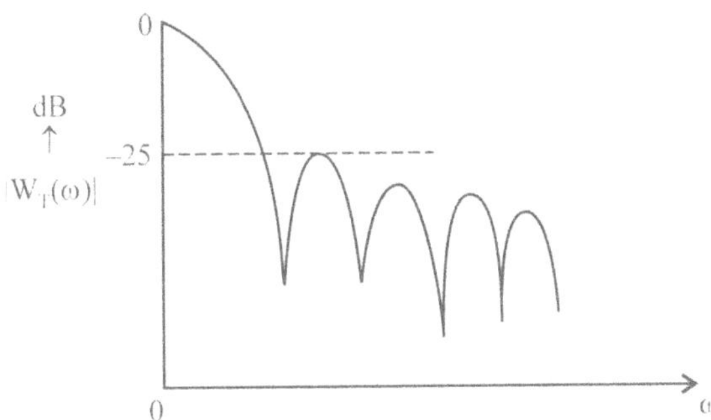

Fig. 5.16 Log magnitude of triangular window.

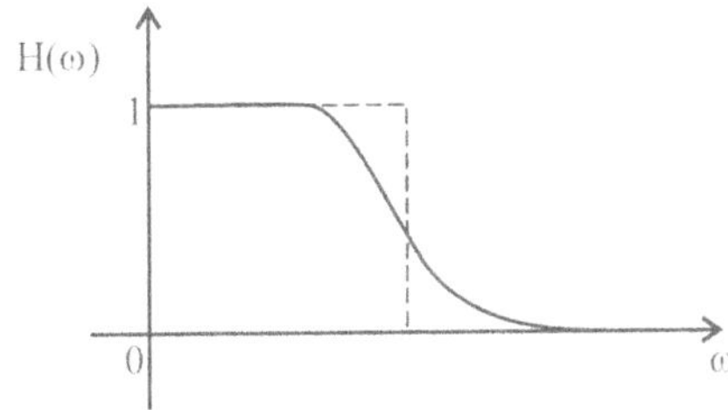

Fig. 5.17 Frequency response of LPF using triangular window.

Observe the Fig 5.16, the side lobe level is smaller i.e., –25 dB than that of rectangular window whose side lobe level is –13 dB. However, the main lobe width is now $\dfrac{8\pi}{N}$, which is twice that of the rectangular window.

From Fig 5.17, it is clear that triangular window produces a monotonic response in both passband and stopband. But it has some disadvantages, when compared with rectangular window.

(i)   Transition region is more

(ii) Attenuation in stopband is less

Due to these disadvantages, this window is not usually a good choice.

5.2.3 Raised Cosine Window

Non-causal window sequence is given by

$$W_\beta(n) = \beta + (1-\beta)\cos\left(\frac{\pi n}{\alpha}\right), \quad -\alpha \le n \le \alpha \qquad\qquad .....(5.2.19a)$$

$$= 0, \qquad\qquad\qquad\qquad \text{otherwise}$$

Causal window sequence can be obtained by replacing n with $n - \alpha$

$$W_\beta(n) = \beta - (1-\beta)\cos\left(\frac{\pi n}{\alpha}\right), \quad 0 \le n \le N-1 \qquad\qquad .....(5.2.19b)$$

$$= 0, \qquad\qquad\qquad\qquad \text{otherwise}$$

where $\alpha = \dfrac{N-1}{2}$

Spectrum of Raised Cosine Window is

$$W_\beta(\omega) = \sum_{n=0}^{N-1}\left[\beta - (1-\beta)\cos\frac{\pi n}{\alpha}\right]e^{-j\omega n}$$

$$W_\beta(\omega) = \beta\sum_{n=0}^{N-1}e^{-j\omega n} - \frac{(1-\beta)}{2}\left[\sum_{n=0}^{N-1}e^{-j\omega n}e^{j\frac{\pi n}{\alpha}} + \sum_{n=0}^{N-1}e^{-j\omega n}e^{-j\frac{\pi n}{\alpha}}\right]$$

$$= \beta \, \frac{1 - e^{-j\omega N}}{1 - e^{-j\omega}} - \frac{(1-\beta)}{2} \, \frac{1 - e^{j\left(\frac{\pi}{\alpha}-\omega\right)N}}{1 - e^{j\left(\frac{\pi}{\alpha}-\omega\right)}} - \frac{(1-\beta)}{2} \, \frac{1 - e^{-j\left(\frac{\pi}{\alpha}+\omega\right)N}}{1 - e^{-j\left(\frac{\pi}{\alpha}+\omega\right)}}$$

$$W_\beta(\omega) = \beta \, \frac{e^{-j\frac{\omega N}{2}}}{e^{-j\omega/2}} \, \frac{\sin\left(\frac{\omega N}{2}\right)}{\sin\left(\frac{\omega}{2}\right)} - \frac{(1-\beta)}{2} \, \frac{e^{j\left(\frac{\pi}{\alpha}-\omega\right)N/2}}{e^{j\left(\frac{\pi}{\alpha}-\omega\right)/2}} \, \frac{\sin\left[\left(\frac{\pi}{\alpha}-\omega\right)N/2\right]}{\sin\left[\left(\frac{\pi}{\alpha}-\omega\right)/2\right]}$$

$$- \frac{(1-\beta)}{2} \, \frac{e^{-j\left(\frac{\pi}{\alpha}+\omega\right)N/2}}{e^{-j\left(\frac{\pi}{\alpha}+\omega\right)/2}} \, \frac{\sin\left[\left(\frac{\pi}{\alpha}+\omega\right)N/2\right]}{\sin\left[\left(\frac{\pi}{\alpha}+\omega\right)/2\right]} \qquad \dots(5.2.20)$$

$$\left|W_\beta(\omega)\right| = \beta \, \frac{\sin\left(\frac{\omega N}{2}\right)}{\sin\left(\frac{\omega}{2}\right)} - \frac{(1-\beta)}{2} \, \frac{\sin\left[\left(\frac{\pi}{\alpha}-\omega\right)N/2\right]}{\sin\left[\left(\frac{\pi}{\alpha}-\omega\right)/2\right]} - \frac{(1-\beta)}{2} \, \frac{\sin\left[\left(\frac{\pi}{\alpha}+\omega\right)N/2\right]}{\sin\left[\left(\frac{\pi}{\alpha}+\omega\right)/2\right]} \qquad \dots(5.2.21)$$

5.2.4 Hanning Window

Non causal Hanning window sequence can be obtained by putting $\beta = 0.5$ in equation 5.2.19a.

$$W_{hn}(n) = 0.5 + 0.5 \, \cos\left(\frac{\pi n}{\alpha}\right), \quad -\alpha \le n \le \alpha \qquad \dots(5.2.22a)$$

$$= 0, \qquad\qquad \text{otherwise}$$

Causal window can be obtained by replacing n with $n - \alpha$

$$W_{hn}(n) = 0.5 - 0.5 \, \cos\left(\frac{\pi n}{\alpha}\right), \quad 0 \le n \le N - 1 \qquad \dots(5.2.22b)$$

$$= 0, \qquad\qquad \text{otherwise}$$

where $\quad \alpha = \dfrac{N-1}{2}$

Spectrum of Hanning window is

$$\left|W_{hn}(\omega)\right| = 0.5 \frac{\sin\left(\dfrac{\omega N}{2}\right)}{\sin\left(\dfrac{\omega}{2}\right)} - 0.25 \frac{\sin\left[\left(\dfrac{\pi}{\alpha}-\omega\right)N/2\right]}{\sin\left[\left(\dfrac{\pi}{\alpha}-\omega\right)/2\right]} - 0.25 \frac{\sin\left[\left(\dfrac{\pi}{\alpha}+\omega\right)N/2\right]}{\sin\left[\left(\dfrac{\pi}{\alpha}+\omega\right)/2\right]} \quad .....(5.2.23)$$

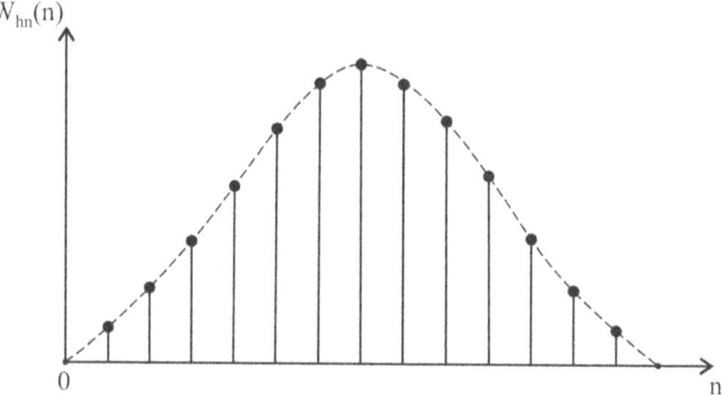

Fig. 5.18 Hanning window.

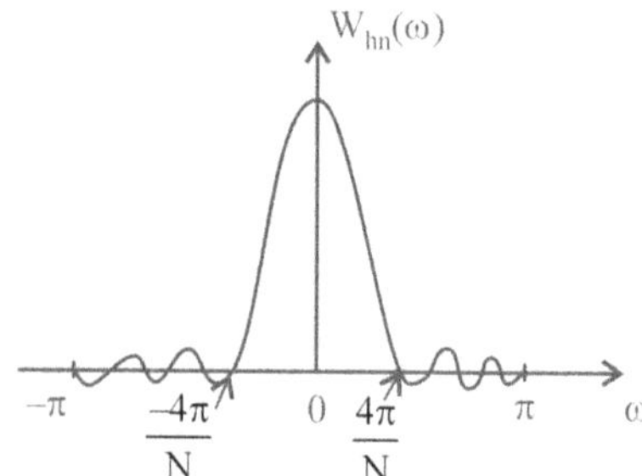

Fig. 5.19 Frequency response of hanning window.

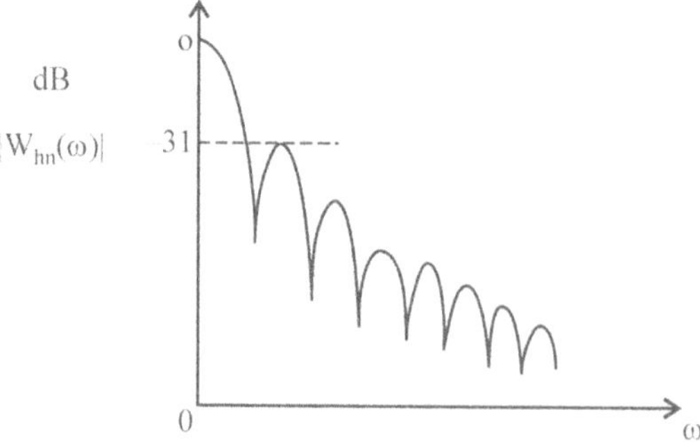

Fig. 5.20 Log magnitude response of hanning window.

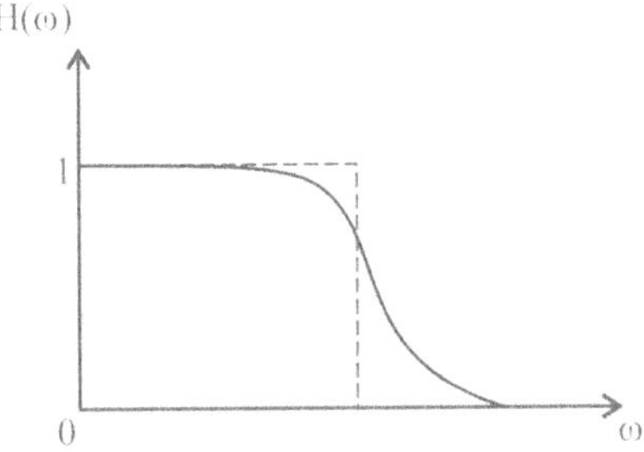

Fig. 5.21 Frequency response of LPF with hanning window.

Hanning window is shown in Fig 5.18. Observe the Fig 5.19, the main lobe width is twice that of rectangular window, due to this transition region of the filter will be doubled and see the Fig 5.20, the level of side lobes is −31 dB, which is very less than the side lobe level of rectangular window. Due to this there will be less number of ripples in both passband and stopband.

5.2.5 Hamming Window

Noncausal Hamming window sequence can be obtained by putting $\beta = 0.54$ in equation (5.2.19a)

$$W_{hm}(n) = 0.54 + 0.46\cos\left(\frac{\pi n}{\alpha}\right), \quad -\alpha \le n \le \alpha \qquad \ldots(5.2.24a)$$

$$= 0, \qquad \text{otherwise}$$

Causal window sequence can be obtained by replacing n with $n - \alpha$ in the above equation.

$$W_{hm}(n) = 0.54 - 0.46\cos\left(\frac{\pi n}{\alpha}\right), \quad 0 \le n \le N - 1 \qquad \ldots(5.2.24b)$$

$$= 0, \qquad \text{otherwise}$$

where $\quad \alpha = \dfrac{N-1}{2}$

Spectrum of Hamming window is

$$|W_{hm}(\omega)| = 0.54\,\frac{\sin\left(\dfrac{\omega N}{2}\right)}{\sin\left(\dfrac{\omega}{2}\right)} - 0.23\,\frac{\sin\left[\left(\dfrac{\pi}{\alpha}-\omega\right)N/2\right]}{\sin\left[\left(\dfrac{\pi}{\alpha}-\omega\right)/2\right]} - 0.23\,\frac{\sin\left[\left(\dfrac{\pi}{\alpha}+\omega\right)N/2\right]}{\sin\left[\left(\dfrac{\pi}{\alpha}+\omega\right)/2\right]} \qquad \ldots(5.2.25)$$

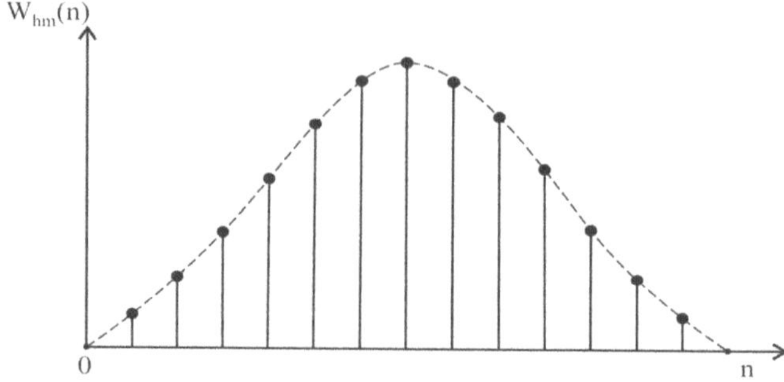

Fig. 5.22 Hamming window.

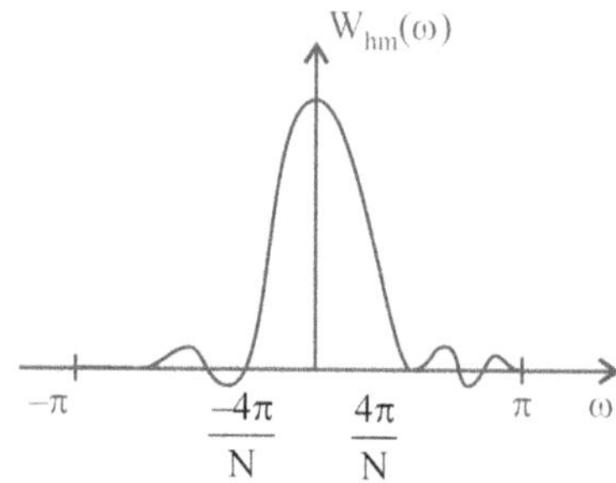

Fig. 5.23 Frequency response of hamming window.

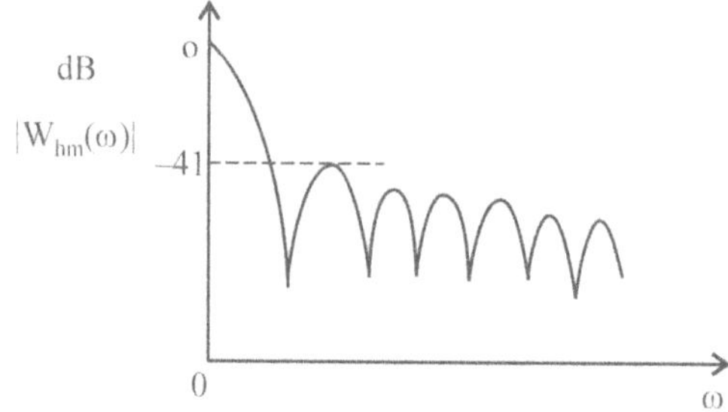

Fig. 5.24 Log magnitude response of hamming window.

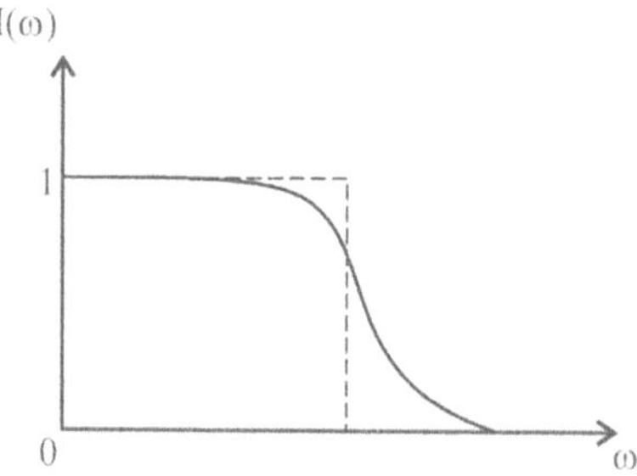

Fig. 5.25 Frequency response of LPF with hamming window.

Hamming window, it's frequency response are shown in Figs 5.22 and 5.23 respectively. Fig 5.25 shows Frequency response of LPF with Hamming window. Observe the Fig 5.24, the side lobe level is –41 dB, which is very very less than that of rectangular window and hanning window. Due to this there will be less number of ripples in both passband and stopband.

Because of these advantages, Hamming window is generally preferred.

5.2.6 Blackman Window

Non-causal Blackman window sequence is

$$W_\beta(n) = 0.42 + 0.5\cos\left(\frac{\pi n}{\alpha}\right) + 0.08\cos\left(\frac{2\pi n}{\alpha}\right), \quad -\alpha \le n \le \alpha \qquad \text{.....(5.2.26a)}$$

$$= 0, \qquad\qquad\qquad\qquad\qquad \text{otherwise}$$

Causal window can be obtained by replacing n with n – α, in the above equation

$$W_\beta(n) = 0.42 - 0.5\cos\left(\frac{\pi n}{\alpha}\right) + 0.08\cos\left(\frac{2\pi n}{\alpha}\right), \quad 0 \le n \le N-1 \qquad \text{.....(5.2.26b)}$$

$$= 0, \qquad\qquad\qquad\qquad\qquad \text{otherwise}$$

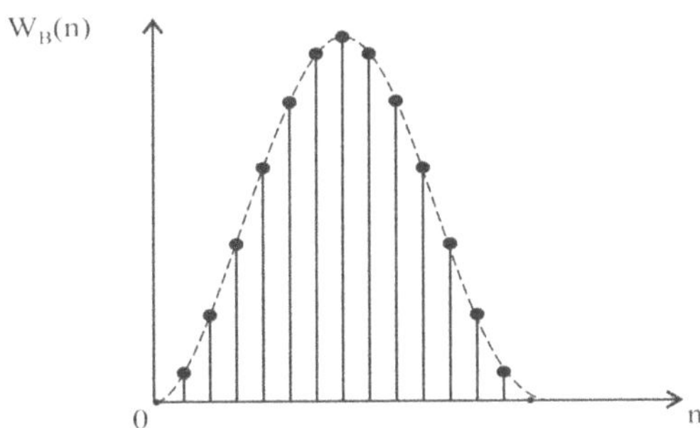

Fig. 5.26 Blackman window.

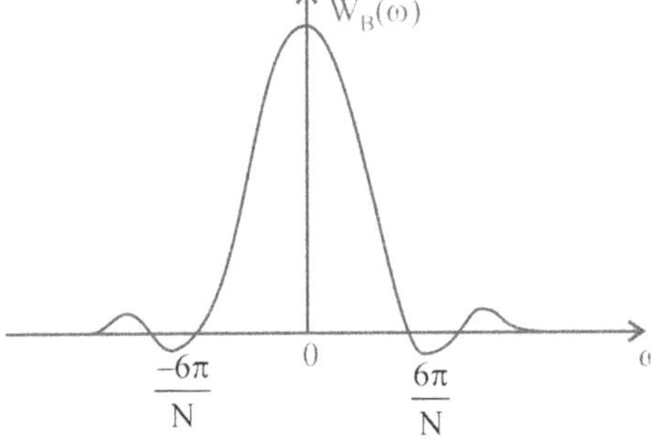

Fig. 5.27 Frequency response of blackman window.

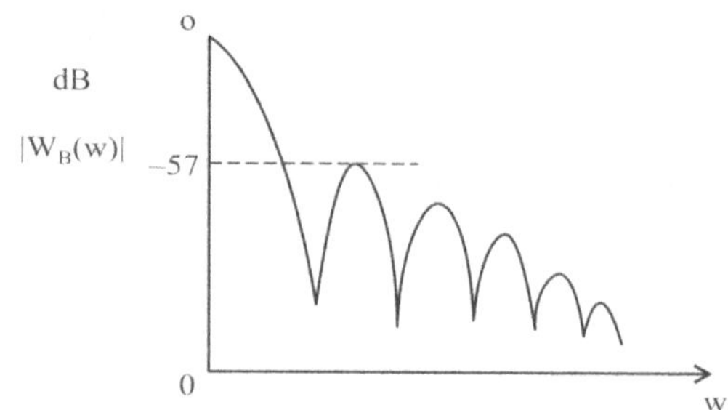

Fig. 5.28 Log magnitude response of blackman window.

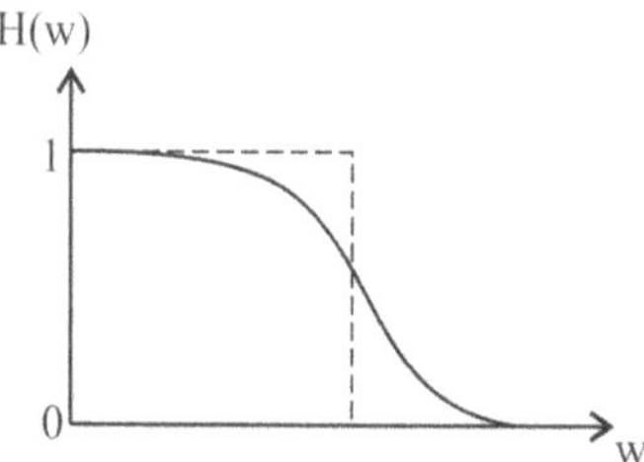

Fig. 5.29 Frequency response of LPF with blackman window.

Blackman window, it's frequency response are shown in Figs 5.26 and 5.27 respectively. Fig 5.29 shows Frequency response of LPF with Blackman window. Due to the additional cosine term, side lobes will be reduced, but the width of main lobe increases to $\dfrac{12\pi}{N}$ when compared with hanning and hamming windows. Observe the Fig. 5.28, the level of side lobes is –57 dB, which is less than the other windows.

5.3 Comparision of IIR and FIR Filters

IIR Filter	FIR Filter
1. All the infinite samples of impulse response are considered.	1. Only N samples of impulse response are considered.
2. IIR filters are not stable always.	2. FIR filters are always stable.
3. Linear phase characteristics cannot be achieved.	3. Linear phase filters can be easily designed.
4. The impulse response cannot be directly converted to digital filter transfer function.	4. The impulse response can be directly converted to digital filter transfer function.
5. The design involves design of analog filter and then transforming analog filter into digital filter.	5. The digital filter can be directly designed to achieve the desired specifications.
6. The specifications include the desired characteristics for magnitude response only.	6. The specifications include the desired characteristics for both magnitude and phase response.
7. For the same order of the filter IIR requires less number of coefficients than FIR.	7. For the same order of the filter FIR requires more number of coefficients than IIR.
8. Memory requirement is less compared to FIR.	8. Memory requirement is more compared to IIR.
9. Execution time is less compared to FIR.	9. Execution time is more compared to IIR.

5.4 Applications of FFT in Spectrum Analysis and Filtering

5.4.1 Applications of FFT in Spectrum Analysis

The important application of DSP methods is in determining the frequency contents of a continuous time signal, in the discrete-time domain known as spectral analysis.

If the continuous time signal $x_a(t)$ is reasonably band limited, the Discrete Time Fourier Transform $X(\omega)$ of its discrete-time equivalent $x(n)$ should provide a good estimate of the spectral properties of $x_a(t)$. However, in most cases, $x_a(t)$ is defined for $-\infty < t < \infty$ and as a result, $x(n)$ is of infinite extent and defined for $-\infty < n < \infty$. Moreover, the DTFT is a continuous function of w and its exact evaluation is in general not feasible.

A more practical approach is that the continuous-time signal $x_a(t)$ is passed through an analog anti aliasing filter before it is sampled to eliminate the effect of aliasing. The output of the filter is then converted into a discrete-time sequence equivalent $x(n)$. The infinite length sequence $x(n)$ is windowed to make it into a finite length sequence $y(n)$ of length N and its DTFT is next evaluated at a set of $R(R \geq N)$ discrete angular frequencies equally spaced in the range $0 \leq w \leq 2\pi$ by computing the R-point DFT of $y(n)$. The DFT is usually computed using FFF algorithm.

The assumptions in the above procedure are:
1. We neglect the effect of aliasing by assuming that the antialiasing filter has been designed appropriately.
2. The A/D converter word length is large enough so that A/D conversion noise is neglected.

To interpret the results of the DFT–based spectral analysis correctly, consider the frequency domain analysis of a sinusoidal sequence.

Consider an infinite-length sinusoidal sequence

$$x(n) = \cos(\omega_0 n + \phi) = \frac{1}{2}\left[e^{j(\omega_0 n + \phi)} + e^{-j(\omega_0 n + \phi)}\right]$$

Its DTFT is

$$X(\omega) = \sum_{n=-\infty}^{\infty} \frac{1}{2}\left[e^{j(\omega_0 n + \phi)} + e^{-j(\omega_0 n + \phi)}\right]e^{-j\omega n}$$

$$= \pi \sum_{n=-\infty}^{\infty} \left(e^{j\phi}\delta(\omega - \omega_0 + 2\pi n) + e^{-j\phi}\delta(\omega + \omega_0 + 2\pi n)\right)$$

DTFT is a periodic function of w with a period 2π. To analyze $x(n)$ in the spectral domain using the DFT, consider the finite length version of the sequence as

$$y(n) = \cos(\omega_0 n + \phi), \qquad 0 \leq n \leq N-1$$

Consider the computation of DFT of length 32, sinusoid of frequency 10 Hz sampled at 64 Hz.

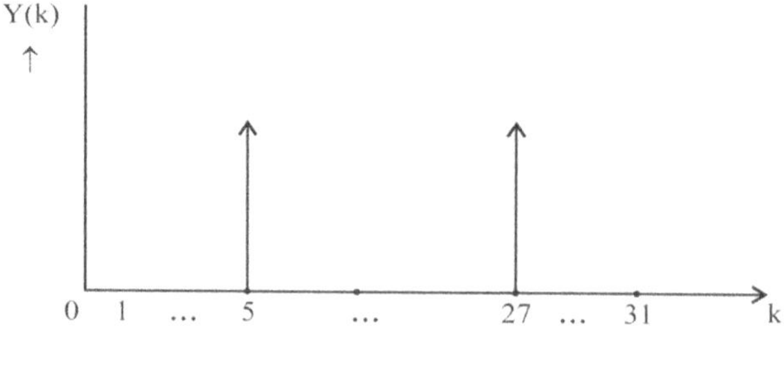

Fig. 5.30

There are two non-zero DFT samples at k = 5 and k = 27 as shown in Fig 5.30.

Consider 32 pt DFT, sinusoid of frequency 11 Hz sampled at 64 Hz.

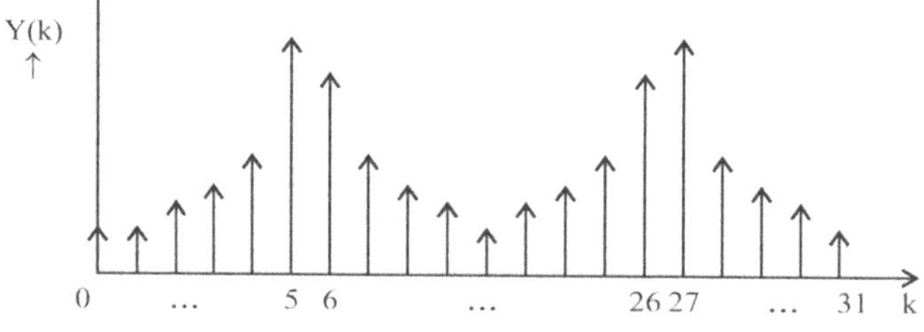

Fig. 5.31

There are two strong parts at k = 5 and 6.

The phenomenon of the spread of energy from a single frequency to many DFT frequency location is called leakage.

The DFT of a length-N sequence y(n) is given by the samples of its DTFT evaluated at $w = 2\pi k/N$, where k = 0, 1, 2, ..., N − 1.

The DTFT of the length 32 sinusoidal sequence of frequency 11 Hz sampled at 64 Hz.

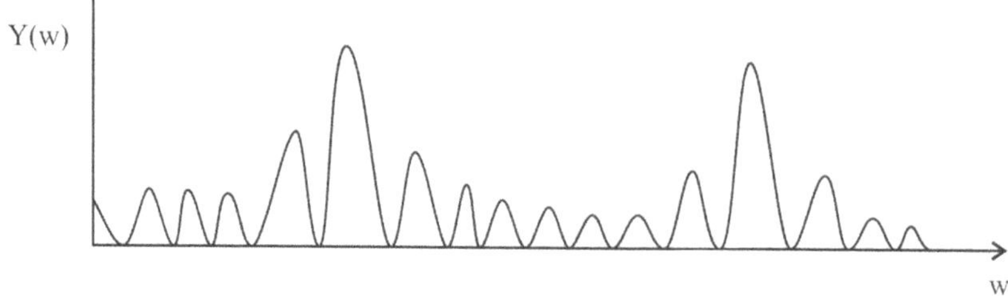

Fig. 5.32

Fig 5.31 is obtained from the frequency samples of Fig 5.30

The sequence y(n) is a windowed version of x(n) using a rectangular window W(n).

$$W(n) = 1, \qquad 0 \le n \le N-1$$
$$= 0, \qquad \text{otherwise}$$

Hence DTFT of y(n) is given by the frequency domain convolution of the DTFT X(w) of x(n) with the DTFT W(ω) of the window ω(n).

$$\therefore \qquad \text{DTFT of y(n)} = \frac{1}{2\pi} \int_{-\pi}^{\pi} X(v)\, W(v-\omega)\, dv$$

where
$$W(\omega) = e^{-j\omega\frac{(N-1)}{2}} \frac{\sin\left(\dfrac{WN}{2}\right)}{\sin\left(\dfrac{W}{2}\right)}$$

substituting X(w)

$$\therefore \qquad \text{DTFT of y(n)} = \frac{1}{2} e^{j\phi}\, W(\omega-\omega_o) + \frac{1}{2} e^{-j\phi}\, W(\omega+\omega_o)$$

So the DTFT of the windowed sequence y(n) is a sum of the frequency shifted and amplitude scaled DTFT W(ω) of the window ω(n) with the amount of frequency shifts given by $\pm\,\omega_0$.

All the samples of DTFT are given by the samples of the side lobes of the DTFT of the window causing the leakage of the frequency components of $\pm\,\omega_0$ to other locations with the amount of leakage determined by the relative amplitude of the main lobe and the side lobes.

Even though the rectangular window has the smallest main lobe width, it has the largest relative side lobe amplitude and so causes considerable leakage. The large amount of leakage results in minor peaks that may be falsely identified as sinusoids.

If we consider hamming window, the leakage has been reduced considerably.

So the performance of DFT-based spectral analysis depends on the type of window being used and its length and the size of the DFT.

To improve frequency resolution, one must use window with a very small main lobe width and to reduce leakage, the window must have a very small relative side lobe level.

The main width can be reduced by increasing the length of the window. Increase in the accuracy of locating the peaks is achieved by increasing the size of DFT.

5.4.2   Applications of FFT in Filtering

Let us assume we have a finite-duration sequence x(n) of length L which excites an FIR filter of length N.

So
$$x(n) = 0, \qquad n < 0 \ \text{and} \ n \geq L$$
$$h(n) = 0, \qquad n < 0 \ \text{and} \ n \geq N$$

where h(n) is the impulse response of the FIR filter.

The output sequence y(n) of the FIR filter can be expressed in time domain as the convolution of x(n) and h(n), i.e.,

$$y(n) = \sum_{k=0}^{N-1} h(k)\, x(n-k) \qquad\qquad .....(5.4.1)$$

Since h(n) and x(n) are finite duration sequences, their convolution is also finite in duration. The duration of y(n) is L+N−1.

The frequency domain equivalent to eqn (5.4.1) is

$$Y(\omega) = X(\omega)\, H(\omega) \qquad\qquad .....(5.4.2)$$

If the sequence y(n) is to be represented in the frequency domain by samples of its spectrum Y(w) at a set of discrete frequencies, the number of distinct samples must equal or exceed L+N−1. Therefore, a DFT of size M ≥ L+N−1 is required to represent y(n) in the frequency domain.

Now if

$$Y(k) = Y(\omega)\Big|_{\omega=2\pi k/M}, \qquad\qquad k = 0, 1, ..., M{-}1$$

$$= X(w)\, H(\omega)\Big|_{w=2\pi k/M}, \qquad k = 0, 1, ..., M{-}1$$

Then
$$Y(k) = X(k)\, H(k), \qquad\qquad k = 0, 1, ..., M{-}1 \qquad .....(5.4.3)$$

where X(k) and H(k) are the N-pt DFTs of the corresponding sequences x(n) and h(n), respectively. Since the sequences x(n) and h(n) have a duration less than M, we simply pad these sequences with zeros to increase their length to M. This increase in the size of the sequences does not alter their spectra X(ω) and H(ω), which are continuous spectra, since the sequences are aperiodic. However, by sampling their spectra at M equally spaced points in frequency (computing the N-pt DFTs using FFT algorithm) we have increased the number of samples that represent these sequences in the frequency domain beyond the minimum number (L, or N, respectively).

Since the M = L+N−1, point DFT of the output sequence y(n) is sufficient to represent y(n) in the frequency domain, it follows that the multiplication of the M-pt DFTs X(k) and H(k), according to eqn (5.4.3) followed by the computation of the M-pt IDFT, using inverse FFT algorithm, yield the sequence y(n).

5.5 Applications of DSP to Speech Processing

A speech signal is made up of non-stationary signals. The speech signal, generated by the excitation of vocal track is composed of two types of basic waveforms.

1. Voiced

2. Unvoiced sounds

A segment of a speech over a small time interval can be considered as a stationary signal and as a result, the DFT of the speech segment can provide a reasonable representation of the frequency domain characteristics of the speech in this time interval.

However in STFT (Short Time Fourier Transform) analysis, the size of the window is critical since a shorter window, developing a wide band spectrogram provides a better time resolution, whereas a longer window developing a narrowband spectrogram results in improved frequency resolution. In order to provide a reasonably good estimate of the changes in the vocal track and the excitation, a wideband spectrogram is preferable. To this end the window is selected to be approximately close to one pitch period, which is adequate for reducing the formants though not adequate to resolve the harmonics of the pitch frequencies. On the otherhand, to resolve the harmonics of the pitch frequencies, a narrowband spectrogram with a window of size of several pitch periods is desirable.

The voiced waveform is considered to be quasi periodic and can be modelled by a sum of finite number of sinusoids.

The lowest frequency of oscillation in this representation is called fundamental frequency or pitch frequency. The unvoiced waveform has no regular fine structure and is more noise like.

5.6 Applications of DSP to Radar Signal Processing

The DFT can be employed for the spectral analysis of a finite length signal composed of sinusoidal components as long as the frequency, amplitude and phase of each sinusoidal component are time invariant and independent of signal length.

If the signal parameters are time varying, they are called non-stationary signals.

$$\textbf{ex:} \qquad x(n) = A\, \cos(w_o n^2)$$

The instantaneous frequency of x(n) is given by ω_0 which is not a constant, but increases linearly with time.

Radar signal is such a non-stationary signal. To describe such signals in the frequency domain using simple DFT of the complete signal will provide misleading results.

To get around the time varying nature of the signal parameters, an alternative approach is to segment the sequence into a set of subsequences of short length with each subsequence centered at uniform intervals of time and its DFT computed separately. If the subsequence length is reasonably small, it can be safely assumed to be stationary for practical purposes.

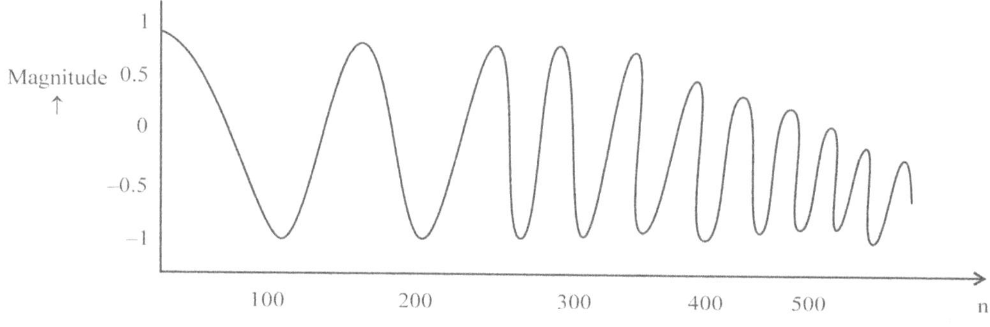

Fig. 5.33

As a result, the frequency domain description of a long sequence is given by a set of short-length DFTs i.e., a time dependent DFT.

To represent a non-stationary signal x(n) in terms of a set of short-length subsequences, we can multiply it with a window w(n) that is stationary with result to time and move the signal through the window.

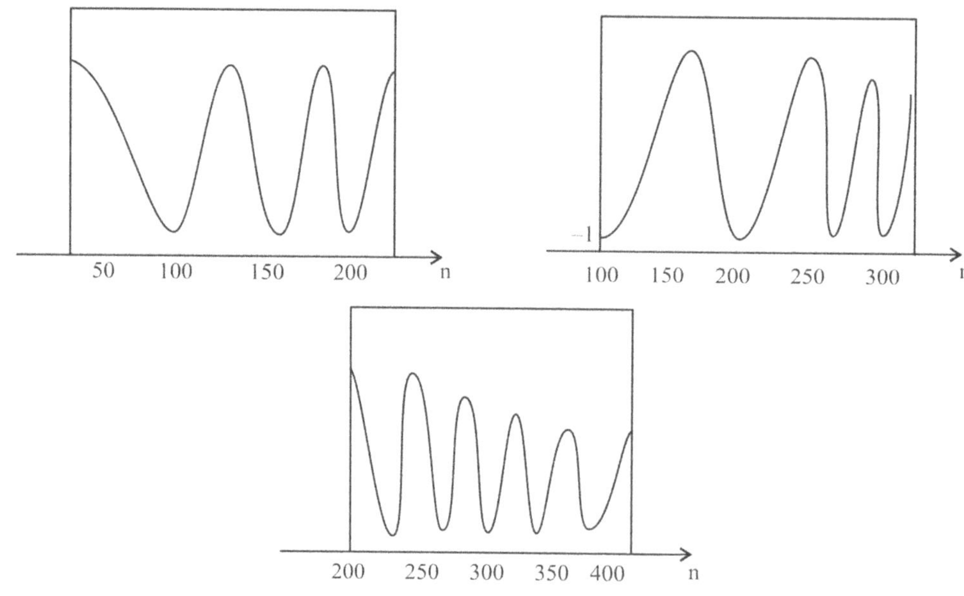

Fig. 5.34

A discrete time Fourier transform of the short sequence obtained by windowing is called short-time fourier transform and is a function of the location of the window relative to the original long sequence and the frequency.

The short time fourier transform (STFT), also known as time dependent fourier transform of a sequence x(n) is defined as

$$X_{STFT}(\omega, n) = \sum_{m=-\infty}^{\infty} x(n-m)\, W(m)\, e^{-j\omega m}$$

where W(n) is a suitably chosen window sequence.

The function of the window sequence is to extract a finite length portion of the sequence x(n) such that the spectral characteristics of the section extracted are approximately stationary over the duration of the window for practical purposes.

If W(n) is 1, then STFT is reduces the conventional DTFT of x(n). Even though the DTFT of x(n) exists under certain well defined conditions, the windowed sequence, being finite in length ensures the existence of STFT for any sequence x(n).

Unlike conventional DTFT, the STFT is a function of two variables. The integer variable time index n and the continuous frequency variable w, and $X_{STFT}(\omega, n)$ is a periodic function of w with period 2π.

In most applications, the magnitude of STFT is considered. The display of the magnitude of STFT is referred to a spectrogram.

Since STFT is a function of two variables, the display of its magnitude would normally require three dimensions. Often it is plotted in two dimensions, with the magnitude represented by the darkness of the plot.

The white area represents zero valued magnitude while the grey areas represent non zero magnitudes, with the largest magnitudes being shown in black.

In STFT display, vertical axis represents the frequency variable (w) and horizontal axis represents the time index n.

In the spectrogram plot, the large-valued DFT samples show up as narrow, nearly black, very short vertical lines, while the other DFT samples show up as grey points. As the instantaneous frequency of the chirp (radar) signal increases linearly, the short black line moves up in vertical direction and eventually, because of aliasing, black line starts moving down in the vertical direction. As a result, the spectrogram of chirp signal appears as a thick line in the form of a triangular shape.

Example 5.1 Design an ideal linear phase bandpass FIR digital filter of length 5 samples using Hamming window for the following specifications.

   (i)   Passband   :   100 Hz to 500 Hz
  (ii) Sampling frequency   :   20000 Hz
 (iii) Passband gain   :   unity

Obtain the impulse response and H(z) of the desired linear phase bandpass FIR digital filter.
[JNTU 2001/S]

Solution:

Given

$$N = 5 \qquad\qquad f_s = 20,000 \text{ Hz}$$

$$\Omega_l = 100 \text{ Hz} \qquad\qquad \Omega_u = 500 \text{ Hz}$$

Digital frequencies are

$$\omega_l = 2\pi \times 100 \times \frac{1}{20,000} = 0.0314 \text{ rad}$$

$$\omega_u = 2\pi \times 500 \times \frac{1}{20,000} = 0.157 \text{ rad}$$

Desired impulse response of bandpass filter is

$$h_d(n) = \frac{1}{\pi(n-\alpha)}\left\{\sin\left[\omega_u(n-\alpha)\right] - \sin\left[\omega_l(n-\alpha)\right]\right\}, \quad n \neq \alpha;$$

$$= \frac{\omega_u - \omega_l}{\pi}; \qquad\qquad\qquad\qquad n = \alpha,$$

where

$$\alpha = \frac{N-1}{2} = \frac{5-1}{2} = 2$$

$\therefore$

$$h_d(0) = \frac{1}{-2\pi}\left\{\sin\left[0.157(-2)\right] - \sin\left[0.0314(-2)\right]\right\}$$

$$= \frac{1}{-2\pi}\left\{-0.314 - (-0.0628)\right\}$$

$$= 0.03998$$

$$h_d(1) = \frac{1}{-\pi}\left\{\sin\left[0.157(-1)\right] - \sin\left[0.0314(-1)\right]\right\}$$

$$= 0.03978$$

$$h_d(2) = \frac{\omega_u - \omega_l}{\pi}, \quad \text{since } n = \alpha$$

$$= \frac{0.157 - 0.0314}{\pi}$$

$$= 0.03998$$

$$h_d(3) = \frac{1}{\pi}\left\{\sin[0.157(1)] - \sin[0.0314(1)]\right\}$$

$$= 0.03978$$

$$h_d(4) = \frac{1}{2\pi}\left\{\sin[0.157(2)] - \sin[0.0314(2)]\right\}$$

$$= 0.03917$$

Hamming window is (causal)

$$W_{hm}(n) = 0.54 - 0.46\cos\left(\frac{\pi n}{\alpha}\right), \quad 0 \le n \le N{-}1$$

where
$$\alpha = \frac{N-1}{2} = \frac{5-1}{2} = 2$$

$$W_{hm}(0) = 0.54 - 0.46\cos(0) = 0.08$$

$$W_{hm}(1) = 0.54 - 0.46\cos\left(\frac{\pi}{2}\right) = 0.54$$

$$W_{hm}(2) = 0.54 - 0.46\cos(\pi) = 1$$

$$W_{hm}(3) = 0.54 - 0.46\cos\left(\frac{3\pi}{2}\right) = 0.54$$

$$W_{hm}(4) = 0.54 - 0.46\cos\left(\frac{4\pi}{2}\right) = 0.08$$

$\therefore$ Impulse response of Linear phase bandpass FIR filter is

$$h(n) = h_d(n)\, W_{hm}(n)$$

$$h(0) = 0.03998\,(0.08) = 0.0032$$

$$h(1) = 0.03978\,(0.54) = 0.0215$$

$$h(2) = 0.03998\,(1) = 0.03998$$

$$h(3) = 0.03978\,(0.54) = 0.0215$$

$$h(4) = 0.03917\,(0.08) = 0.0032$$

$\therefore$ $h(n) = \{0.0032,\ 0.0215,\ 0.03998,\ 0.0215,\ 0.0032\}$

H(z) of the filter is

$$H(z) = 0.0032 + 0.0215\, z^{-1} + 0.03998\, z^{-2} + 0.0215\, z^{-3} + 0.0032\, z^{-4}$$

Example 5.2: Design a linear phase FIR digital filter corresponding to a differentiator with a cut-off frequency of 500 Hz. Take 5 samples and use Hamming window. Obtain H(z) of the filter. [JNTU 1998]

Solution:

Given $\qquad \Omega_c = 500$ Hz

$$\omega_c = 2\pi \times 500 = 1000\,\pi \text{ rad/sec}$$

Frequency response of discrete-time differentiator with linear phase is

$$H_d(\omega) = j\omega\, e^{-j\omega\alpha}, \quad -\pi < \omega < \pi$$

where $\qquad \alpha = \dfrac{N-1}{2} = \dfrac{5-1}{2} = 2$

If we draw frequency response, we can find that the filter coefficients are antisymmetric about $n = \alpha$.

The desired impulse response is

$$h_d(n) = \frac{1}{2\pi}\left[\int_{-\pi}^{-\omega_c} j\omega\, e^{-j\omega\alpha} e^{j\omega n}\, d\omega + \int_{\omega_c}^{\pi} j\omega\, e^{-j\omega\alpha} e^{j\omega n}\, d\omega \right]$$

(since differentiator is highpass filter)

$$h_d(n) = \left\{ \frac{j}{2\pi}\left[\frac{\omega}{j(n-\alpha)} e^{j\omega(n-\alpha)} \right]_{-\pi}^{-\omega_c} + \left[\frac{j}{(n-\alpha)^2} e^{j\omega(n-\alpha)} \right]_{-\pi}^{-\omega_c} \right.$$

$$\left. + \left[\frac{\omega}{j(n-\alpha)} e^{j\omega(n-\alpha)} \right]_{\omega_c}^{\pi} + \left[\frac{j}{(n-\alpha)^2} e^{j\omega(n-\alpha)} \right]_{\omega_c}^{\pi} \right\}$$

$$= \frac{j}{2\pi}\left\{ \left[\frac{-\omega_c}{j(n-\alpha)} e^{-j\omega_c(n-\alpha)} + \frac{\pi}{j(n-\alpha)} e^{-j\pi(n-\alpha)} + \frac{j}{(n-\alpha)^2} e^{-j\omega_c(n-\alpha)} - \frac{j}{(n-\alpha)^2} e^{-j\pi(n-\alpha)} \right] \right.$$

$$\left. + \left[\frac{\pi}{j(n-\alpha)} e^{j\pi(n-\alpha)} - \frac{W_c}{j(n-\alpha)} e^{j\omega_c(n-\alpha)} + \frac{j}{(n-\alpha)^2} e^{j\pi(n-\alpha)} - \frac{j}{(n-\alpha)^2} e^{j\omega_c(n-\alpha)} \right] \right\}$$

$$= \frac{j}{2\pi}\left\{ \frac{-\omega_c}{j(n-\alpha)}\left[e^{j\omega_c(n-\alpha)} + e^{-j\omega_c(n-\alpha)}\right] + \frac{\pi}{j(n-\alpha)}\left[e^{j\pi(n-\alpha)} + e^{-j\pi(n-\alpha)}\right] \right.$$

$$\left. - \frac{j}{(n-\alpha)^2}\left[e^{j\omega_c(n-\alpha)} - e^{-j\omega_c(n-\alpha)}\right] + \frac{j}{(n-\alpha)^2}\left[e^{j\pi(n-\alpha)} - e^{-j\pi(n-\alpha)}\right] \right\}$$

$$= \frac{-\omega_c}{\pi(n-\alpha)}\cos\left[\omega_c(n-\alpha)\right] + \frac{1}{(n-\alpha)}\cos\left[\pi(n-\alpha)\right] + \frac{j}{\pi(n-\alpha)^2}\sin\left[\omega_c(n-\alpha)\right]$$

$$- \frac{j}{\pi(n-\alpha)^2}\sin\left[\pi(n-\alpha)\right]$$

Since α is an integer, $n - \alpha$ is also an integer.

$\therefore \ \sin[\pi(n-\alpha)] = 0$

$$\therefore \ h_d(n) = \frac{\cos\left[\pi(n-\alpha)\right]}{(n-\alpha)} - \frac{\omega_c}{\pi(n-\alpha)}\cos\left[\omega_c(n-\alpha)\right] + \frac{j}{\pi(n-\alpha)^2}\sin\left[\omega_c(n-\alpha)\right]$$

$N = 5$,

$$h_d(0) = \frac{\cos(-2\pi)}{-2\pi} - \frac{1000\pi}{-2\pi}\cos(-2000\pi) + \frac{j}{4\pi}\sin(-2000\pi)$$

$$= -0.1592 + 500 = 499.8$$

$$h_d(1) = \frac{\cos(-\pi)}{-\pi} - \frac{1000\pi}{-\pi}\cos(-1000\pi) + \frac{j}{\pi}\sin(-1000\pi)$$

$$= 0.3183 + 1000 = 1000.3183$$

$$h_d(2) = \frac{1}{2\pi}\left[\int_{-\pi}^{-\omega_c} j\omega \, d\omega + \int_{w_c}^{\pi} j\omega \, d\omega\right], \qquad \text{since } n = \alpha$$

$$= \frac{j}{2\pi}\left\{\left[\frac{\omega^2}{2}\right]_{-\pi}^{-\omega_c} + \left[\frac{\omega^2}{2}\right]_{\omega_c}^{\pi}\right\}$$

$$= \frac{j}{2\pi}\left[\frac{\omega_c^2}{2} - \frac{\pi^2}{2} + \frac{\pi^2}{2} - \frac{\omega_c^2}{2}\right] = 0$$

$$h_d(3) = \frac{\cos\pi}{\pi} - \frac{100\pi}{\pi}\cos(1000\pi) + \frac{j}{\pi}\sin(1000\pi)$$

$$= -0.3183 - 1000 = -1000.3183$$

$$h_d(4) = \frac{\cos(2\pi)}{2\pi} - \frac{1000\pi}{2\pi}\cos(2000\pi) + \frac{j}{4\pi}\sin(2000\pi)$$

$$= 0.1592 - 500 = -499.8$$

Causal Hamming window is

$$\omega_{hm}(n) = 0.54 - 0.46 \, \cos\left(\frac{\pi n}{\alpha}\right), \qquad\qquad 0 \le n \le N-1$$

where $\alpha = \dfrac{N-1}{2} = 2$

$$\omega_{hm}(0) = 0.54 - 0.46 \, \cos(0) = 0.08$$

$$\omega_{hm}(1) = 0.54 - 0.46 \, \cos\left(\frac{\pi}{2}\right) = 0.54$$

$$\omega_{hm}(2) = 0.54 - 0.46 \, \cos(\pi) = 1$$

$$\omega_{hm}(3) = 0.54 - 0.46 \, \cos\left(\frac{3\pi}{2}\right) = 0.54$$

$$\omega_{hm}(4) = 0.54 - 0.46 \, \cos(2\pi) = 0.08$$

Impulse response coefficients of digital differentiator are

$$h(n) = h_d(n) \, W_{hm}(n)$$
$$h(0) = h_d(0) \, W_{hm}(0) = 499.8 \,(0.08) = 39.984$$
$$h(1) = h_d(1) \, W_{hm}(1) = 1000.3183 \,(0.54) = 540.17$$
$$h(2) = h_d(2) \, W_{hm}(2) = 0$$
$$h(3) = h_d(3) \, W_{hm}(3) = -1000.3198 \,(0.54) = -540.17$$
$$h(4) = h_d(4) \, W_{hm}(4) = -499.8 \,(0.08) = -39.984$$

$\therefore \qquad h(n) = \{39.984, 540.17, 0, -540.17, -39.984\}$

$$H(z) = 39.984 + 540.17 \, z^{-1} - 540.17 \, z^{-3} - 39.984 \, z^{-4}$$

Example 5.3: Design an ideal differentiator with frequency response

$$H(\omega) = j\omega \qquad -\pi \le \omega \le \pi$$

Using Hamming window with $N = 7$

Solution:

Here $\alpha = 0$, since $e^{-j\omega\alpha} = 1$

If we draw frequency response, we can find that the filter coefficients are antisymmetric about $n = \alpha = 0$, satisfying $h_d(n) = -h_d(-n)$.

$\therefore$

$$h_d(n) = \frac{1}{2\pi} \int_{-\pi}^{\pi} j\omega\, e^{j\omega n}\, d\omega$$

$$= \frac{j}{2\pi}\left\{\left[\frac{1}{jn}\omega e^{j\omega n}\right]_{-\pi}^{\pi} + \left[\frac{j}{n^2}e^{j\omega n}\right]_{-\pi}^{\pi}\right\}$$

$$= \frac{j}{2\pi}\left\{\frac{\pi}{jn}e^{j\pi n} + \frac{\pi}{jn}e^{-j\pi n} + \frac{j}{n^2}e^{j\pi n} - \frac{j}{n^2}e^{-j\pi n}\right\}$$

$$= \frac{1}{n}\cos \pi n - \frac{j}{\pi n^2}\sin \pi n \qquad (\sin \pi n = 0, \text{ for all integer}$$

values of n).

$$h_d(n) = \frac{\cos \pi n}{n}$$

Given $\qquad N = 7$

Since $\alpha = 0$, we get noncausal $h_d(n)$ i.e., from $-3 \le n \le 3$ and $h_d(n) = -h_d(-n)$

$n = 0;\qquad h_d(0) = -h_d(0) \rightarrow h_d(0) = 0$

(or)

$$h_d(0) = \frac{1}{2\pi}\int_{-\pi}^{\pi} jw\, dw = \frac{j}{2\pi}\left[\frac{w^2}{2}\right]_{-\pi}^{\pi} = 0$$

$n = 1;\qquad h_d(1) = -h_d(-1) = \dfrac{\cos \pi(+1)}{(+1)} = -1$

$n = 2;\qquad h_d(2) = -h_d(-2) = \dfrac{\cos \pi(+2)}{(+2)} = +\dfrac{1}{2}$

$n = 3;\qquad h_d(3) = -h_d(-3) = \dfrac{\cos \pi(+3)}{(+3)} = -\dfrac{1}{3}$

$\therefore$ Noncausal $h_d(n)$ is

$$h_d(n) = \left\{+\frac{1}{3}, -\frac{1}{2}, +1, \underset{\uparrow}{0}, -1, +\frac{1}{2}, -\frac{1}{3}\right\}$$

Causal $h_d(n)$ can be obtained by adding $\left(\dfrac{N-1}{2}\right)$ i.e., $\dfrac{7-1}{2} = 3$ to indices of $h_d(n)$.

$\therefore$ causal $\quad h_d(n) = \left\{ \dfrac{1}{3}, -\dfrac{1}{2}, 1, 0, -1, \dfrac{1}{2}, -\dfrac{1}{3} \right\}$

Causal Hamming window is

$$W_{hm}(n) = 0.54 - 0.46 \cos\left(\dfrac{\pi n}{\alpha}\right), \quad 0 \le n \le N-1$$

where $\quad \alpha = \dfrac{N-1}{2} = \dfrac{7-1}{2} = 3$

$$W_{hm}(0) = 0.54 - 0.46 \cos(0) = 0.08$$

$$W_{hm}(1) = 0.54 - 0.46 \cos\left(\dfrac{\pi}{3}\right) = 0.31$$

$$W_{hm}(2) = 0.54 - 0.46 \cos\left(\dfrac{2\pi}{3}\right) = 0.77$$

$$W_{hm}(3) = 0.54 - 0.46 \cos\left(\dfrac{3\pi}{3}\right) = 1$$

$$W_{hm}(4) = 0.54 - 0.46 \cos\left(\dfrac{4\pi}{3}\right) = 0.77$$

$$W_{hm}(5) = 0.54 - 0.46 \cos\left(\dfrac{5\pi}{3}\right) = 0.31$$

$$W_{hm}(6) = 0.54 - 0.46 \cos\left(\dfrac{6\pi}{3}\right) = 0.08$$

Causal impulse response of ideal differentiator is

$$h(n) = h_d(n)\, W_{hm}(n)$$

$$h(0) = h_d(0)\, W_{hm}(0) = +\dfrac{1}{3}(0.08) = 0.0267$$

$$h(1) = h_d(1)\, W_{hm}(1) = -\dfrac{1}{2}(0.31) = -0.155$$

$$h(2) = h_d(2)\, W_{hm}(2) = 1(0.77) = 0.77$$

$$h(3) = h_d(3)\ W_{hm}(3) = 0\ (1)\ = 0$$

$$h(4) = h_d(4)\ W_{hm}(4) = -1(0.77)\ = -\,0.77$$

$$h(5) = h_d(5)\ W_{hm}(5) = \frac{1}{2}(0.31) = 0.155$$

$$h(6) = h_d(6)\ W_{hm}(6) = -\frac{1}{3}(0.08)\ = -\,0.0267$$

$$\therefore \qquad h(n) = \{0.0267,\ -0.155,\ 0.77,\ 0,\ -0.77,\ 0.155,\ -0.0267\}$$

Example 5.4 Design an ideal lowpass filter with a frequency response

$$H_d(\omega) = 1, \qquad -\frac{\pi}{2} \le \omega \le \frac{\pi}{2}$$

$$= 0, \qquad \frac{\pi}{2} \le |\omega| \le \pi$$

Using Hanning window with $N = 7$

Solution:

Given
$$H_d(\omega) = 1, \qquad -\frac{\pi}{2} \le \omega \le \frac{\pi}{2}$$

$$= 0, \qquad \frac{\pi}{2} \le |\omega| \le \pi$$

$$e^{-j\omega\alpha} = 1 \qquad \rightarrow \qquad \alpha = 0$$

If we draw frequency response, we can find that the filter coefficients are symmetric about $n = \alpha = 0$, satisfying $h_d(n) = h_d(-n)$

$$\therefore \qquad h_d(n) = \frac{1}{2\pi} \int_{-\pi/2}^{\pi/2} 1\ e^{j\omega n}\ d\omega$$

$$= \frac{1}{2\pi} \left[\frac{e^{j\omega n}}{jn} \right]_{-\pi/2}^{\pi/2}$$

$$= \frac{1}{2\pi} \left[\frac{e^{j\frac{\pi}{2}n}}{jn} - \frac{e^{-j\frac{\pi}{2}n}}{jn} \right]$$

$$h_d(n) = \frac{1}{\pi n} \sin\left(\frac{\pi}{2}n\right)$$

Given N = 7,

$h_d(n)$ is noncausal, since it is symmetric about origin.

$$n = 0; \qquad h_d(0) = \frac{1}{2\pi}\int_{-\pi/2}^{\pi/2} d\omega = \frac{1}{2\pi}\big[w\big]_{-\pi/2}^{\pi/2} = \frac{1}{2}$$

$$n = 1; \qquad h_d(1) = h_d(-1) = \frac{\sin\left(\dfrac{\pi}{2}\right)}{\pi} = 0.3183$$

$$n = 2; \qquad h_d(2) = h_d(-2) = \frac{\sin\left(\dfrac{2\pi}{2}\right)}{2\pi} = 0$$

$$n = 3; \qquad h_d(3) = h_d(-3) = \frac{\sin\left(\dfrac{3\pi}{2}\right)}{3\pi} = -0.1061$$

$\therefore$ Noncausal $h_d(n)$ is

$$h_d(n) = \{-0.1061,\ 0,\ 0.3183,\ 0.5,\ 0.3183,\ 0,\ -0.1061\}$$

Causal $h_d(n)$ can be obtained by adding $\left(\dfrac{N-1}{2}\right)$ i.e., 3 to the indices of $h_d(n)$.

$\therefore$ Causal $h_d(n)$ is

$$h_d(n) = \{-0.1061, 0, 0.3183, 0.5, 0.3183, 0, -0.1061\}$$

Causal hanning window is

$$W_{hn}(n) = 0.5 - 0.5\ \cos\left(\frac{\pi n}{\alpha}\right), \qquad 0 \le n \le N-1$$

where $\alpha = \dfrac{N-1}{2} = \dfrac{7-1}{2} = 3$

$$W_{hn}(0) = 0.5 - 0.5\ \cos(0) = 0$$

$$W_{hn}(1) = 0.5 - 0.5\ \cos\left(\frac{\pi}{3}\right) = 0.25$$

$$W_{hn}(2) = 0.5 - 0.5\ \cos\left(\frac{2\pi}{3}\right) = 0.75$$

$$W_{hn}(3) = 0.5 - 0.5 \ \cos(\pi) = 1$$

$$W_{hn}(4) = 0.5 - 0.5 \ \cos\left(\frac{4\pi}{3}\right) = 0.75$$

$$W_{hn}(5) = 0.5 - 0.5 \ \cos\left(\frac{5\pi}{3}\right) = 0.25$$

$$W_{hn}(6) = 0.5 - 0.5 \ \cos\left(\frac{6\pi}{3}\right) = 0$$

Causal impulse response of ideal lowpass filter is

$$h(n) = h_d(n) \ W_{hn}(n)$$

$$h(0) = h_d(0) \ W_{hn}(0) = -0.1061(0) = 0$$

$$h(1) = h_d(1) \ W_{hn}(1) = 0 \ (0.25) = 0$$

$$h(2) = h_d(2) \ W_{hn}(2) = 0.3183 \ (0.75) = 0.2387$$

$$h(3) = h_d(3) \ W_{hn}(3) = 0.5 \ (1) = 0.5$$

$$h(4) = h_d(4) \ W_{hn}(4) = 0.3183 \ (0.75) = 0.2387$$

$$h(5) = h_d(5) \ W_{hn}(5) = 0 \ (0.25) = 0$$

$$h(6) = h_d(6) \ W_{hn}(6) = -0.1061 \ (0) = 0$$

$$h(n) = \{0, \ 0, \ 0.2387, \ 0.5, \ 0.2387, \ 0, \ 0\}$$

Example 5.5: A linear phase FIR digital lowpass filter is desired with the following specifications

(i)   Cut-off frequency = 100 Hz

(ii)   Sampling frequency = 1 kHz

(iii) Impulse duration = 10 milliseconds

Compute and sketch the impulse response of the desired FIR digital filter using Hamming window.   [JNTU 2000]

Solution:

Given $\Omega_c = 100$ Hz $f_s = 1$ kHz

$\therefore$ $\omega_c = 2\pi \times 100 \times \dfrac{1}{1 \times 10^3} = 0.6283$ rad

Impulse duration is 10 milliseconds.

i.e.,   at 0th millisecond, we get 1 sample

at 1 millisecond, we get 2 samples

at 2 milliseconds, we get 3 samples

$$\vdots$$

at 10 milliseconds, we get 11 samples

$$\therefore \qquad N = 11$$

Linear phase FIR digital lowpass filter desired frequency response is

$$H_d(\omega) = 1\, e^{-j\omega\alpha}, \qquad -\omega_c \le |\omega| \le \omega_c$$

$$= 0, \qquad\qquad \text{otherwise}$$

where $\quad \alpha = \dfrac{N-1}{2} = \dfrac{11-1}{2} = 5$

Corresponding desired impulse response, which is symmetry about $\alpha = \dfrac{N-1}{2} = 5$, satisfying the condition $h_d(n) = h_d(N-1-n)$

$$h_d(n) = \frac{1}{2\pi} \int\limits_{-\omega_c}^{\omega_c} e^{-j\omega\alpha} e^{j\omega n}\, d\omega$$

$$= \frac{\sin\omega_c(n-\alpha)}{\pi(n-\alpha)}, \qquad n \ne \alpha$$

$$= \frac{\omega_c}{\pi}, \qquad\qquad n = \alpha$$

$$\therefore \qquad h_d(0) = \frac{\sin\left[0.6283(-5)\right]}{\pi(-5)} = 0.000005898 = h_d(10)$$

$$h_d(1) = \frac{\sin\left[0.6283(-4)\right]}{\pi(-4)} = 0.0468 = h_d(9)$$

$$h_d(2) = \frac{\sin\left[0.6283(-3)\right]}{\pi(-3)} = 0.1009 = h_d(8)$$

$$h_d(3) = \frac{\sin\left[0.6283(-2)\right]}{\pi(-2)} = 0.1514 = h_d(7)$$

$$h_d(4) = \frac{\sin\left[0.6283(-1)\right]}{\pi(-1)} = 0.1871 = h_d(6)$$

$$h_d(5) = \frac{0.6283}{\pi} = 0.199994$$

$$h_d(6) = \frac{\sin\left[0.6283(1)\right]}{\pi(1)} = 0.1871$$

Causal Hamming window is, it also satisfies $W_{hm}(n) = W_{hm}(N-1-n)$

$$W_{hm}(n) = 0.54 - 0.46\cos\left(\frac{\pi n}{\alpha}\right), \qquad 0 \leq n \leq N-1$$

$$W_{hm}(0) = W_{hm}(10) = 0.54 - 0.46\cos(0) = 0.08$$

$$W_{hm}(1) = W_{hm}(9) = 0.54 - 0.46\cos\left(\frac{\pi}{5}\right) = 0.1679$$

$$W_{hm}(2) = W_{hm}(8) = 0.54 - 0.46\cos\left(\frac{2\pi}{5}\right) = 0.3979$$

$$W_{hm}(3) = W_{hm}(7) = 0.54 - 0.46\cos\left(\frac{3\pi}{5}\right) = 0.6822$$

$$W_{hm}(4) = W_{hm}(6) = 0.54 - 0.46\cos\left(\frac{4\pi}{5}\right) = 0.9125$$

$$W_{hm}(5) = 0.54 - 0.46\cos(\pi) = 1$$

Impulse response of liner phase FIR digital lowpass filter is, it satisfies $h(n) = h(N-1-n)$

$$h(n) = h_d(n)W_{hm}(n)$$

$$h(0) = h(10) = h_d(0)W_{hm}(0) = 0.000005898(0.08)$$

$$\approx 0$$

$$h(1) = h(9) = h_d(1)W_{hm}(1) = 0.0468(0.1679) = 0.00786$$

$$h(2) = h(8) = h_d(2)W_{hm}(2) = 0.1009(0.3979) = 0.0402$$

$$h(3) = h(7) = h_d(3)W_{hm}(3) = 0.1514(0.6822) = 0.1033$$

$$h(4) = h(6) = h_d(4)W_{hm}(4) = 0.1871(0.9125) = 0.1707$$

$$h(5) = h_d(5)W_{hm}(5) = 0.199994(1) \approx 0.2$$

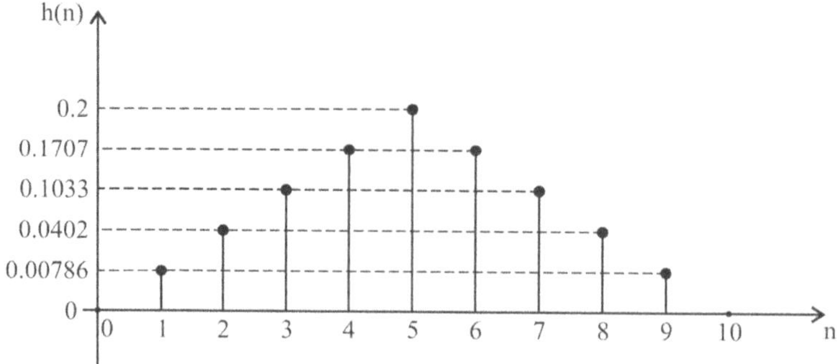

Example 5.6: An FIR linear phase filter has the following impulse response

$$h(n) = \begin{cases} 1, & 0 \le n \le 4 \\ 0, & \text{otherwise} \end{cases}$$

Use Barlett's window and compute the impulse response of the filter. Find its magnitude and phase response as a function of frequency. [JNTU 2000]

Solution:

Given

$$h(n) = \begin{cases} 1, & 0 \le n \le 4 \\ 0, & \text{otherwise} \end{cases}$$

Barlett's window (causal)

$$W_T(n) = 1 - \frac{|n - \alpha|}{\alpha}, \qquad 0 \le n \le N - 1$$

where $\quad \alpha = \dfrac{N-1}{2} \quad N = 5; \quad \alpha = \dfrac{5-1}{2} = 2$

$$W_T(0) = 1 - \frac{|0 - 2|}{2} = 0 = W_T(4)$$

$$W_T(1) = 1 - \frac{|1 - 2|}{2} = 0.5 = W_T(3)$$

$$W_T(2) = 1 - \frac{|2 - 2|}{2} = 1$$

$\therefore$ Impulse response of FIR filter with Barlett window is

$$h_{FIR}(n) = h(n) W_T(n)$$

$$h_{FIR}(0) = h_{FIR}(4) = h(0) W_T(0) = 1(0) = 0$$

$$h_{FIR}(1) = h_{FIR}(3) = h(1) W_T(1) = 1(0.5) = 0.5$$

$$h_{FIR}(2) = h(2) W_T(2) = 1(1) = 1$$

$$h_{FIR}(n) = \{0, \ 0.5, \ 1, \ 0.5, \ 0\}$$

$$H(z) = 0.5z^{-1} + z^{-2} + 0.5z^{-3}$$

Frequency response is

$$H(\omega) = 0.5\ e^{-j\omega} + e^{-j2\omega} + 0.5\ e^{-j3\omega}$$

$$H(\omega) = e^{-j2\omega}\left[1 + \frac{e^{j\omega} + e^{-j\omega}}{2}\right]$$

$$= e^{-j2\omega}\left[1 + \cos\ \omega\right]$$

Magnitude response is

$$\left|H(\omega)\right| = 1 + \cos\ \omega$$

Phase response is

$$\underline{/H(\omega)} = -2\ \omega$$

Example 5.7: An FIR digital filter has the following impulse response

$$h(n) = 1, \qquad 0 \le n \le 4$$

$$= 0, \qquad \text{otherwise}$$

Use hamming window and compute the impulse response of FIR digital filter with linear phase. Sketch its magnitude and phase response of a function of frequency for $\omega \le \pi.$ [JNTU 2000/s]

Solution: Given $\qquad h(n) = \begin{cases} 1, & 0 \le n \le 4 \\ 0, & \text{otherwise} \end{cases}$

Causal hamming window is

$$W_{hm}(n) = 0.54 - 0.46\ \cos\left(\frac{\pi n}{\alpha}\right), \qquad 0 \le n \le N-1$$

where $\qquad \alpha = \dfrac{N-1}{2};\ N = 5$

$$\alpha = \frac{5-1}{2} = 2$$

$$W_{hm}(0) = W_{hm}(4) = 0.54 - 0.46 \cos(0)$$

$$= 0.08$$

$$W_{hm}(1) = W_{hm}(3) = 0.54 - 0.46 \cos\left(\frac{\pi}{2}\right)$$

$$= 0.54$$

$$W_{hm}(2) = 0.54 - 0.46 \cos(\pi) = 1$$

Impulse response of FIR filter is

$$h_{FIR}(n) = h(n)\,W_{hm}(n)$$

$$h_{FIR}(0) = h_{FIR}(4) = h(0)\,W_{hm}(0) = 0.08$$

$$h_{FIR}(1) = h_{FIR}(3) = h(1)\,W_{hm}(1) = 0.54$$

$$h_{FIR}(2) = h(2)\,W_{hm}(2) = 1$$

$$h_{FIR}(n) = \{0.08,\ 0.54,\ 1,\ 0.54,\ 0.08\}$$

$$H(Z) = 0.08 + 0.54z^{-1} + z^{-2} + 0.54z^{-3} + 0.08z^{-4}$$

Frequency response is

$$H(\omega) = 0.08 + 0.54\,e^{-j\omega} + e^{-j2\omega} + 0.54\,e^{-j3\omega} + 0.08\,e^{-j4\omega}$$

$$H(\omega) = e^{-j2\omega}\left[0.08\left(e^{j2\omega} + e^{-j2\omega}\right) + 0.54\left(e^{j\omega} + e^{-j\omega}\right) + 1\right]$$

$$= e^{-j2\omega}\left[0.16 \cos(2\omega) + 1.08 \cos(\omega) + 1\right]$$

Magnitude response is

$$|H(\omega)| = 1 + 1.08 \cos(\omega) + 0.16 \cos(2\omega)$$

Phase response is

$$\angle H(\omega) = \theta(\omega) = -2\omega$$

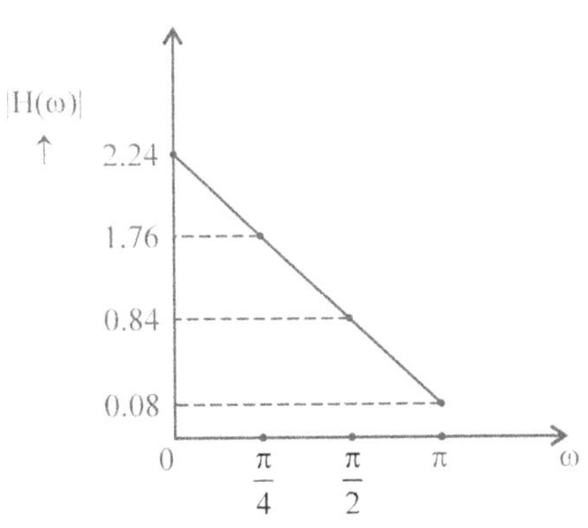

Magnitude response

Phase response

Example 5.8: Design a band pass filter to pass frequencies in the range $1 - 2$ radians/sec using hanning window $N = 5$ 　　　　　　　　　　　　　　　　　[JNTU 2002]

Solution: Given $\qquad \omega_L = 1 \text{ rad/sec}$

$$\omega_U = 2 \text{ rad/sec}$$

$$N = 5$$

$h_d(n)$ of linear phase band pass filter is

$$h_d(n) = \frac{\sin \omega_u(n-\alpha) - \sin \omega_L(n-\alpha)}{\pi(n-\alpha)}$$

where

$$\alpha = \frac{N-1}{2} = \frac{5-1}{2} = 2$$

$$h_d(0) = \frac{\sin 2(-2) - \sin 1(-2)}{\pi(-2)} = -0.2652$$

$$h_d(1) = \frac{\sin 2(-1) - \sin 1(-1)}{\pi(-1)} = 0.02159$$

$$h_d(2) = \frac{\omega_U - \omega_L}{\pi}; \qquad \because \ n = \alpha$$

$$= \frac{2-1}{\pi} = 0.3183$$

$$h_d(3) = \frac{\sin 2(1) - \sin 1(1)}{\pi(1)} = 0.02159$$

$$h_d(4) = \frac{\sin 2(2) - \sin 1(2)}{\pi(2)} = -0.2652$$

Hanning window is

$$W_{hn}(n) = 0.5 - 0.5 \cos\left(\frac{\pi n}{\alpha}\right), \quad 0 \le n \le N-1$$

where

$$\alpha = \frac{N-1}{2} = \frac{5-1}{2} = 2$$

$$W_{hn}(0) = 0.5 - 0.5 \cos(0) = 0 = W_{hn}(4)$$

$$W_{hn}(1) = 0.5 - 0.5 \cos\left(\frac{\pi}{2}\right) = 0.5 = W_{hn}(3)$$

$$W_{hn}(2) = 0.5 - 0.5 \cos(\pi) = 1$$

$\therefore$ Impulse response of Band pass filter is

$$h(n) = h_d(n) W_{hn}(n)$$

$$h(0) = h(4) = h_d(0) W_{hn}(0) = (-0.2652)(0) = 0$$

$$h(1) = h(3) = h_d(1) W_{hn}(1) = 0.02159(0.5) = 0.010795$$

$$h(2) = h_d(2) W_{hn}(2) = 0.3183(1) = 0.3183$$

$\therefore \qquad h(n) = \{0,\ 0.010795,\ 0.3183,\ 0.010795,\ 0\}$

Example 5.9: Design an ideal Hilbert transformer having frequency response

$$H(\omega) = j, \quad -\pi \le \omega \le 0$$

$$= -j, \quad 0 \le \omega \le \pi$$

Using blackman window with $N = 7$

Solution:

$$h_d(n) = \frac{1}{2\pi}\left[\int_{-\pi}^{0} j\, e^{j\omega n} d\omega + \int_{0}^{\pi} -j\, e^{j\omega n} d\omega\right]$$

$$= \frac{1 - \cos \pi n}{\pi n}$$

From the given frequency response $e^{-j\omega\alpha} = 1 \to \alpha = 0$

If we draw frequency response, we can find that filter coefficients are antisymmetrical about $n = \alpha = 0$, satisfying $h_d(n) = -h_d(-n)$

$$n = 0;\ h_d(0) = \frac{j}{2\pi}\left[\int_{-\pi}^{0} d\omega - \int_{0}^{\pi} d\omega\right]$$

$$h_d(0) = \frac{j}{2\pi}\left\{[\omega]_{-\pi}^{0} - [\omega]_{0}^{\pi}\right\}$$

$$= \frac{j}{2\pi}[0 + \pi - \pi + 0] = 0$$

$$n = 1;\ h_d(1) = -h_d(-1) = \frac{1 - \cos(-\pi)}{-\pi} = -0.6366$$

$$n = 2;\ h_d(2) = -h_d(-2) = \frac{1 - \cos(-2\pi)}{-2\pi} = 0$$

$$n = 3;\ h_d(3) = -h_d(-3) = \frac{1 - \cos(-3\pi)}{-3\pi} = -0.2122$$

Non causal $h_d(n) = \{0.2122,\ 0,\ 0.6366,\ \underset{\uparrow}{0},\ -0.6366, 0,\ -0.2122\}$

causal $h_d(n) = \{\underset{\uparrow}{0.2122},\ 0,\ 0.6366,\ 0,\ -0.6366,\ 0,\ -0.2122\}$

Causal blackman window is

$$W_B(n) = 0.42 - 0.5\cos\left(\frac{\pi n}{\alpha}\right) + 0.08\cos\left(\frac{2\pi n}{\alpha}\right),\quad 0 \le n \le N-1$$

where

$$\alpha = \frac{N-1}{2} = \frac{7-1}{2} = 3$$

$$W_B(0) = 0.42 - 0.5\cos(0) + 0.08\cos(0) = 0$$

$$W_B(1) = 0.42 - 0.5\cos\left(\frac{\pi}{3}\right) + 0.08\cos\left(\frac{2\pi}{3}\right) = 0.13$$

$$W_B(2) = 0.42 - 0.5\cos\left(\frac{2\pi}{3}\right) + 0.08\cos\left(\frac{4\pi}{3}\right) = 0.63$$

$$W_B(3) = 0.42 - 0.5\cos(\pi) + 0.08\cos(2\pi) = 1$$

$$W_B(4) = 0.42 - 0.5\cos\left(\frac{4\pi}{3}\right) + 0.08\cos\left(\frac{8\pi}{3}\right) = 0.63$$

$$W_B(5) = 0.42 - 0.5\cos\left(\frac{5\pi}{3}\right) + 0.08\cos\left(\frac{10\pi}{3}\right) = 0.13$$

$$W_B(6) = 0.42 - 0.5\cos(2\pi) + 0.08\cos(4\pi) = 0$$

Impulse response of ideal Hilbert transformer is

$$h(n) = h_d(n) W_B(n)$$

$$h(0) = h(6) = h_d(0) W_B(0) = 0.2122(0) = 0$$

$$h(1) = h(5) = h_d(1) W_B(1) = 0(0.13) = 0$$

$$h(2) = h(4) = h_d(2) W_B(2) = 0.6366(0.63) = 0.401$$

$$h(3) = h_d(3) W_B(3) = 0(1) = 0$$

$$h(n) = \{0,\ 0,\ 0.401,\ 0,\ 0.401,\ 0,\ 0\}$$

Example 5.10: A filter is specified by its unit sample response $h(n) = -\dfrac{1}{3}\delta(n-1) + \dfrac{1}{3}\delta(n) + \dfrac{1}{3}\delta(n-2)$. Verify whether it is a linear phase filter or not.

Solution: Given

$$h(n) = -\frac{1}{3}\delta(n-1) + \frac{1}{3}\delta(n) + \frac{1}{3}\delta(n-2)$$

$$H(Z) = -\frac{1}{3}z^{-1} + \frac{1}{3} + \frac{1}{3}z^{-2}$$

Frequency response is

$$H(\omega) = -\frac{1}{3}e^{-j\omega} + \frac{1}{3} + \frac{1}{3}e^{-j2\omega}$$

$$= e^{-j\omega}\left[-\frac{1}{3} + \frac{1}{3}e^{j\omega} + \frac{1}{3}e^{-j\omega} \right]$$

$$= e^{-j\omega}\left[\frac{2}{3}\cos\omega - \frac{1}{3} \right]$$

The phase is i.e., $\angle H(\omega) = -\omega$, which is linear

So, the system is a linear phase system

Review Questions

1. For sharp cutoff filters FIR requires _________ coefficients when compared with IIR
   - (a) Same
   - (b) More
   - (c) Less
   - (d) Can't say   *Ans:* [b]
2. Memory requirement for FIR filter implementation is _________ when compared with IIR
   - (a) Same
   - (b) Less
   - (c) More
   - (d) Can't say   *Ans:* [c]

3. Execution time of FIR filter is __________ when compared with IIR
   - (a) Same
   - (b) More
   - (c) Less
   - (d) Can't say   *Ans:* [b]

4. Narrow transition band FIR filter implementation is __________ when compared with its IIR counter part
   - (a) Costlier
   - (b) Cheaper
   - (c) Equally cost
   - (d) Can't say   *Ans:* [a]

5. Analog filters can be directly transformed into equivalent digital filters using
   - (a) FIR
   - (b) IIR
   - (c) Both (a) and (b)
   - (d) Neither (a) nor (b)   *Ans:* [b]

6. Linear phase response is possible with
   - (a) FIR
   - (b) IIR
   - (c) Both (a) and (b)
   - (d) Neither (a) nor (b)   *Ans:* [a]

7. __________ filters are always stable
   - (a) FIR
   - (b) IIR
   - (c) Both (a) and (b)
   - (d) Neither (a) nor (b)   *Ans:* [a]

8. __________ of the filter provides a useful measure of how the filter modifies the phase characteristics of the signal
   - (a) Phase delay
   - (b) Group delay
   - (c) Either (a) or (b)
   - (d) Neither (c) or (b)   *Ans:* [c]

9. The amount of time delay each frequency component of the signal suffers in going through the filter is called
   - (a) Phase delay
   - (b) Group delay
   - (c) Either (a) or (b)
   - (d) Neither (a) or (b)   *Ans:* [a]

10. The average time delay the composite signal suffers at each frequency is called
    - (a) Phase delay
    - (b) Group delay
    - (c) Either (a) or (b)
    - (d) Neither (a) or (b)   *Ans:* [b]

11. The phase delay in terms of phase response $\theta(\omega)$ is
    - (a) $\dfrac{-\theta(\omega)}{\omega}$
    - (b) $\dfrac{-d\theta(\omega)}{d\omega}$
    - (c) $-\omega\theta(\omega)$
    - (d) $\dfrac{-\omega d\,\theta(\omega)}{d\omega}$ *Ans:* [a]

12. The group delay in terms of phase response $\theta(\omega)$ is

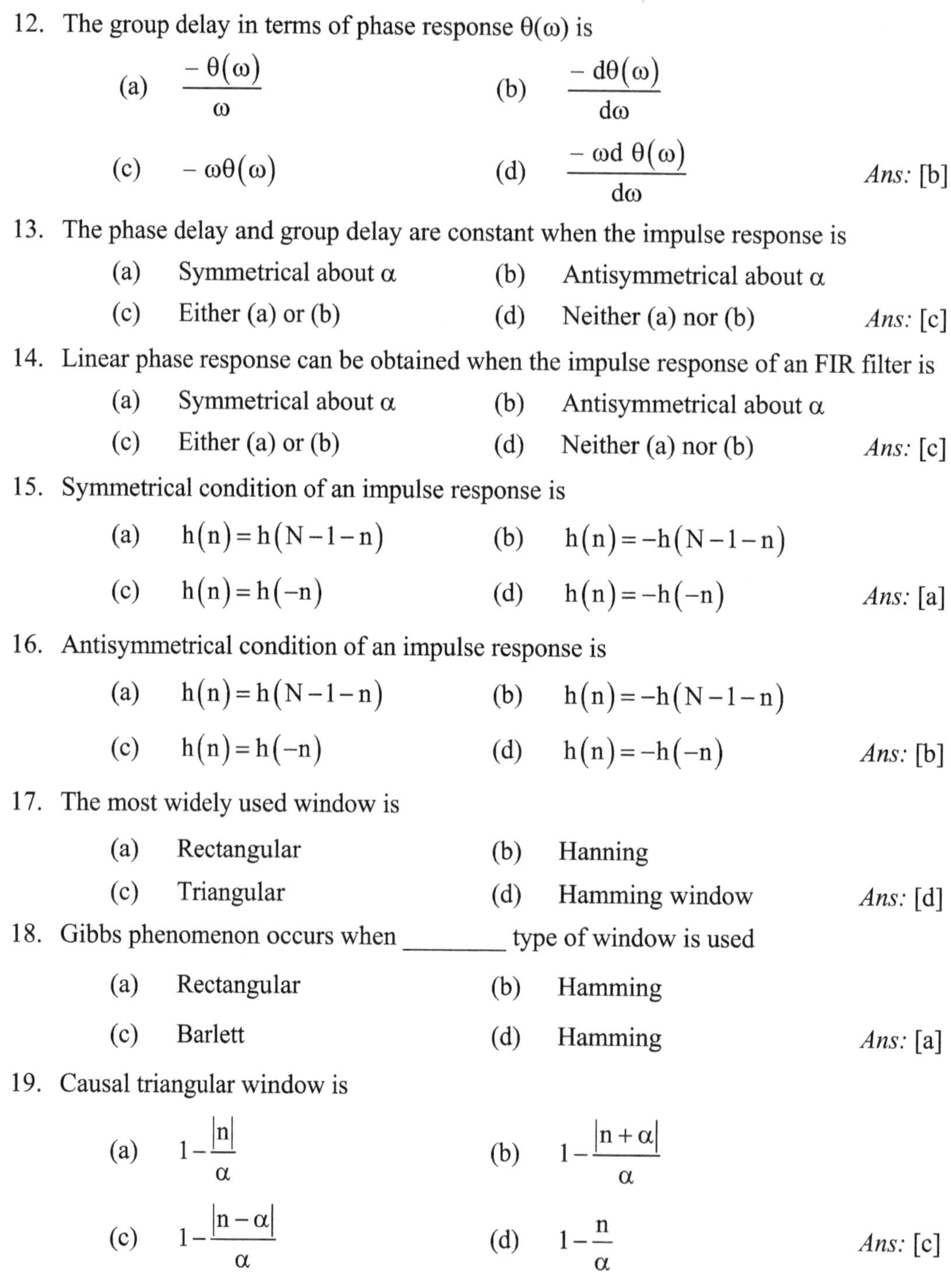

   (a) $\dfrac{-\theta(\omega)}{\omega}$ (b) $\dfrac{-d\theta(\omega)}{d\omega}$

   (c) $-\omega\theta(\omega)$ (d) $\dfrac{-\omega d\,\theta(\omega)}{d\omega}$ *Ans:* [b]

13. The phase delay and group delay are constant when the impulse response is

   (a) Symmetrical about α (b) Antisymmetrical about α

   (c) Either (a) or (b)   (d) Neither (a) nor (b)   *Ans:* [c]

14. Linear phase response can be obtained when the impulse response of an FIR filter is

   (a) Symmetrical about α (b) Antisymmetrical about α

   (c) Either (a) or (b)   (d) Neither (a) nor (b)   *Ans:* [c]

15. Symmetrical condition of an impulse response is

   (a) $h(n)=h(N-1-n)$ (b) $h(n)=-h(N-1-n)$

   (c) $h(n)=h(-n)$ (d) $h(n)=-h(-n)$ *Ans:* [a]

16. Antisymmetrical condition of an impulse response is

   (a) $h(n)=h(N-1-n)$ (b) $h(n)=-h(N-1-n)$

   (c) $h(n)=h(-n)$ (d) $h(n)=-h(-n)$ *Ans:* [b]

17. The most widely used window is

   (a) Rectangular   (b) Hanning

   (c) Triangular   (d) Hamming window   *Ans:* [d]

18. Gibbs phenomenon occurs when _________ type of window is used

   (a) Rectangular   (b) Hamming

   (c) Barlett   (d) Hamming   *Ans:* [a]

19. Causal triangular window is

   (a) $1-\dfrac{|n|}{\alpha}$ (b) $1-\dfrac{|n+\alpha|}{\alpha}$

   (c) $1-\dfrac{|n-\alpha|}{\alpha}$ (d) $1-\dfrac{n}{\alpha}$ *Ans:* [c]

20. Non causal triangular window is

(a) $1 - \dfrac{|n|}{\alpha}$ (b) $1 - \dfrac{|n + \alpha|}{\alpha}$

(c) $1 - \dfrac{|n - \alpha|}{\alpha}$ (d) $1 - \dfrac{n}{\alpha}$ *Ans:* [a]

21. Side lobe level of triangular window is _________ when compared with that of rectangular window

(a) Less   (b) More

(c) Equal   (d) Can't say   *Ans:* [a]

Chapter 6

Multirate Digital Signal Processing

6.1 Introduction

Multirate digital signal processing is a digital signal processing system that uses signals with different sampling frequencies. Multirate digital signal processing often uses sample rate conversion to convert from one sampling frequency to another sampling frequency. Sample rate conversion may be done either in the analog domain or in digital domain, in analog domain it requires D/A conversion then A/D conversion, which introduce errors and noise into the signal, therefore Sample rate conversion is done in digital domain that uses decimation to decrease the sampling rate, interpolation to increase the sampling rate. Multirate systems have gained popularity since the early 1980s and they are commonly used for audio and video processing, communications systems, and transform analysis etc. In most applications multirate systems are used to improve the performance, or for increased computational efficiency.

6.2 Decimation or Down Sampling

Decimation removes samples from a signal as shown in Fig. 6.1. Decimation can therefore only down sample the signal by an integer factor:

$$\frac{F_s}{F_s^{new}} = M > 1 \quad \text{so that} \quad F_s > F_s^{new}$$

where M is an integer, F_s is the old sampling rate (number of samples per second) and F_s^{new} is the new sampling rate.

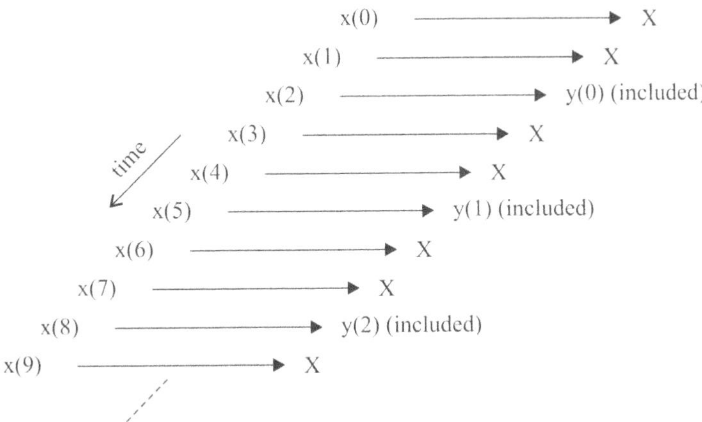

Fig. 6.1 Decimation of a general sequence.

From the above analysis, Decimation can be regarded as the discrete-time counterpart of sampling. Whereas in sampling we start with a continuous-time signal x(t) and convert it into a sequence of samples x(n), in decimation we start with a discrete-time signal x(n) and convert it into another discrete-time signal y(n), which consists of sub-samples of x(n). Thus, the formal definition of M-fold decimation, or down-sampling, is defined by Equation 6.1.1. In decimation, the sampling rate is reduced from F_s to F_s/M by discarding M – 1 samples for every M samples in the original sequence.

$$y(n) = v(nM) = \sum_{k=-\infty}^{k=\infty} h(k)x(nM - k) \qquad \ldots\ldots(6.2.1)$$

The sampling theorem states that the highest frequency in a signal should be less than half the sampling frequency.

A digital anti-aliasing filter has to be applied to remove frequencies higher than:

$$F_{cf} = \frac{F_s^{new}}{2}$$

So in digital frequency the cut-off frequency is: $\Omega_{cf} = \frac{\Omega_s^{new}}{2} = \frac{2\pi \dfrac{F_s^{new}}{F_s}}{2} = \pi \dfrac{F_s^{new}}{F_s} < \pi$

as $F_s^{new} < F_s$

This means that the signal has to be filtered in the digital domain before decimation. The block diagram notation of the decimation process is depicted in Fig. 6.2. An anti-aliasing digital filter precedes the down-sampler to prevent aliasing from occurring, due to the lower sampling rate.

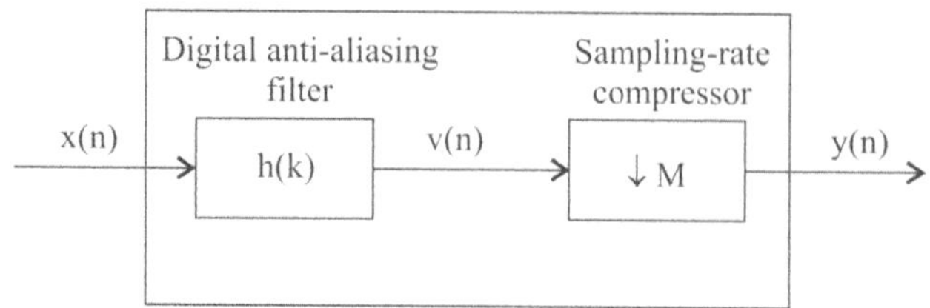

Fig. 6.2 Block diagram notation of decimation, by a factor of M.

In Fig. 6.3 below, it illustrates the concept of 3-fold decimation i.e., M = 3. Here, the samples of x(n) corresponding to n = …, –2, 1, 4,… and n = …, –1, 2, 5,… are lost in the decimation process. In general, the samples of x(n) corresponding to n ≠ kM, where k is an integer, are discarded in M-fold decimation. In Figure 6.2 (b), it shows samples of the decimated signal y(n) spaced three times wider than the samples of x(n). This is not a coincidence. In real time, the decimated signal appears at a slower rate than that of the original signal by a factor of M. If the sampling frequency of x(n) is F_s, then that of y(n) is F_s/M.

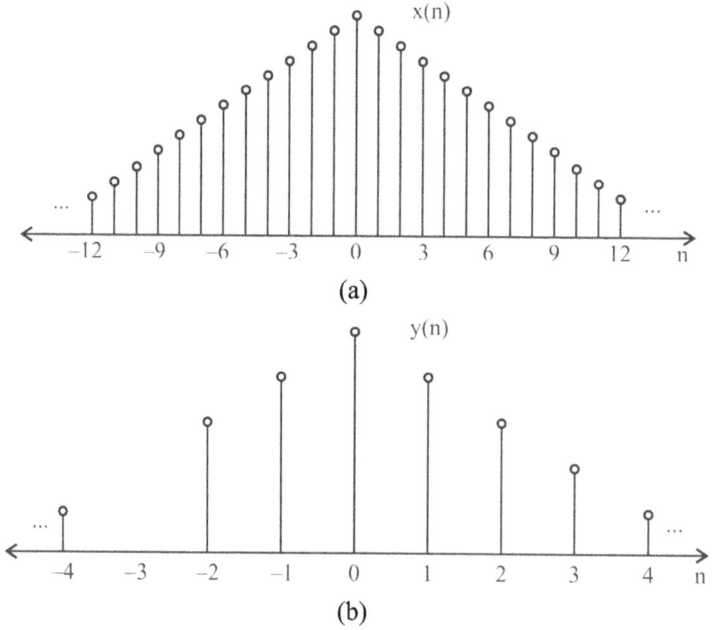

Fig. 6.3 Decimation of a discrete-time signal by a factor of 3.

6.3 Interpolation or Up Sampling

Interpolation increases the sampling frequency by estimating the value of the signal between samples as shown in Fig.6.4.

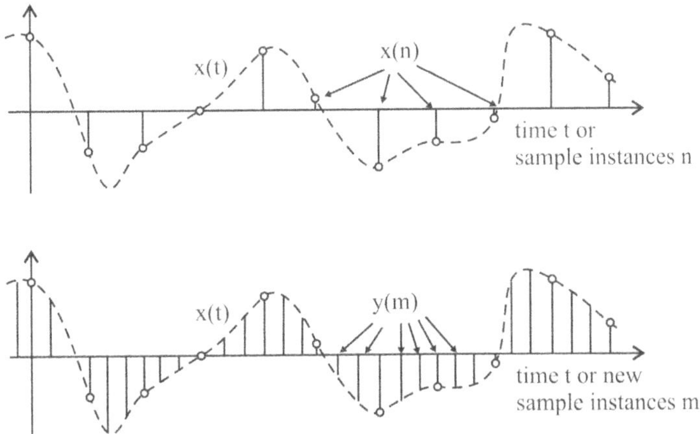

Fig. 6.4 Interpolation of a general sequence.

The new sampling frequency is greater than the old sampling frequency: $F_s^{new} > F_s$ where F_s is the old sampling frequency and F_s^{new} is the new sampling frequency. Also, the new sampling frequency has to be an integer multiple of the original sampling frequency:

$$\frac{F_s^{new}}{F_s} = L > 1 \qquad \text{where L is an integer.}$$

From the above analysis we can say that Interpolation is the exact opposite of decimation. It is an information preserving operation, in that all samples of x(n) are present in the expanded signal y(n). The mathematical definition of L-fold interpolation is defined by Equation (6.1.2) and the block diagram notation is depicted in Fig. 6.5. Interpolation works by inserting (L–1) zero-valued samples for each input sample. The sampling rate therefore increases from F_s to LF_s. With reference to Fig. 6.5, the expansion process is followed by a unique digital low-pass filter called an anti-imaging filter. Although the expansion process does not cause aliasing in the interpolated signal, it does however yield undesirable replicas in the signal's frequency spectrum. We shall see how this special filter, in Section 6.4, is necessary to remove these replicas from the frequency spectrum.

$$y(n) = L \sum_{k=-\infty}^{k=\infty} h(k)w(n-k) \qquad \dots\dots(6.3.1)$$

where
$$w(n) = \begin{cases} x(n/L), & \text{if } n/L \text{ is an integer} \\ 0, & \text{if } n/L \text{ is non-integer} \end{cases}$$

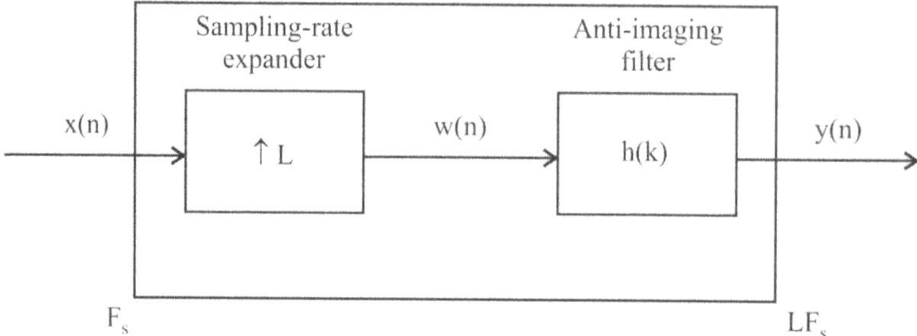

Fig. 6.5 Block diagram notation of interpolation, by a factor of L.

A common interpolation approach is zero filling based interpolation as shown in Fig.6.6(a).

There are two stages:

1. zero filling
2. low pass filtering

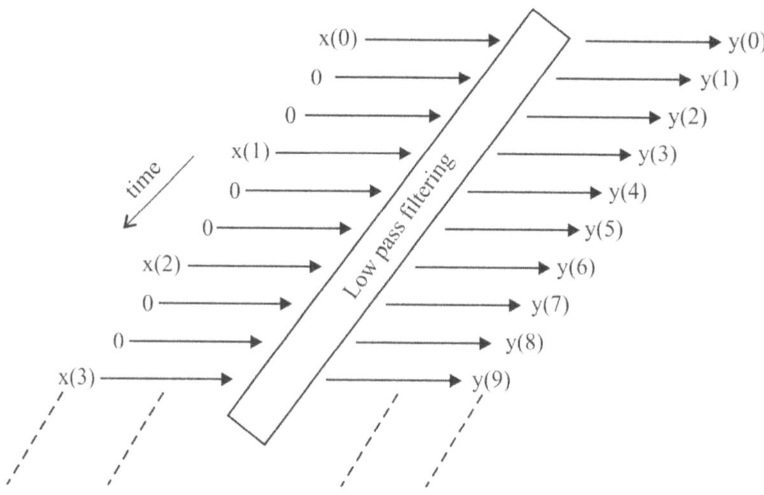

Fig. 6.6(a) Zero filling based interpolation.

In Fig. 6.6(b) below, it depicts 3-fold interpolation of the signal x(n) i.e., L=3. The insertion of zeros(two zero samples are inserted between each original sample) effectively attenuates the signal by L, so the output of the anti-imaging filter must be multiplied by L, to maintain the same signal magnitude.

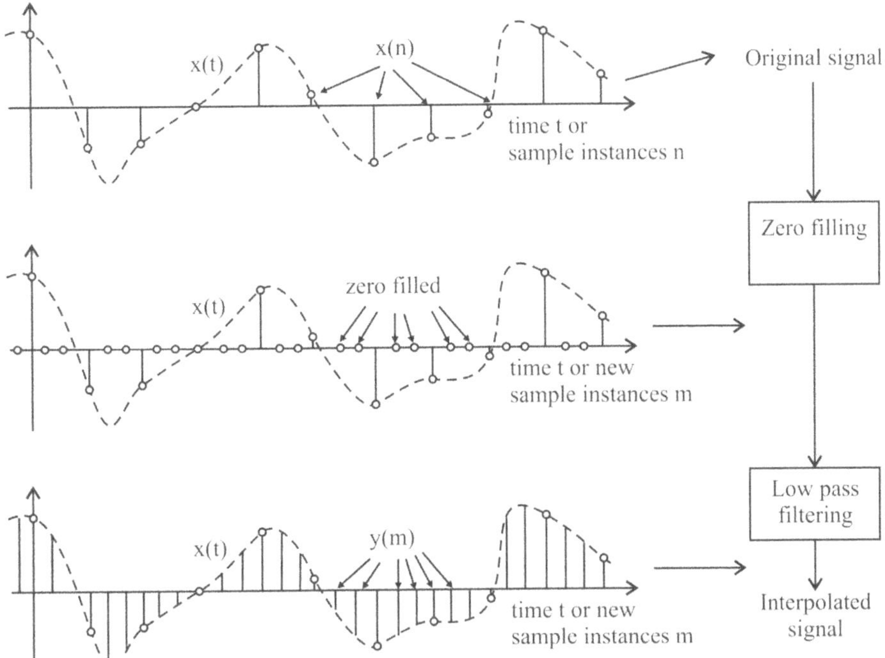

Fig. 6.6(b) Interpolation of a discrete-time signal by a factor of 3.

6.4 Frequency Transforms of Decimated and Expanded Sequences

The analysis of decimation and expansion is better understood by assessing their respective frequency spectra using the Fourier transform.

6.4.1 Decimation

The implications of aliasing caused by decimation are very similar to those in the case of sampling a continuous-time signal. In general, if the Fourier transform of a signal, $X(\omega)$, occupies the entire bandwidth from $(-\pi, \pi)$, then the Fourier transform of the decimated signal, $X_{(\downarrow M)}(\omega)$, will be aliased. This is due to the superposition of the M shifted and frequency-scaled transforms. This is illustrated in Fig. 6.7 below, which shows the aliasing phenomenon for $M = 3$.

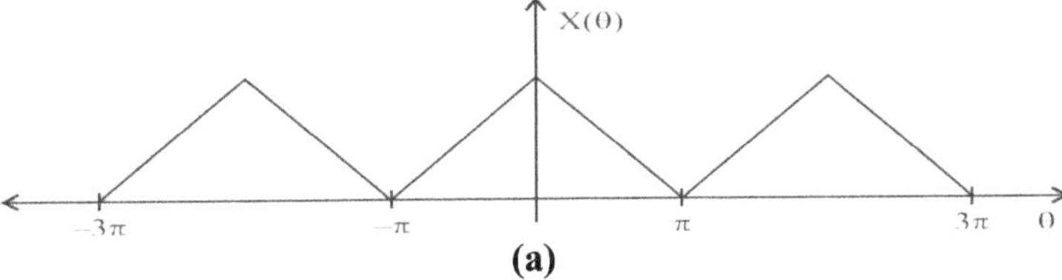

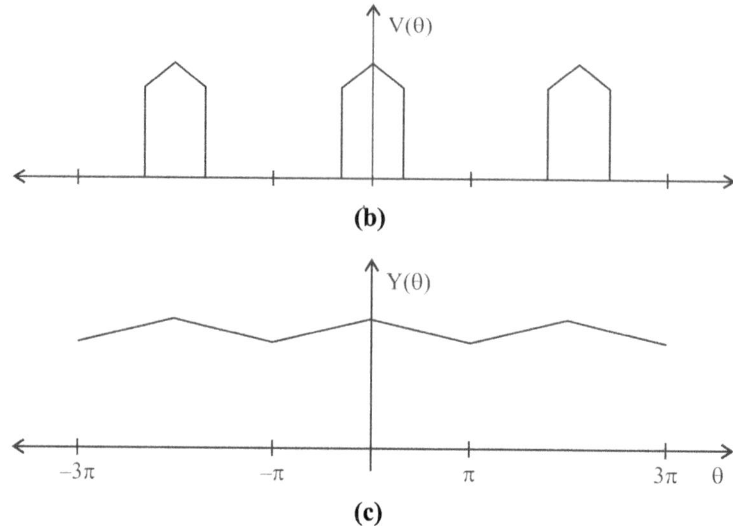

(b)

(c)

Fig. 6.7 Aliasing caused by decimation; (a) Fourier transform of the original signal;
(b) After decimation filtering; (c) Fourier transform of the decimated signal.

In Fig. 6.7 (a) it shows the Fourier transform of the original signal. Part (b) shows the signal after lowpass filtering. In Figure 6.7(c), it depicts the expanded spectrum after decimation.

6.4.2 Expansion

The effect of expansion on a signal in the frequency domain is illustrated in Fig. 6.8 below. Part (a) shows the Fourier transform of the original signal; part (b) illustrates the Fourier transform of the signal with zeros added $W(\omega)$; and part (c) shows the Fourier transform of the signal after the interpolation filter. It is clearly visible that the shape of the Fourier transform is compressed by a factor L in the frequency axis and is also repeated L times in the range of $(-\pi, \pi)$. Despite the compression of the signal in the frequency axis, the shape of the Fourier transform is still preserved, confirming that expansion does not lead to aliasing. These replicas are removed by a digital low-pass filter called an anti-imaging filter, as indicated in Fig. 6.5.

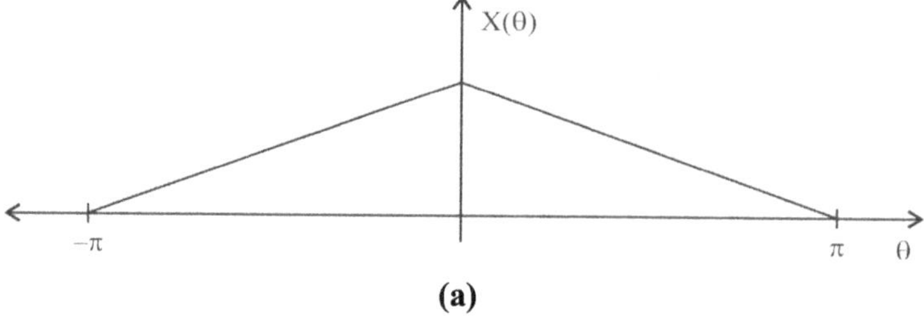

(a)

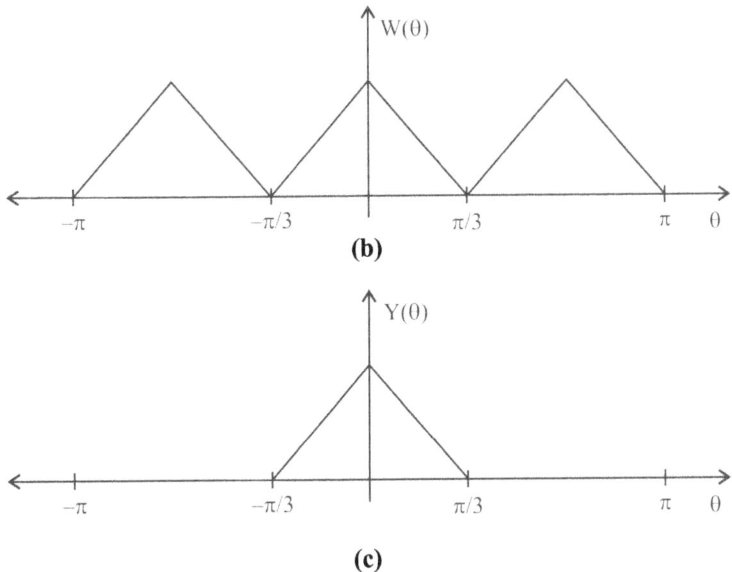

Fig. 6.8 Expansion in the frequency domain of the original signal (a) Fourier transform of the original signal (b) Fourier transform of the signal with zeros added (c) Fourier transform of the signal after the interpolation filter.

6.5 Sampling-Rate Conversion

A common use of multirate signal processing is for sampling-rate conversion. Suppose a digital signal $x(n)$ is sampled at an interval T_1, and we wish to obtain a signal $y(n)$ sampled at an interval T_2. Then the techniques of decimation and interpolation enable this operation, providing the ratio T_1/T_2 is a rational number i.e., L/M. Sampling-rate conversion can be accomplished by L-fold expansion, followed by low-pass filtering and then M-fold decimation, as depicted in Fig. 6.9. It is important to emphasis that the interpolation should be performed first and decimation second, to preserve the desired spectral characteristics of $x(n)$. Furthermore by cascading the two in this manner, both of the filters can be combined into one single low-pass filter.

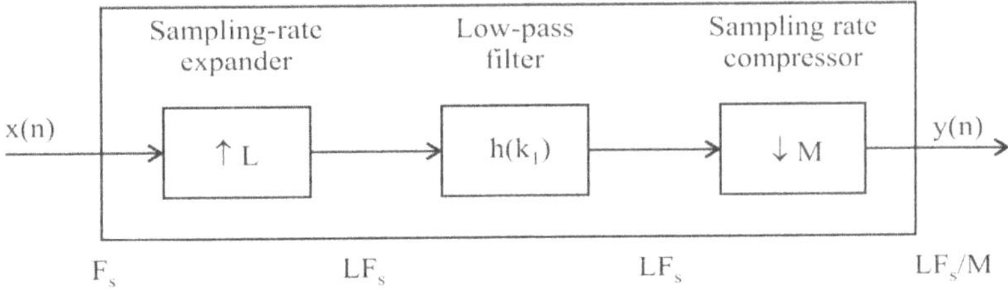

Fig. 6.9 Sampling-rate conversion by expansion, filtering and decimation.

Sampling-rate conversion would take place when data from a CD(which stores music at 44.1 KHz) is transferred onto a Professional music recording device that processes audio at 48 KHz. Here the sampling-rate is increased from 44.1 kHz to 48 kHz. Transfer of the music to or from the CD player and the professional audio device using decimation only or interpolation only are not possible because

$$\frac{L}{M} = \frac{48}{44.1} = \frac{160}{147} = 1.08844$$

Which is not an integer.

Non-integer sample rate conversion can be obtained by combining decimation and interpolation, similar to finding a common denominator in fractions.

Step1: Find common (integer) factor of the two sample rates, L

Step2: Interpolate (upsample) by this common factor L

Step3: Decimate (downsample) to the new sample rate F_s^{new} by downsampling by an integer factor M.

Then the Sample rate conversion is $\dfrac{L}{M} = \dfrac{F_s^{new}}{F_s}$

Example 6.1

Get audio from 44.1 kHz sampled source (CD player) and transfer to professional audio processor requiring 48 kHz sample rate.

Solution:

This process requires up sampling to 48 kHz from 44.1 kHz as shown in Fig.6.10.

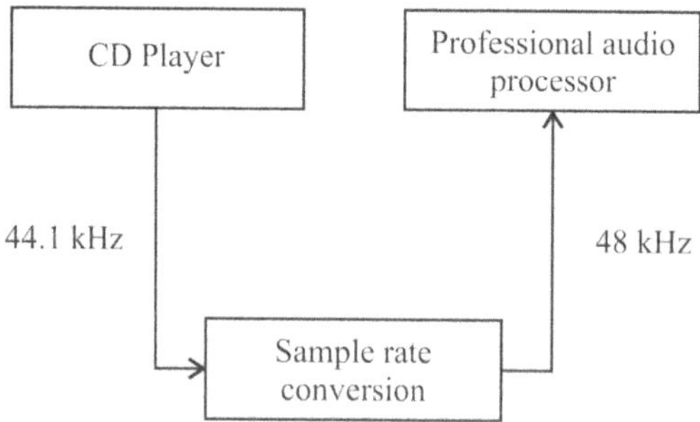

Fig. 6.10 Upsampling to 48 kHz from 44.1 kHz.

Step1: Common integer factor of the two sample rates is L=160, i.e.,

$$\frac{L}{M} = \frac{F_s^{new}}{F_s} = \frac{48}{44.1} = \frac{160}{147}$$

Step 2: So interpolate (upsample) by factor L by inserting 159 zeros for each sample in 44.1kHz CD player signal to obtain $(44.1 \times 160) = 7056$ kHz, then low pass filtering.

Step 3: Then decimate (downsample) from 7056 kHz to 48 kHz by removing 146 sample in every

147 (M = L × 44.1 kHz/48 kHz) from the upsampled signal (after applying anti-aliasing low pass filter).

The resulting sample rate conversion is:

$$\frac{L}{M} = \frac{160}{147} = 1.088$$

Which is the same as

$$\frac{F_s^{new}}{F_s} = \frac{48\text{kHz}}{44.1\text{kHz}} = 1.088.$$

Non-Integer Sample rate Conversion may be optimized as follows:

There are ×2 low pass filters (low pass filtering and anti-aliasing filtering) for non-integer sample rate conversion as shown in Fig.6.11.

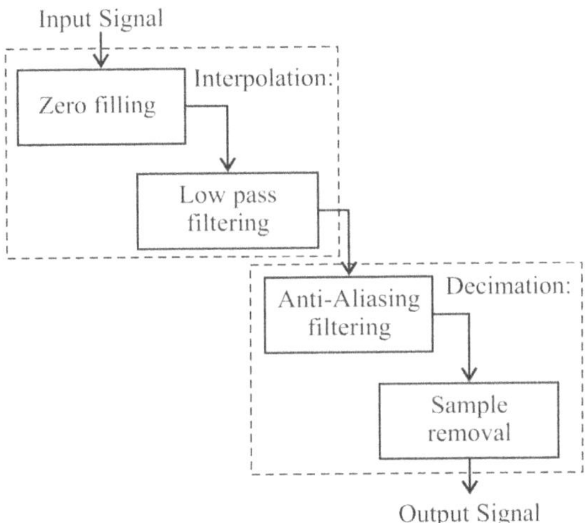

Fig. 6.11 Block diagram of non-integer sample rate conversion.

The interpolation low pass filter and the anti-aliasing filter for the decimation stage can be combined with a cut-off frequency equal to the lower of the two filters' cut-off frequencies as shown in Fig.6.12.

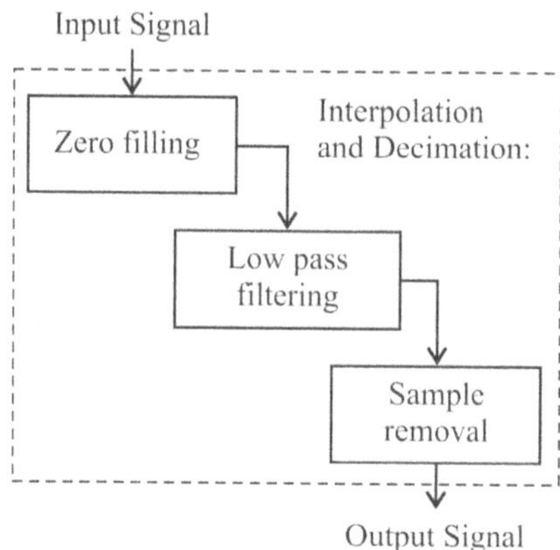

Fig. 6.12 Optimization of non-integer sample rate conversion.

6.6 Multistage Approach

When the sampling-rate changes are large, it is often better to perform the operation in multiple stages, where $M_i(L_i)$, an integer, is the factor for the stage i.

$$M = M_1M_2...M_I \text{ or } L = L_1L_2...L_I$$

An example of the multistage approach for decimation is shown in Fig. 6.13. The multistage approach allows a significant relaxation of the anti-alias and anti-imaging filters, with a consequent reduction in the filter complexity. The optimum number of stages is one that leads to the least computational effort in terms of either the multiplications per second (MPS), or the total storage requirement (TSR).

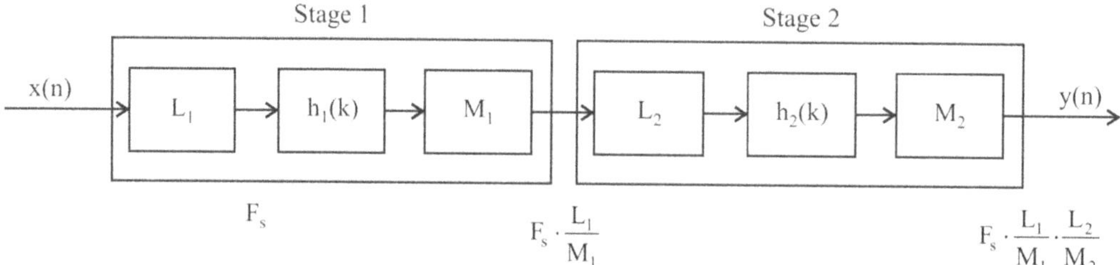

Fig. 6.13 Multistage approach for the decimation process.

In real world applications sample rate conversion converts a sampling frequency to another sampling frequency that is:

Very much greater $\left(F_s^{new} \gg F_s\right)$ or very much smaller $F_s^{new} \ll F_s$ than the original signal sampling frequency.

The problem with the above process is best explained by the following example.

Example 6.2

A signal x(n), sampled at 4.096 kHz has to be decimated to 128 Hz. There should be an antialiasing filter:

- that rejects frequencies above 64Hz,
- with a stopband ripple, $\delta_s \approx 0.001$,
- and a passband ripple of $\delta_p \approx 0.001$,
- The transition width should be $f_{tw} = 4Hz$,
- So that frequencies below 60Hz are kept.

Solution:

A Blackman window can achieve a stop band ripple 75 dB and passband ripple of 0.0014dB.

This can be compared with the requirements of this antialiasing filter of $\delta_s \approx 0.001$, which is $-20\log(0.001) = 60dB$ and a passband ripple $\delta_p \approx 0.001$, or $20\log(1 + 0.001) = 0.0087dB$.

According to the low pass FIR filter design, the number of filter coefficients for a Blackman window will then be:

$$N = 5.98 \times \frac{f_s}{f_{tw}} = 5.98 \times 4096 / 4 = 6123.5$$

So the number of filter coefficients is very high.

Multiple stages for decimation (or interpolation) can reduce the number of filter coefficients in the filter specifications. The signal can be decimated more than once, using a gradual change in sampling frequency

Conventional decimation is as shown in Fig.6.2.

Decimation in mutliple stages (3 stages) is as shown in Fig.6.14

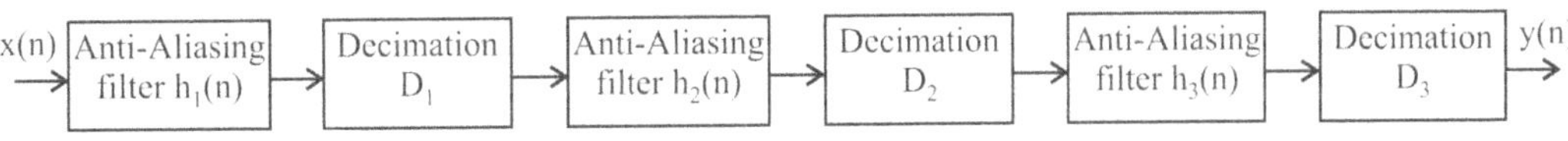

Fig. 6.14 Decimation in multiple stages.

The multi-stage sample rate conversion decimation values, D_i:

$$\frac{F_s}{F_s^{new}} = D = D_1 \times D_2 \times D_3 \ldots \times D_k = \prod_{i=1}^{k} D_i$$

Where all D_i are integers.

So for three stages, $k = 3$ and $D = D_1 \times D_2 \times D_3$.

Example 6.3

The earlier problem can now be implemented using 2 decimation stages. Find out how many filter coefficients are necessary for a 2 stage decimation process.

Solution:

The original sampling frequency $F_s = 4.096$ kHz and the new (decimated signal) should have a sampling frequency of $F_s^{new} = 128\,\text{Hz}$. Multistage decimation with 2 stages requires that:

$$\frac{F_s}{F_s^{new}} = D = \frac{4096}{128} = 32 = D_1 \times D_2$$

The multistage decimation values can therefore be $D_1 = 8$ and $D_2 = 4$, creating an intermediate signal with sampling frequency: $F_s^{(1)} = F_s / 8 = 512\,\text{Hz}$.

The transition width can be longer with this higher sampling rate.

We can keep the same passband frequency(60Hz).

The transition width can go up to half the sampling rate:

$$F_{tw}^{(1)} \leq \frac{512\,\text{Hz}}{2} - 60\,\text{Hz} = 196\,\text{Hz}$$

The number of Blackman filter coefficients for this stage is:

$$N_1 = 5.98 \times \frac{F_s^{(1)}}{F_{tw}^{(1)}} = 5.98 \times \frac{512}{196} = 16, \text{(rounded up to integer value)}.$$

So $N_1 = 16$ filter coefficients are required for the first decimation stage.

The intermediate signal sampled at 512 Hz is to be decimated by a factor of 4 to 128 Hz for the second stage:

$$F_s^{new} = F_s^{(2)} = \frac{512\,\text{Hz}}{4} = 128\,\text{Hz}.$$

The transition width for this (final) stage can then be:

$$F_{tw}^{(2)} = \frac{128\,\text{Hz}}{2} - 60\,\text{Hz} = 4\,\text{Hz}$$

So that the number of Blackman filter coefficients for this stage is

$$N_2 = 5.98 \times \frac{F_s^{(2)}}{F_{tw}^{(2)}} = 5.98 \times \frac{128}{4} = 192.$$

192 filter coefficients are required for this final stage. The combined filter coefficients for the two stages is

$$N_1 + N_2 = 16 + 192 = 208,$$

which is considerably less than the original non-multistage decimation antialiasing filter requiring $N = 6124$ coefficients.

6.7 Polyphase Filters

Potential computational savings can be made within the process of decimation, interpolation and sampling-rate conversion. Polyphase filters is the name given to certain realisations of multirate filtering operations, which facilitate computational savings in both hardware and software.

As an example, the combined low-pass filter in the sampling-rate converter, as illustrated in Fig. 6.9, can be re-drawn as a realisation structure. In principle, the simplest realisation of the low-pass filter is the direct-form FIR structure, as depicted in Fig. 6.15. However, this type of structure is very inefficient owing to the interpolation process, which introduces $(L - 1)$ zeros between consecutive points in the signal. If L is large, then the majority of the signal components fed into the FIR filter are zero. As a result, most of the multiplications and additions are zero i.e., many pointless calculations. Furthermore, the decimation process itself implies that only one out of every M output samples is required at the output of the sampling-rate converter. Consequently, only one out of every M possible values at the output of the filter needs to be computed. This type of structure therefore, leads to much inefficiency during the process of sampling-rate conversion. Fig. 6.16. It takes into account that after the interpolation process the signal consists of $(L-1)$ zero coefficients, and the decimation process implies that only one out of every M samples is required at the output of the converter. To make the scheme more efficient, the low-pass filter in Fig. 6.15 is replaced by a bank of filters arranged in parallel, as illustrated in the efficient realisation. The sampling-rate conversion process is undertaken by the multiplexer at the output by selecting every MT/L samples. In this example, the efficient realisation is illustrated for a signal which is interpolated by $L = 3$ and decimated by $M = 2$ samples.

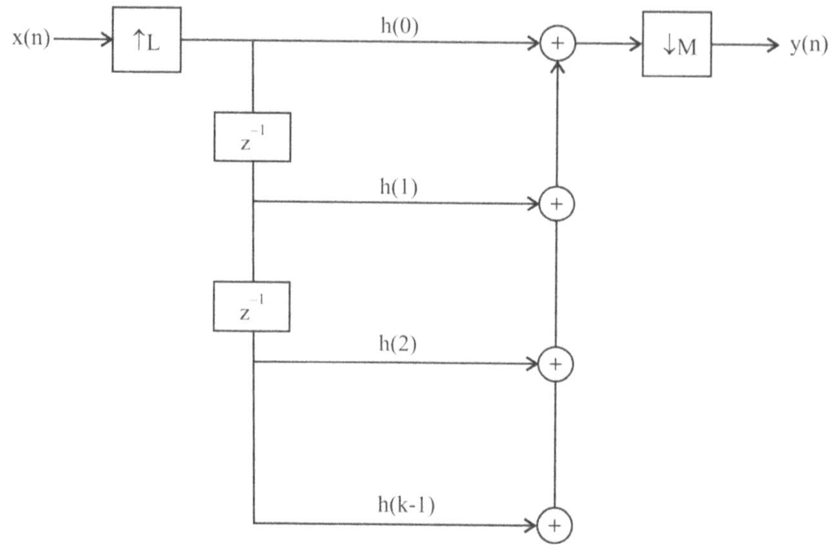

Fig. 6.15 Realisation structure of sampling-rate conversion.

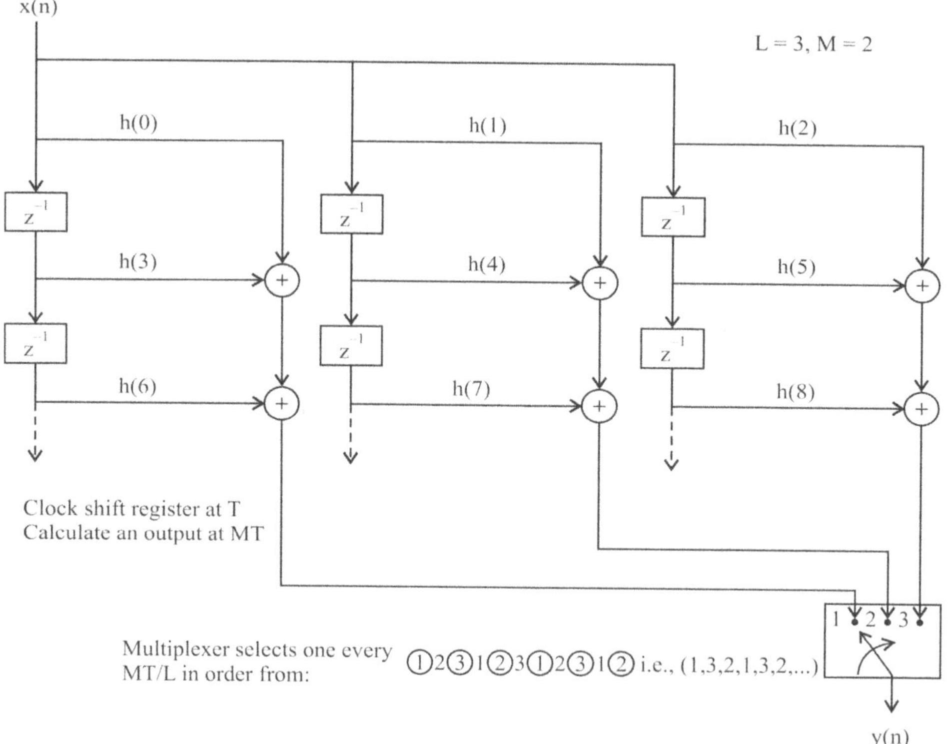

Fig. 6.16 Efficient realisation structure for sampling-rate conversion.

6.8 Applications of Multirate DSP

Multirate systems are used in a CD player when the music signal is converted from digital into analogue (DAC). Digital data (16-bit words) are read from the disk at a sampling rate of 44.1 kHz. If this data were converted directly into an analogue signal, image frequency bands centred on multiples of the sampling-rate would occur, causing amplifier overload, and distortion in the music signal. To protect against this, a common technique called *oversampling* is often implemented nowadays in all CD players and in most digital processing systems of music signals. Fig. 6.17 below illustrates a basic block diagram of a CD player and how oversampling is utilised. It is customary to oversample (or expand) the digital signal by a factor of x8, followed by an interpolation filter to remove the image frequencies. The sampling rate of the resulting signal is now increased up to 352.8 kHz. The digital signal is then converted into an analogue waveform by passing it through a 14-bit DAC. Then the output from this device is passed through an analogue low-pass filter before it is sent to the speakers.

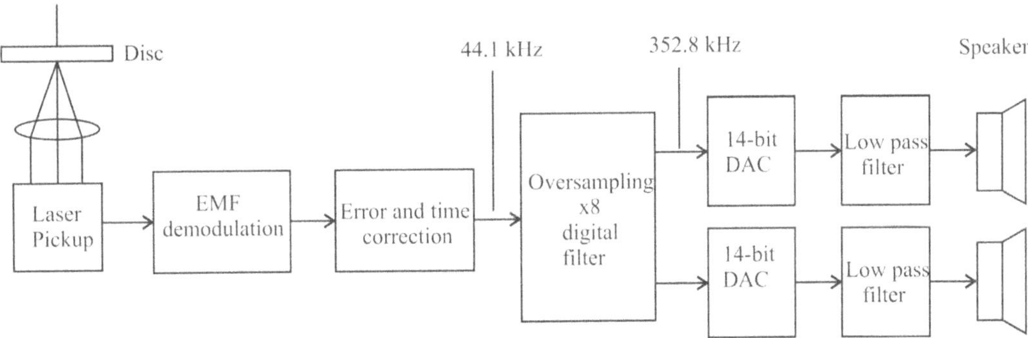

Fig. 6.17 Digital to analogue conversion for a CD player using x8 oversampling.

Fig. 6.18 illustrates the procedure of converting a digital waveform into an analogue signal in a CD player using x8 oversampling. As an example, Fig.6.18 (a) illustrates a 20 kHz sinusoidal signal sampled at 44.1 kHz, denoted by x(n). The six samples of the signal represent the waveform over two periods. If the signal x(n) was converted directly into an analogue waveform, it would be very hard to exactly reconstruct the 20 kHz signal from this diagram. Now, Fig.6.18 (b) shows x(n) with an x8 interpolation, denoted by y(n). Fig.6.18 (c) shows the analogue signal y(t), reconstructed from the digital signal y(n) by passing it through a DAC. Finally, Fig.6.18 (d) shows the waveform of z(t), which is obtained by passing the signal y(t) through an analogue low-pass filter.

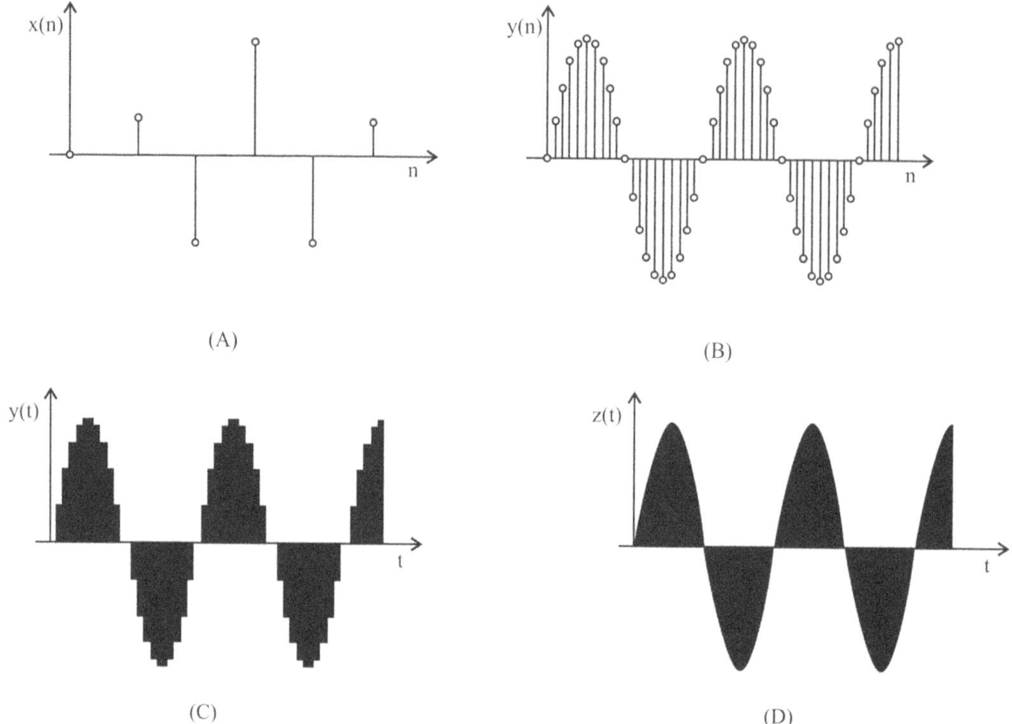

Fig. 6.18 Illustration of oversampling in CD music signal reconstruction.

The effect of oversampling also has some other desirable features. Firstly, it causes the image frequencies to be much higher and therefore easier to filter out. The anti-alias filter specification can therefore be very much relaxed i.e. the cut-off frequency of the filter for the previous example increases from (44.1/2) = 22.05 kHz to (44.1x8/2) = 176.4 kHz after the interpolation.

One other attractive feature about oversampling is the effect of reducing the noise *power spectral density*, by spreading the noise power over a larger bandwidth. This is illustrated in Fig. 6.19 and mathematically defined below by Equation (6.3).

$$\text{Noise power spectral density} = \frac{\text{Total power}}{\text{Bandwidth}} \qquad .....(6.8.1)$$

For both sequences, the *total noise power* (shaded area in Fig. 6.19) remains the same. However, as the bandwidth is increased by a factor of x8 because of the interpolation process, it causes the level of the noise power spectral density to decrease by a factor of x8, over the whole range of the bandwidth.

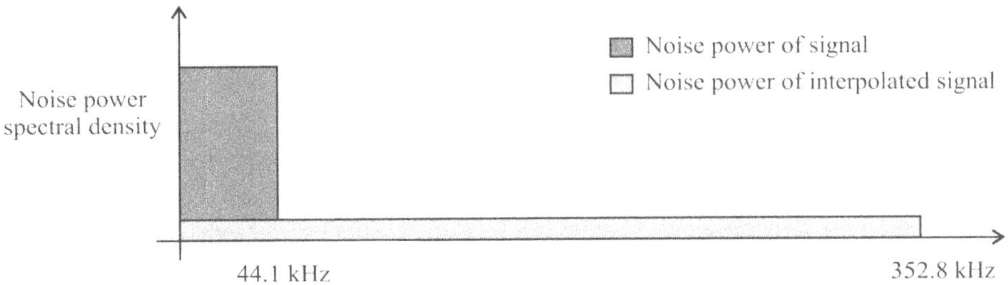

Fig. 6.19 Illustration of noise power spectral density reduction due to oversampling.

As a consequence of the reduction in the noise power spectral density, it means that the level of tolerable noise can be increased by a factor of 8. In terms of the quantisation noise power, q^2, it means that it can now be 8 times greater (or the quantisation step size, q, can be increased by $\sqrt{8}$). This ultimately means that a reduction in the number of bits for the DAC is possible. In general, the reduction in the number of bits for the DAC process is given by Equation (6.1.4) below.

$$\text{DAC bit reduction} = \frac{1}{2}\log_2^{\text{(oversample factor)}} \qquad ....(6.8.2)$$

For the previous example, the DAC bit reduction owing to the x8 oversample factor is $1/2\log_2(8) = 1.5$ bits.

There are in fact more sophisticated oversampled ADCs and DACs that use various feedback paths within the system to move most of the quantisation noise into a high frequency out-of-band region. Substantially larger savings in the number of bits can then be made, even to one bit only.

Example 6.4: Develop an expression for the output y(n) as a function of the input x(n) for the multirate structure of Figure 6.20.

$$x(n) \longrightarrow \boxed{\uparrow 5} \longrightarrow \boxed{\downarrow 10} \longrightarrow \boxed{\uparrow 2} \longrightarrow y(n)$$

Fig. 6.20

Solution:

$$x(n) \longrightarrow \boxed{\uparrow 5} \longrightarrow \boxed{\downarrow 10} \longrightarrow \boxed{\uparrow 2} \longrightarrow y(n) \equiv$$

$$x(n) \longrightarrow \boxed{\uparrow 5} \longrightarrow \boxed{\downarrow 5} \longrightarrow \boxed{\downarrow 2} \longrightarrow \boxed{\uparrow 2} \longrightarrow y(n) \equiv x(n) \longrightarrow \boxed{\downarrow 2} \xrightarrow{x_1(n)} \boxed{\uparrow 2} \longrightarrow y(n)$$

Hence, $x(n) = x(2n)$ and $y(n) = \begin{cases} x_1(n/2), & \text{for } n=2r \\ 0, & \text{otherwise} \end{cases} = \begin{cases} x(n), & \text{for } n=2r \\ 0, & \text{otherwise} \end{cases}$

therefore

$$y(n) = \begin{cases} x(n), & \text{for } n=2r \\ 0, & \text{otherwise} \end{cases}$$

Example 6.5 Consider the multirate structure of Fig.6.21(a) where $H_0(z)$, $H_1(z)$, and $H_2(z)$ are, respectively, ideal zero-phase real-coefficient lowpass, bandpass, and highpass filters with frequency responses as indicated in Fig.6.21(b). If the input is a real sequence with a discrete-time Fourier transform as shown in Fig.6.21(c), sketch the discrete-time Fourier transforms of the outputs $y_0(n)$, $y_1(n)$, and $y_2(n)$.

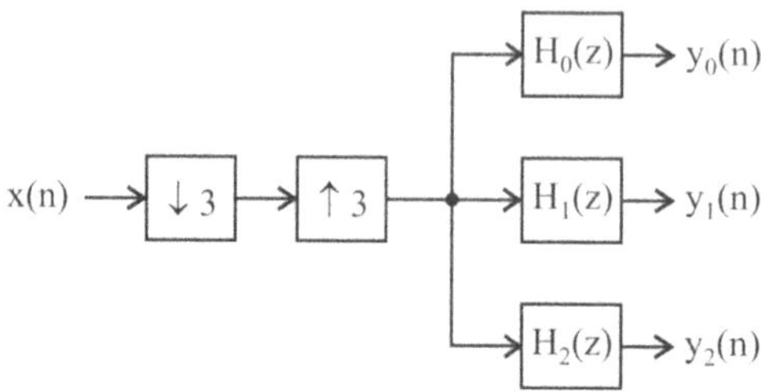

Fig. 6.21(a)

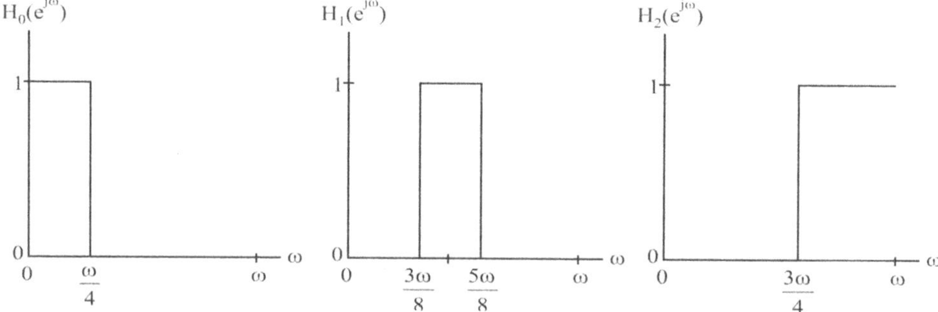

Fig. 6.21(b)

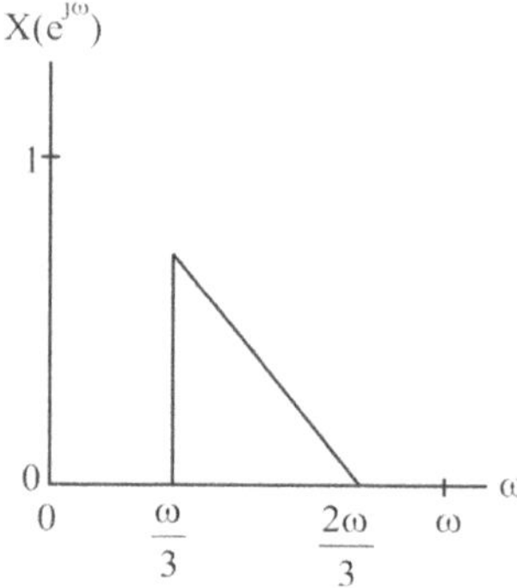

Fig. 6.21(c)

Solution:

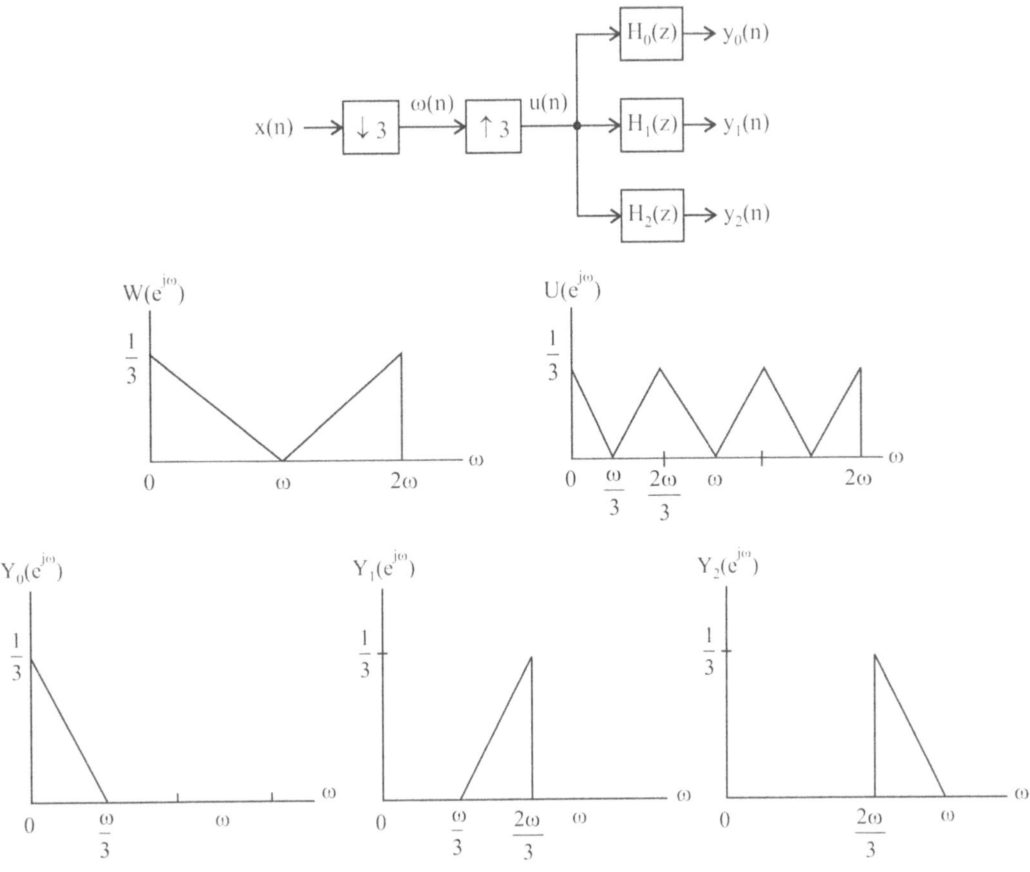

Example 6.6: Develop a computationally efficient realization of a factor-of-4 interpolator employing a length-16 linear-phase FIR filter.

Solution:

A computationally efficient realization of the factor-of-4 interpolator

is obtained by applying a 4-branch polyphase decomposition to H(z):

$$H(z) = E_0(z^4) + z^{-1}E_1(z^4) + z^{-2}E_2(z^4) + z^{-3}E_3(z^4)$$

and then moving the down-sampler through the polyphase filters resulting in

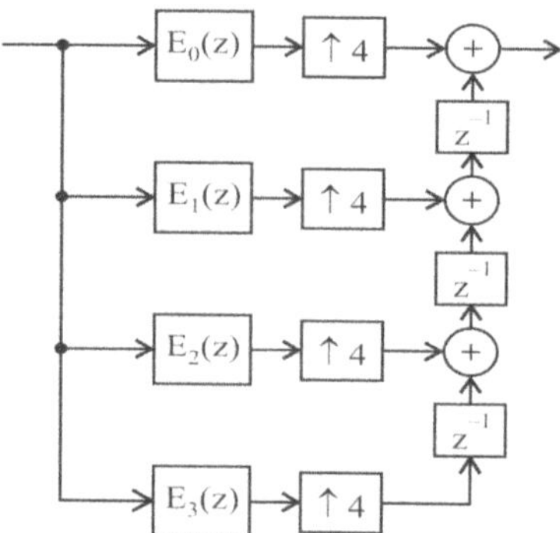

Further reduction in computational complexity is achieved by sharing common multipliers if H(z) is a linear-phase FIR filter. For example, for a length-16 Type II FIR transfer function a computationally efficient factor-of-4 interpolator structure based on the above equation is as shown below:

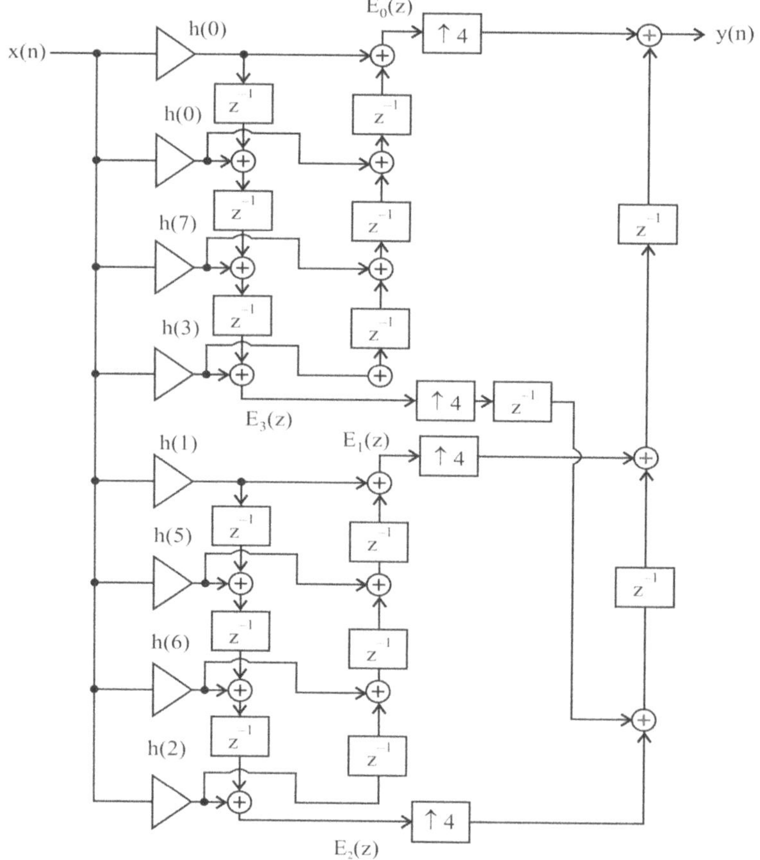

Review Questions and Answers

1. What does multirate mean?

Multirate simply means "multiple sampling rates". A multirate DSP system uses multiple sampling rates within the system. Whenever a signal at one rate has to be used by a system that expects a different rate, the rate has to be increased or decreased, and some processing is required to do so. Therefore "Multirate DSP" really refers to the art or science of *changing* sampling rates.

2. Why should we do multirate DSP?

The most immediate reason is when you need to pass data between two systems which use incompatible sampling rates. For example, professional audio systems use 48 kHz rate, but consumer CD players use 44.1 kHz; when audio professionals transfer their recorded music to CDs, they need to do a rate conversion.

But the most common reason is that multirate DSP can greatly increase processing efficiency (even by orders of magnitude!), which reduces DSP system cost. *This makes the subject of multirate DSP vital to all professional DSP practitioners.*

3. What are the categories of multirate?

Multirate consists of:

1. **Decimation:** To decrease the sampling rate,

2. **Interpolation:** To increase the sampling rate, or,

3. **Resampling:** To combine decimation and interpolation in order to change the sampling rate by a fractional value that can be expressed as a ratio. For example, to resample by a factor of 1.5, you just interpolate by a factor of 3 then decimate by a factor of 2 (to change the sampling rate by a factor of $3/2 = 1.5$.)

4. What are "decimation" and "downsampling"?

Loosely speaking, "decimation" is the process of reducing the sampling rate. In practice, this usually implies lowpass-filtering a signal, then throwing away some of its samples.

"Downsampling" is a more specific term which refers to just the process of throwing away samples, without the lowpass filtering operation. Throughout this text book, though, we'll just use the term "decimation" loosely, sometimes to mean "downsampling".

5. What is the "decimation factor"?

The decimation factor is simply the ratio of the input rate to the output rate. It is usually symbolized by "M", so input rate/output rate = M.

Tip: You can remember that "M" is the symbol for decimation factor by thinking of "deci-M-ation". (Exercise for the student: which letter is used as the symbol for interpo-L-ation factor?)

6. **Why should we decimate?**

The most immediate reason to decimate is simply to reduce the sampling rate at the output of one system so a system operating at a lower sampling rate can input the signal. But a much more common motivation for decimation is to reduce the cost of processing: the calculation and/or memory required to implement a DSP system generally is proportional to the sampling rate, so the use of a lower sampling rate usually results in a cheaper implementation.

Almost anything you do to/with the signal can be done with fewer operations at a lower sample rate, and the workload is almost always reduced by more than a factor of M.

For example, if you double the sample rate, an equivalent filter will require four times as many operations to implement. This is because both amount of data (per second) and the length of the filter increase by two, so convolution goes up by four. Thus, if you can halve the sample rate, you can decrease the work load by a factor of four. I guess you could say that if you reduce the sample rate by M, the workload for a filter goes down to $(1/M)^2$.

7. **Is there a restriction on decimation factors we can use?**

Yes. Decimation involves throwing away samples, so you can only decimate by integer factors; you cannot decimate by fractional factors. (However, you can do interpolation prior to decimation to achieve an overall rational factor, for example, "4/5"; see Resampling questions.)

8. **Which signals can be downsampled?**

A signal can be downsampled (without doing any filtering) whenever it is "oversampled", that is, when a sampling rate was used that was greater than the Nyquist criteria required. Specifically, the signal's highest frequency must be less than half the post-decimation sampling rate. (This just boils down to applying the Nyquist criteria to the input signal, relative to the new sampling rate.)

In most cases, though, you'll end up lowpass-filtering your signal prior to downsampling, in order to enforce the Nyquist criteria at the post-decimation rate. For example, suppose you have a signal sampled at a rate of 30 kHz, whose highest frequency component is 10 kHz (which is less than the Nyquist frequency of 15 kHz). If you wish to reduce the sampling rate by a factor of three to 10 kHz, you must ensure that you have no components greater than 5 kHz, which is the Nyquist frequency for the reduced rate. However, since the original signal has components up to 10 kHz, you must lowpass-filter the signal prior to downsampling to remove all components above 5 kHz so that no aliasing will occur when downsampling.

This combined operation of filtering and downsampling is called decimation.

9. **What happens if we violate the Nyquist criteria in downsampling or decimating?**

You get aliasing--just as with other cases of violating the Nyquist criteria. (Aliasing is a type of distortion which cannot be corrected once it occurs.)

10. **Can we decimate in multiple stages?**

Yes, so long as the decimation factor, M, is not a prime number. For example, to decimate by a factor of 15, you could decimate by 5, then decimate by 3. The more prime factors M has, the more choices you have. For example you could decimate by a factor of 24 using:

- one stage: 24

- two stages: 6 and 4, or 8 and 3

- three stages: 4, 3, and 2

- four stages: 3, 2, 2, and 2

11. **Why should we use multiple stages in decimation?**

If you are simply downsampling (that is, throwing away samples without filtering), there's no benefit. But in the more common case of decimating (combining filtering and downsampling), the computational and memory requirements of the filters can usually be reduced by using multiple stages.

12. **How do we figure out the optimum number of stages, and the decimation factor at each stage?**

That's a tough one. There isn't a simple answer to this one: the answer varies depending on many things, so if you really want to find the optimum, you have to evaluate the resource requirements of each possibility.

However, here are a couple of rules of thumb which may help narrow down the choices:

- Using two or three stages is usually optimal or near-optimal.

- Decimate in order from the largest to smallest factor. In other words, use the largest factor at the highest sampling rate. For example, when decimating by a factor of 60 in three stages, decimate by 5, then by 4, then by 3.

13. **How do decimation is implemented?**

Decimation consists of the processes of lowpass filtering, followed by downsampling.

To implement the filtering part, you can use either FIR or IIR filters.

To implement the downsampling part (by a downsampling factor of "M") simply keep every M^{th} sample, and throw away the M-1 samples in between. For example, to decimate by 4, keep every fourth sample, and throw three out of every four samples away.

14. If we are going to throw away most of the lowpass filter's outputs, why bother to calculate them in the first place?

You may be onto something. In the case of FIR filters, any output is a function only of the past inputs (because there is no feedback). Therefore, you only have to calculate outputs which will be used.

For IIR filters, you still have to do part or all of the filter calculation for each input, even when the corresponding output won't be used. (Depending on the filter topology used, certain feed-forward parts of the calculation can be omitted.). The reason is that outputs you do use are affected by the feedback from the outputs you don't use.

The fact that only the outputs which will be used have to be calculated explains why decimating filters are almost always implemented using FIR filters!

15. What computational savings do we gain by using a FIR decimator?

Since you compute only one of every M outputs, you save M-1 operations per output, or an overall "savings" of $(M - 1)/M$. Therefore, the larger the decimation factor is, the larger the savings, percentage-wise.

A simple way to think of the amount of computation required to implement a FIR decimator is that it is equal to the computation required for a non-decimating N-tap filter operating at the output rate.

16. How much memory savings do we gain by using a FIR decimator?

None. You still have to store every input sample in the FIR's delay line, so the memory requirement is the same size as for a non-decimated FIR having the same number of taps.

17. How do a FIR decimator is designed?

Just use your favorite FIR design method. The design criteria are:

1. The passband lower frequency is zero; the passband upper frequency is whatever information bandwidth you want to preserve after decimating. The passband ripple is whatever your application can tolerate.

2. The stopband lower frequency is half the output rate minus the passband upper frequency. The stopband attenuation is set according to whatever aliasing your application can stand. (Note that there will always be aliasing

in a decimator, but you just reduce it to a negligible value with the decimating filter.)

3. As with any FIR, the number of taps is whatever is required to meet the passband and stopband specifications.

18. How do a FIR decimator is implemented?

A decimating FIR is actually the same as a regular FIR, except that you shift M samples into the delay line for each output you calculate. More specifically:

1. Store M samples in the delay line.

2. Calculate the decimated output as the sum-of-products of the delay line values and the filter coefficients.

3. Shift the delay line by M places to make room for the inputs of the next decimation.

Also, just as with ordinary FIRs, circular buffers can be used to eliminate the requirement to literally shift the data in the delay line.

19. How do a FIR decimator is tested?

You can test a decimating FIR in most of the ways you might test an ordinary FIR:

1. A special case of a decimator is an "ordinary" FIR. When given a value of "1" for M, a decimator should act exactly like an ordinary FIR. You can then do impulse, step, and sine tests on it just like you can on an ordinary FIR.

2. If you put in a sine whose frequency is within the decimator's passband, the output should be distortion-free (once the filter reaches steady-state), and the frequency of the output should be the same as the frequency of the input, in terms of absolute Hz.

3. You also can extend the "impulse response" test used for ordinary FIRs by using a "fat impulse", consisting of M consecutive "1" samples followed by a series of "0" samples. In that case, if the decimator has been implemented correctly, the output will not be the literal FIR filter coefficients, but will be the sum of every subset of M coefficients.

4. You can use a step response test. Given a unity-valued step input, the output should be the sum of the FIR coefficients once the filter has reached steady state.

20. What are "upsampling" and "interpolation"?

"Upsampling" is the process of inserting zero-valued samples between original samples to increase the sampling rate. (This is called "zero-stuffing".) Upsampling adds to the original signal undesired spectral images which are centered on multiples of the original sampling rate.

"Interpolation", in the DSP sense, is the process of upsampling followed by filtering. (The filtering removes the undesired spectral images.) As a linear process, the DSP sense of interpolation is somewhat different from the "math" sense of interpolation, but the result is conceptually similar: to create "in-between" samples from the original samples. The result is as if you had just originally sampled your signal at the higher rate.

21. Why should we interpolate?

The primary reason to interpolate is simply to increase the sampling rate at the output of one system so that another system operating at a higher sampling rate can input the signal.

22. What is the "interpolation factor"?

The interpolation factor is simply the ratio of the output rate to the input rate. It is usually symbolized by "L", so output rate / input rate = L.

Tip: You can remember that "L" is the symbol for interpolation factor by thinking of "interpo-L-ation".

23. Is there a restriction on interpolation factors we can use?

Yes. Since interpolation relies on zero-stuffing you can only interpolate by integer factors; you cannot interpolate by fractional factors. (However, you can combine interpolation and decimation to achieve an overall rational factor, for example, 4/5; see resampling related questions.)

24. Which signals can be interpolated?

All there is no restriction.

25. When interpolating, do we always need to do filtering?

Yes. Otherwise, you're doing upsampling.

26. Do we always need to do interpolation (upsampling followed by filtering) or can we get by with doing just upsampling?

Upsampling adds undesired spectral images to the signal at multiples of the original sampling rate, so unless you remove those by filtering, the upsampled signal is not the same as the original: it's distorted.

Some applications may be able to tolerate that, for example, if the images get removed later by an analog filter, but in most applications you will have to remove the undesired images via digital filtering. Therefore, interpolation is far more common that upsampling alone.

27. Can we interpolate in multiple stages?

Yes, so long as the interpolation ratio, L, is not a prime number. For example, to interpolate by a factor of 15, you could interpolate by 3 then interpolate by 5. The more factors L has, the more choices you have. For example you could interpolate by 16 in:

- one stage: 16

- two stages: 4 and 4

- three stages: 2, 2, and 4

- four stages: 2, 2, 2, and 2

28. Why should we use multiple stages in interpolation?

Just as with decimation, the computational and memory requirements of interpolation filtering can often be reduced by using multiple stages.

29. How do we figure out the optimum number of stages, and the interpolation ratio at each stage?

There isn't a simple answer to this one: the answer varies depending on many things. However, here are a couple of rules of thumb:

- Using two or three stages is usually optimal or near-optimal.

- Interpolate in order of the smallest to largest factors. For example, when interpolating by a factor of 60 in three stages, interpolate by 3, then by 4, then by 5. (Use the largest ratio on the highest rate.)

30. How do interpolation is implemented?

Interpolation always consists of two processes:

1. Inserting L-1 zero-valued samples between each pair of input samples. This operation is called "zero stuffing".

2. Lowpass-filtering the result.

The result (assuming an ideal interpolation filter) is a signal at L times the original sampling rate which has the same spectrum over the input Nyquist (0 to Fs/2) range, and with zero spectral content above the original Fs/2.

31. How the implementation of interpolation works?

1. The zero-stuffing creates a higher-rate signal whose spectrum is the same as the original over the original bandwidth, but has images of the original spectrum centered on multiples of the original sampling rate.

2. The lowpass filtering eliminates the images.

32. Why do interpolation by zero-stuffing? Doesn't it make more sense to create the additional samples by just copying the original samples?

This idea is appealing because, intuitively, this "stairstep" output seems more similar to the original than the zero-stuffed version. But in this case, intuition leads us down the garden path. This process causes a "zero-order hold" distortion in the original passband, and still creates undesired images (see below).

Although these effects could be un-done by filtering, it turns out that zero-stuffing approach is not only more "correct", it actually reduces the amount of computation required to implement a FIR interpolation filter. Therefore, interpolation is always done via zero-stuffing.

33. How does zero-stuffing reduce computation of the interpolation filter?

The output of a FIR filter is the sum each coefficient multiplied by each corresponding input sample. In the case of a FIR interpolation filter, some of the input samples are stuffed zeros. Each stuffed zero gets multiplied by a coefficient and summed with the others. However, this adding-and-summing processing has no effect when the data sample is zero--which we know in advance will be the case for L-1 out of each L input samples of a FIR interpolation filter. So why bother to calculate these taps?

The net result is that to interpolate by a factor of L, you calculate L outputs for each input using L different "sub-filters" derived from your original filter.

34. Give an example of a FIR interpolator?

Here's an example of a 12-tap FIR filter that implements interpolation by a factor of four. The coefficients are h_0-h_{11}, and three data samples, x_0-x_2 (with the newest, x_2, on the left) have made their way into the filter's delay line:

h_0	h_1	h_2	h_3	h_4	h_5	h_6	h_7	h_8	h_9	h_{10}	h_{11}		Result
x_2	0	0	0	x_1	0	0	0	x_0	0	0	0		$x_2 \cdot h_0 + x_1 \cdot h_4 + x_0 \cdot h_8$
0	x_2	0	0	0	x_1	0	0	0	x_0	0	0		$x_2 \cdot h_1 + x_1 \cdot h_5 + x_0 \cdot h_9$
0	0	x_2	0	0	0	x_1	0	0	0	x_0	0		$x_2 \cdot h_2 + x_1 \cdot h_6 + x_0 \cdot h_{10}$
0	0	0	x_2	0	0	0	x_1	0	0	0	x_0		$x_2 \cdot h_3 + x_1 \cdot h_7 + x_0 \cdot h_{11}$

35. What is generalized from the example given in 6.33?

The table suggests the following general observations about FIR interpolators:

- Since the interpolation ratio is four (L = 4), there are four "sub-filters" (whose coefficient sets are marked here with matching colors.) These sub-filters are officially called "polyphase filters".

- For each input, we calculate L outputs by doing L basic FIR calculations, each using a different set of coefficients.

- The number of taps per polyphase filter is 3, or, expressed as a formula: Npoly = Ntotal /L.

- The coefficients of each polyphase filter can be determined by skipping every Lth coefficient, starting at coefficients 0 through L-1, to calculate corresponding outputs 0 through $L - 1$.

- Alternatively, if you rearranged your coefficients in advance in "scrambled" order like this: h_0, h_4, h_8, h_1, h_5, h_9, h_2, h_6, h_{10}, h_3, h_7, h_{11}, then you could just step through them in order.

- We have hinted here at the fact that N should be a multiple of L. This isn't absolutely necessary, but if N isn't a multiple of L, the added complication of using a non-multiple of L often isn't worth it. So if the minimum number of taps that *your* filter specification requires doesn't happen to be a multiple of L, your best bet is usually to just increase N to the next multiple of L. You can do this either by adding some zero-valued coefficients onto the end of the filter, or by re-designing the filter using the larger N value.

36. What computational savings do we gain by using a FIR interpolator?

Since each output is calculated using only N/L coefficients (rather than N coefficients), you get an overall computational "savings" of (N - N/L) per output .

A simple way to think of the amount of computation required to implement a FIR interpolator is that it is equal to the computation required for a non-interpolating N-tap filter operating at the input rate. In effect, you have to calculate L filters using N/L taps each, so that's N total taps calculated per input.

37. How much memory savings do we gain by using a FIR interpolator?

Compared to the straight-forward implementation of interpolation by upsampling the signal by stuffing it with L-1 zeros , then filtering it, you save memory by a factor of (L-1)/L. In other words, you don't have to store L-1 zero-stuffed "upsamples" per actual input sample.

38. How do a FIR interpolator is implemented?

An interpolating FIR is actually the same as a regular FIR, except that, for each input, you calculate L outputs per input using L polyphase filters, each having N/L taps. More specifically:

1. Store a sample in the delay line. (The size of the delay line is N/L.)

2. For each of L polyphase coefficient sets, calculate an output as the sum-of-products of the delay line values and the filter coefficients.

3. Shift the delay line by one to make room for the next input.

Also, just as with ordinary FIRs, circular buffers can be used to eliminate the requirement to literally shift the data in the delay line.

39. How do a FIR interpolator is tested?

You can test an interpolating FIR in most of the ways you might test an ordinary FIR:

1. A special case of an interpolator is an ordinary FIR. When given a value of 1 for L, an interpolator should act exactly like an ordinary FIR. You can then do impulse, step, and sine tests on it just like you can on an ordinary FIR.

2. If you put in a sine whose frequency is within the interpolator's passband, the output should be distortion-free (once the filter reaches steady state), and the frequency of the output should be the same as the frequency of the input, in terms of absolute Hz.

3. You can use a step response test. Given a unity-valued step input, every group of L outputs should be the same as the sums of the coefficients of the L individual polyphase filters, once the filter has reached steady state.

40. What is "resampling"?

"Resampling" means combining interpolation and decimation to change the sampling rate by a rational factor.

41. Why should we resample?

Resampling is usually done to interface two systems which have different sampling rates. If the ratio of two system's rates happens to be an integer, decimation or interpolation can be used to change the sampling rate (depending on whether the rate is being decreased or increased); otherwise, interpolation and decimation must be used together to change the rate

A practical and well-known example results from the fact that professional audio equipment uses a sampling rate of 48 kHz, but consumer audio equipment uses a rate of 44.1 kHz. Therefore, to transfer music from a professional recording to a CD, the sampling rate must be changed by a factor of:

$(44100 / 48000) = (441 / 480) = (147 / 160)$

There are no common factors in 147 and 160, so we must stop factoring at that point. Therefore, in this example, we would interpolate by a factor of 147 then decimate by a factor of 160.

42. What is the "resampling factor"?

The interpolation factor is simply the ratio of the output rate to the input rate. Given that the interpolation factor is L and the decimation factor is M, the resampling factor is L / M. In the above example, the resampling factor is 147 / 160 = 0.91875

43. Is there a restriction on the resampling factor we can use?

Yes. As always, the Nyquist criteria must be met relative to the resulting output sampling rate, or aliasing will result. In other words, the output rate cannot be less than twice the highest frequency (of interest) of the input signal.

44. When resampling, do we always need to a filter?

Yes. Since resampling includes interpolation, you need an interpolation filter. Otherwise, the images created by the zero-stuffing part of interpolation will remain, and the interpolated signal will not be "the same" as the original.

Likewise, since resampling includes decimation, you seemingly need a decimation filter. Or do you? Since the interpolation filter is in-line with the decimation filter, you could just combine the two filters by convolving their coefficients into a single filter to use for decimation. Better yet, since both are lowpass filters, just use whichever filter has the lowest cutoff frequency as the interpolation filter.

45. Can we resample in multiple stages?

Yes, but there are a couple of restrictions:

- If either the interpolation or decimation factors are prime numbers, you won't be able to decompose those parts of the resampler into stages.

- You must preserve the Nyquist criteria at each stage or else aliasing will result. That is, no stage can have an output rate which is less than twice the highest frequency of interest.

46. Why should we use multiple stages in resampling?

Just as with interpolation and decimation, the computational and/or memory requirements of the resampling filtering can sometimes be greatly reduced by using multiple stages.

47. How do resampling is implemented?

The straight-forward implementation of resampling is to do interpolation by a factor of L, then decimation by a factor of M. (You must do it in that order; otherwise, the decimator would remove part of the desired signal--which the interpolator could not restore.)

48. Is that straight-forward implementation efficient?

No. The problem is that for resampling factors close to 1.0, the interpolation factor can be quite large. For example, in the case described above of changing from the sampling rate from 48 kHz to 44.1 kHz, the ratio is only 0.91875, yet the interpolation factor is 147!

Also, you are filtering the signal twice: once in the interpolator and once in the decimator. However, one of the filters has a larger bandwidth than the other, so the larger-bandwidth filter is redundant.

49. How to overcome the problem mentioned in 6.47?

Just combine the computational and memory advantages that FIR interpolator and decimator implementations can provide.

First, let's briefly review what makes FIR interpolation and decimation efficient:

- When interpolating by a factor of L, you only have to actually calculate 1/L of the FIR taps per interpolator output.
- When decimating by a factor of M, you only have to calculate one output for every M decimator inputs.

So, combining these ideas, we will calculate only the outputs we actually need, using only a subset of the interpolation coefficients to calculate each output. That makes it possible to efficiently implement even FIR resamplers which have large interpolation and/or decimation factors.

The tricky part is figuring out which polyphase filters to apply to which inputs, to calculate the desired outputs, as a function of L and M. There are various ways of doing that, but they're all beyond our scope here.

50. How do a FIR resampler is tested?

1. The most obvious method is to put in a sine whose frequency is within the resampler's passband. If an undistorted sine comes out, that's a good sign. Note, however, that there will typically be a "ramp up" at the beginning of the sine, due to the filter's delay line filling with samples. Therefore, if you analyze the spectral content of the sine, be sure to skip past the ramp-up portion.

2. Depending on the resampling factor, resampling can be thought of as a general case of other types of multirate filtering. It can be:
   - Interpolation: The interpolation factor, L, is greater than one, and the decimation factor, M, is one.
   - Decimation: The interpolation factor, L, is one, but the decimation factor, M, is greater than one.
   - "Ordinary" Filtering: The interpolation and decimation factors, L and M, are both one.
   - Therefore, if you successfully test it with all these cases using the methods appropriate for each case, it probably is correct.

Index

A

Adders or summing element, 13
Analog Filter Approximations, 191
Analog Spectral Transformations, 213
Analog-to-digital, 2
Antisymmetric (odd) signals, 9
Aperiodic signals, 9

B

Band reject filter, 189, 191
Band Reject FIR Filter, 305
Bandpass filter, 189, 190
Bandpass FIR Filter, 304
BIBO stable is, 162
Bilinear Transformation method, 216
Bilinear Transformation, 225
Bit Reversal, 94
Blackman window, 299, 313
Butterworth Approximation, 193

C

Cascade Form Realization, 175
Cascade form, 168, 175
Causal systems, 14, 16
Chebyshev Approximation, 199
Circular Convolution, 37, 52, 69

D

DAC, 359
Decimation in Time (DIT) Algorithm, 85
Decimation, 344, 349
Decimation-in-Frequency (DIF) FFT Algorithm, 97
Delay elements, 13
Design of FIR Filters, 298
DFT of Delayed Sequence, 50
Diagonal method, 63, 75
Difference equations, 22
Differentiation in Z-domain, 126
Digital signal, 2
Digital Spectral Transformation, 230
Digital systems, 3
Digital-to-analog, 2
Direct form I, 168
Direct form II (or) Canonic form, 171
Direct Method, 60, 71
Discrete Fourier Series (DFS), 32
Discrete Fourier Transform (DFT), 39
Discrete–Time Fourier Series (DTFS), 32
Discrete-time Fourier Transform (DTFT), 42
Discrete-time signals, 3
Down Sampling, 344
Duality, 34

O

P

R

S

T